Inference, Asymptotics, and Applications

Selected Papers of Ib Michael Skovgaard, with Introductions by his Colleagues

Inference, Asymptotics, and Applications

Selected Papers of Ib Michael Skovgaard, with Introductions by his Colleagues

Editors

Nancy Reid
University of Toronto, Canada

Torben Martinussen
University of Copenhagen, Denmark

World Scientific

EW JERSEY • LONDON • SINGAPORE • BEIJING • SHANGHAI • HONG KONG • TAIPEI • CHENNAI • TOKYO

Published by

World Scientific Publishing Co. Pte. Ltd.

5 Toh Tuck Link, Singapore 596224

USA office: 27 Warren Street, Suite 401-402, Hackensack, NJ 07601

UK office: 57 Shelton Street, Covent Garden, London WC2H 9HE

Library of Congress Cataloging-in-Publication Data

Names: Skovgaard, Ib M. | Reid, N., editor. | Martinussen, Torben., editor.

Title: Inference, asymptotics, and applications : selected papers of Ib Michael Skovgaard, with introductions by his colleagues / edited by Nancy Reid (University of Toronto, Canada), Torben Martinussen (University of Copenhagen, Denmark).

Description: New Jersey : World Scientific, 2017. | Includes bibliographical references.

Identifiers: LCCN 2017001100 | ISBN 9789813207875 (hardcover : alk. paper)

Subjects: LCSH: Mathematical statistics--Asymptotic theory. | Approximation theory. | Nonparametric statistics. | Stochastic processes.

Classification: LCC QA276 .S5555 2017 | DDC 519.5/4--dc23

LC record available at https://lccn.loc.gov/2017001100

British Library Cataloguing-in-Publication Data

A catalogue record for this book is available from the British Library.

Printed in Singapore

To Ib Michael Skovgaard, with admiration and respect

Preface

We are pleased to present this volume of selected papers by our colleague Ib Michael Skovgaard. Through different routes we have both come to admire the exceptional clarity of his writing and the depth of understanding with which he addressed many very difficult problems. Our shared conviction that his work could profitably be re-read by students and researchers interested in the theory and application of statistics, and students of writing style, prompted us to undertake the compiling of this book. We were very happy that all the invited contributors responded quickly and positively, and you will read many expressions of admiration in the introductions to the papers.

We would like to express our appreciation to Rochelle Kronzek and Tan Rok Ting at World Scientific Press for their efforts in bringing the book to fruition. A special thanks is owed to Ruoyong Xu of the University of Toronto for her prompt and careful work in preparing the final version. NR's research is partially supported by the Natural Sciences and Engineering Research Council of Canada.

Nancy Reid and Torben Martinussen, October 2016

Introduction

Ib Skovgaard obtained his PhD degree in statistics from the Department of Mathematical Statistics at the University of Copenhagen in 1982 with Steffen Lauritzen as supervisor. Steffen's task was very easy and just involved discussing things from time to time. Generally, the department was a very lively and stimulating research environment, and Ib also discussed his work with several staff members. In 1992 Ib received the higher dr. scient. (Sc.D.) degree from the University of Copenhagen. This degree was based on the monograph "Analytic Statistical Models", which was published by the Institute of Mathematical Statistics.

In 1978, after he had received his cand. stat. degree (M.Sc. degree in statistics), Ib worked for a short while in Danske Bank, where he developed highly original applications of Brownian motion to banking. This work has never been published — perhaps because of its usefulness to the bank.

From 1981 Ib was employed by the Royal Veterinary and Agricultural University in Copenhagen, first as assistant and associate professor, and from 1996 as full professor. He was Head of the Department of Mathematics, 1986–1989, and Dean of the Faculty of Basic Science and Food Science, 1989–1991. In 2007 the Royal Veterinary and Agricultural University merged with the University of Copenhagen, and in 2012 part of the statistics group merged with the Statistics Section of the Department of Mathematical Sciences. Thus Ib returned to his old department, which had in the meantime merged with the Department of Mathematics. In 2014 he retired from his professor position.

In his theoretical research, Ib made seminal contributions to the vigorous developments in neo-Fisherian statistical theory and asymptotic theory that took place in the 1980s and 1990s. In particular, he worked on higher-order asymptotics and saddlepoint methods, but also on likelihood inference and principles of statistical inference. Biostatistical applications and development of appropriate statistical methods for this work have played a prominent role in Ib's research over the years. Application areas include food science, automatic alignment of electropherograms, statistical genetics and bioinformatics projects.

Ib is a brilliant teacher, always attentive to the students and their needs and competencies. Apart from traditional lecture based statistics courses, Ib was also active in developing e-learning courses before this became mainstream, as well as courses based on project work, and PhD courses for young researchers who use statistics in their thesis work. Moreover, he has held a huge number of consultancy meetings with students and researchers throughout the years.

During the 1970s Mats Rudemo had established an environment for statistics at the Royal Veterinary and Agricultural University. Ib quickly became an important player in this work, and he and Mats consolidated and developed the environment over the years. The statistics group grew steadily over the years, and in 2005 Ib became officially the Head of the Statistics Group — in practice he had this role long before 2005. The concept was clear: tight connections to and close interaction with other scientists at the Royal Veterinary and Agricultural University — both with respect to teaching and research — combined with theoretical statistical research of high standard and personal integrity. This combination is not as easy to establish and maintain as it may appear, and the good reputation of statistics and statisticians among students and researchers in the agricultural and veterinary sciences at University of Copenhagen is to a large extent due to Ib's efforts.

Many young Danish statisticians have had their first position in Ib's group, arriving with a theoretical background and little experience with applied statistics. For these young researchers Ib has been an invaluable mentor. Always friendly, positive and accommodating, he generously shared his enormous knowledge and experience. None of the young statisticians were ever turned down by Ib or met with arrogant answers — no matter how stupid they later realised that their questions had been.

Ib is a keen and highly skilled chess player. On one occasion, he visited the University of Aarhus as an external examiner. It was a Friday afternoon, and some students were drinking beer and playing chess. Ib, with his usual modest and understated behaviour, seemed like an easy target and was challenged by the best player among the students, who thought of himself as the best chess player in town. After only a few moves, the student was visibly shaken, and it did not take long for Ib to win the game. In his retirement, Ib has again taken up a serious interest in chess.

Helle Sørensen and Michael Sørensen

I had the pleasure to be Ib's PhD student from 1991–1994 and after that colleague with Ib ten more years at the Royal Veterinary and Agricultural University of Denmark ("KVL"). My PhD work was on the topic of what has become known as sensometrics — the statistical analysis of human sensory and consumer data. This field is rooted in food science, one of the major areas of research at KVL at that time. It exemplifies explicitly how Ib engaged himself in and contributed to many of the main strategic research areas of KVL, the only university of its kind in Denmark. The role and position of the field of statistics at a university like this, and hence in Denmark, is by no means assured. I have no doubt that the strong position of and respect for the statistics group at KVL was founded by two persons: Mats Rudemo and Ib Skovgaard. Without their combination of formal statistical professionalism, true interest in the integration of this into all the relevant bio, food- and agri-sciences and their open and strong personalities, statistics within these fields in Denmark would not have been the same.

Ib has been supervising and mentoring me for many years in a way that I can myself only strive to be able to live up to in my own endeavours. He was always open, positive and supportive, and never short of time or attention. I am grateful for the way that I was welcomed as a young PhD student in 1991 moving to the capital of Denmark for the first time in my life. And I have enjoyed the fantastic working environment of the statistics group and its closest friends at KVL for many years, very much due to Ib's naturally friendly personality. On top of all the other activities, Ib was also a passionate runner, an interest that we shared at that time. Among other things we participated from the department in the DHL Relay. It will come as no surprise that Ib was outrunning all the rest of us.

Per Bruun Brockhoff

I joined the Royal Veterinary and Agricultural University in 1997 starting there as an Assistant Professor, and I worked there for about ten years. All those years Ib was head of the Statistics Group. I soon learned that Ib was immensely helpful. He always took time to answer all kind of questions. He had a quite spacious office with room for both an ordinary desk and a meeting table. If one came to his office to seek his advice he would typically be sitting at his desk working on something but he would then rise and walk to the meeting table in this way showing his willingness to help. Ib is a great statistician and mathematician as well. Numerous applied statistical projects have benefitted from his great insight into which specific analysis to apply in a given practical situation. Ib's own research area turned out to be higher order asymptotics but I had the feeling that it could just as well had been any other area because of his fantastic mathematical skills. I have benefitted from this on many occasions and I particularly remember a situation where I worked together with Thomas Scheike on a Cox model with time-dependent regression coefficents. We had developed an estimator of the cumulative coefficients and we were pretty sure that the estimator was consistent and asymptotically normal but had difficulties in showing it. We asked Ib for assistance and soon he delivered a very elegant proof of both conjectures. To me this is very typical, you can ask Ib about almost anything and he will be able to put you in a situation where you are better off. Ib is also the kind of person who almost never complains. One of the first courses I was responsible for had about 150 students and it ended with a four hour written exam that I subsequently had to grade. While grading all these exams I felt a little sorry for myself and at some point complained to Ib about this boring task. It turned out that he was also grading exams, but the course he was responsible for had around 400 students. It took him almost a month to grade all these exams and I never heard him complain. I was so impressed about this, that whenever I have to do less interesting jobs it still helps me to remind myself about this event.

Torben Martinussen

Ib and I are of the same generation of statistical scientists — I met him at Stanford in the early 1980s, and he visited Vancouver around the same time. We were both lucky enough to be part of the 'asymptotics wave' in inference that made great strides throughout the 1980s and 1990s, and often met at conferences and workshops. I had the honour of being the foreign examiner of his dr. scient., and attending the formal ceremonies for that. I was expected, I think, to challenge his dissertation research, but I'm afraid I was unable to find much to challenge!

It is a huge, but often unstated, joy of academic life that one makes friends around the world, and has opportunities, with luck, to see these friends regularly, if not often. I am very grateful to count Ib Skovgaard among my academic friends, and enjoyed immensely contributing to this volume in his honour.

Nancy Reid

From left to right: Masafumi Akahira, Art Owen, Murray Aitkin, Richard Smith, Nancy Reid, Ib, at the 47th Session of the International Statistical Institute, Paris, 1989.

Ib and his nephew Martin.

Contents

Chapter 1

Edgeworth expansions of the distributions of maximum likelihood estimators in the general (Non I.I.D.) case

Introduction by Rahul Mukerjee

Indian Institute of Management Calcutta

1.1 Background

Edgeworth expansion is a major pillar in the foundation of the modern higher order asymptotic theory of statistics. It is particularly useful in theoretical calculations, with application to such diverse areas as power properties of tests, Bayesian asymptotics, bootstrap methods, empirical likelihood, and so on. *Kendall's Advanced Theory of Statistics*, Volume 1 (Stuart and Ord, 1987, p. 223) attributed both the Gram-Charlier series of Type A and its Edgeworth form to P.L. Chebyshev and L.H.F. Oppermann. Edgeworth (1904) derived the expansion from a different perspective via the theory of errors and claimed that his series, based on cumulants, gives a better representation of density functions than the Gram-Charlier series of Type A, which involves moments. A heuristic explanation for this claim, when comparable numbers of terms are retained in both series, can be based on the fast tapering nature of cumulants in many applications, such as with the standardized sum of independent, identically distributed random variables, a property that moments do not share (Stuart and Ord, 1987, p. 225). The point just noted leads to deep theoretical questions on the validity of a formal Edgeworth expansion, that is, one where only a pre-assigned number of terms are retained and approximate cumulants obtained by the formal delta method replace the exact ones when the latter are intractable. For example, if such a formal expansion is integrated to approximate a distribution function, then what is the order of the error of approximation, and under precisely what conditions? An informative discussion on this and related issues is available in Wallace (1958), where earlier references can also be found.

In a major breakthrough, Bhattacharya and Ghosh (1978) established the validity of the formal Edgeworth expansion in the case of independent, identically distributed observations, under appropriate moment conditions, for statistics that are functions of sample averages and applied the findings to obtain asymptotic expansions of distributions of minimum contrast estimators of a vector parameter. See their paper for references to related earlier work in the 1970s in more specialized settings and sometimes under more restrictive assumptions. Further references to work immediately preceding the paper that is being commented on are available in the paper itself and Skovgaard (1986).

1.2 Comments on the paper

The first sentence of the introduction aptly encapsulates the contributions of the paper: "The purpose of the present paper is to derive simple sufficient conditions for the validity of an Edgeworth expansion of the maximum likelihood estimator in the non independent, identically distributed case, and to compute the quantities needed for this expansion". Indeed, post Bhattacharya and Ghosh (1978), there was a compelling need to explore the non i.i.d. case in view of numerous potential applications, and this landmark paper by Skovgaard makes a very significant advance on this difficult problem in full rigour. The introduction tells the reader clearly about the advantages and limitations of the Edgeworth expansion, such as nice performance in theoretical calculations at the cost of numerical accuracy near the tails. It also contains a few useful notes of caution on the use of Edgeworth expansions.

§2 of the paper introduces the notation and regularity conditions (Assumptions 2.1). The notation appears in a coordinate-free form to facilitate the presentation and proofs of results in the multivariate case. For possible ease in application, coordinate versions of the results are also given at the end of §§3, 4 and 5, where the summation convention is followed. The regularity conditions are not hard to verify and hold under wide generality. In fact, as Remark 2.2 notes, some of these conditions, including one on differentiation inside integrals, are not really necessary but still included to avoid technical difficulties.

The key results of the paper appear in §3. With reference to a sequence of experiments indexed by $n \in N$, let $\hat{\theta}_n$ be the maximizer of the likelihood function in a neighborhood of θ_0, the true value of a vector parameter θ. Note that $\hat{\theta}_n$ coincides with the maximum likelihood estimator when the likelihood equation has a unique solution. Also, let $S_1, \ldots, S_p$ be the centered derivatives of the log-likelihood at θ_0, up to order p. Theorem 3.5, the main result of this section, shows the uniqueness of $\hat{\theta}_n$ and obtains an Edgeworth expansion of its distribution, precisely indicating the order of approximation. This is done under the assumptions that (a) the joint distribution of $S_1, \ldots, S_p$ admits an Edgeworth expansion up to the same order of approximation (Assumptions 3.1), and (b) the derivatives of $\hat{\theta}_n$ with respect to $S_1, \ldots, S_p$ are sufficiently well behaved (Assumptions 3.2). These assumptions suggest that the results in Skovgaard (1981a) on transformation of Edgeworth expansion should be useful in proving Theorem 3.5. As seen later in the proof, this is, indeed, the case. The rest of §3 dwells on a stochastic expansion of $\hat{\theta}_n$ (Remark 3.6) and the consequent calculation of approximate cumulants (Remark 3.7) appearing in the Edgeworth expansion in Theorem 3.5. These details as well as the coordinate versions of the findings shown at the end of the section will endear the paper to readers interested in applying the results to specific situations. This section also presents a result (Corollary 3.10) on the Edgeworth expansion of a sufficiently well behaved function of $\hat{\theta}_n$. The results in Skovgaard (1981a) again facilitate its proof.

§4 relates the results of the previous section to exponential family models and shows how then the calculation of approximate cumulants can be simplified. §5 specializes to the commonly used non-linear regression model with normally distributed errors. It is shown that then the conditions of Theorem 3.5 can be considerably simplified (Lemma 5.2).

Finally, the proofs are presented in §6. As indicated above, the proofs draw on the results in Skovgaard (1981a), but reaching the stage where these results can be used calls for an intricate analysis. The proofs display the high level of depth, rigor and sophistication that is the hallmark of Ib's papers.

1.3 Impact

A JSTOR search reveals that this influential work of Skovgaard has received 31 citations in papers on a wide range of topics such as nonlinear regression (Hougaard, 1985; Schmidt and Zwanzig, 1986), saddlepoint methods (Reid, 1988), confidence bands (Sun *et al.*, 2000), strongly dependent processes

(Lieberman *et al.*, 2003), and so on. Given its depth, rigor and generality, this masterpiece will definitely continue to inspire serious researchers in years to come, with possible applications to semiparametric settings (cf. Linton, 1996) and Bayesian asymptotics, among other areas.

Scand J Statist 8: 227–236, 1981

Edgeworth Expansions of the Distributions of Maximum Likelihood Estimators in the General (Non I.I.D.) Case

IB M. SKOVGAARD

University of Copenhagen

Received June 1980, in final form March 1981

ABSTRACT. In this paper we use the method described in Skovgaard (1981) to derive Edgeworth expansions of the distributions of maximum likelihood estimators in the general (non i.i.d.) case. Comparatively simple sufficient conditions for the validity of the expansion are derived and further simplification obtained in the non-linear normal regression models. A precise description of how to compute the expansion is given, and the first four terms of the corresponding stochastic expansion are given in an explicit form. In case of a smooth hypothesis of an exponential family, also an explicit version of the approximate cumulants, needed to compute the first four terms of the Edgeworth expansion, is given. It is shown that corresponding results for functions of the estimator are easily derived from the original expansion.

Key words: approximate cumulants, Edgeworth expansion, exponential family models, maximum likelihood estimator, non-linear normal regression

1. Introduction

The purpose of the present paper is to derive simple sufficient conditions for the validity of an Edgeworth expansion of the maximum likelihood estimator in the non i.i.d. case, and to compute the quantities needed for this expansion. The expansion is obtained by formally calculating moments of a stochastic expansion which is a Taylor series expansion of the maximum likelihood estimator.

The proof of the main theorem is based on the results of Skovgaard (1981), the method being similar to the one used in Bhattacharya & Ghosh (1978). In the i.i.d. case the moments used in the expansions may be identified with those given in a number of papers, e.g. Shenton & Bowman (1977).

The reason, that we are exclusively dealing with expansions of the Edgeworth type, is mainly their nice performance in theoretical calculations. In particular, this class of expansions is closed under a large class of transformations; see Bhattacharya & Ghosh (1978) and Skovgaard (1981). As a result explicit expansions can be computed in great generality and these form a good basis for investigations of asymptotic properties. An alternative approach is the saddlepoint approximation (see Daniels (1954)), which undoubtedly gives better numerical results especially in the tails of the distribution. The saddlepoint approximation depends however on more complicated features of the underlying distribution (than the moments) and is not as easy to handle in theoretical calculations, although Barndorff-Nielsen & Cox (1979) have shown, that it may be applied to many problems of statistical relevance.

A few words of warning concerning the use of Edgeworth expansions should be given. First, the expansions are typically *not convergent*, and it is therefore *not safe to include many terms*. Second, the absolute error is uniformly bounded over the whole range of the distribution, but the *relative error* (of the density) *is usually unbounded*. A specific phenomenon related to this is, that the approximating measure, which is a signed measure, may give *negative tail probabilities*. To prevent such misuses of the expansion one should inspect the individual terms. If these are *rapidly decreasing in magnitude*, this indicates a good approximation, although a precise result of this kind is hardly obtainable.

In Section 2 we present the notation and the regularity conditions used in this paper. Section 3 contains the main results as Theorem 3.5 and Corollary 3.10 proving the validity of Edgeworth expansions of the maximum likelihood estimator and functions of this. Also a method of obtaining the stochastic expansion of the MLE is described. In Section 4 we relate these results to smooth hypotheses in exponential families, since in this (very common) case, the cumulants can be given in a more explicit form.

In Section 5 we consider an important class of models, namely the non-linear regression model with normally distributed errors. Here the conditions of Theorem 3.5 may be replaced by one simple condition, and the results also simplifies considerably.

Section 6 contains the proofs of the various results. To make the results easier to apply we give coordinate versions of these at the end of Sections 3, 4 and 5.

2. Notation and basic assumptions

Let V, W be finite dimensional Euclidean vector spaces. If v_1, v_2 belongs to V, then $\langle v_1, v_2\rangle$ denotes their inner product and

$$\|v_1\| = \langle v_1, v_1\rangle^{\frac{1}{2}} \tag{2.1}$$

$$(v_1'^m) = (v_1, ..., v_1) \in V^m, \quad m \in \mathbf{N} \tag{2.2}$$

$\mathcal{B}(V)$ is the Borel system on V and $\mathcal{B}_k$ that on $\mathbf{R}^k$. Hom (V, W), the class of linear mappings of V into W, and $B_p(V, W)$, the class of p-linear, *symmetric* mappings of V^p into W, are in the natural way given the structure of normed vector spaces, where the norm of $A \in B_p(V, W)$ is given by

$$\|A\| = \sup \{\|A(v'^p)\| \mid \|v\| \leq 1\}. \tag{2.3}$$

We shall use the usual isomorphisms between vector spaces, e.g. $B_p(V, W) \simeq B_{p-1}(V, \text{Hom}(V, W))$ without distinguishing between these. If $A \in \text{Hom}(V, W)$, then A^* denotes its adjoint, i.e.

$$\langle A(v), w\rangle = \langle v, A^*(w)\rangle, \quad v \in V, w \in W. \tag{2.4}$$

$C^p(V, W)$ denotes the class of p times continuously differentiable functions of V into W. The pth differential of $f \in C^p(V, W)$ at v_0 is the function in $B_p(V, W)$ given by

$$D^p f(v_0)(v'^p) = \frac{d^p}{dh^p} f(v_0 + hv)\Big|_{h=0}, \quad v, v_0 \in V, h \in \mathbf{R} \tag{2.5}$$

Cumulants and moments of a distribution on V will, if they exist, be considered as multilinear, symmetric forms on V, e.g. the pth moment, μ_p, of a random vector X in V is given by

$$\mu_p(v'^p) = E\{\langle v, X\rangle^p\}, \quad v \in V \tag{2.6}$$

where $E\{...\}$ denotes expectation.

The normal density on V with mean zero and variance equal to the inner product on V will be denoted ϕ (V is understood), and $\phi_{\mu, \Sigma}$ will denote the normal density with mean μ und variance Σ.

The Cramér-Edgeworth polynomials $(\tilde{P}_r)$ are as usual defined by the formal identity

$$\sum_{r=0}^{\infty} u^r \tilde{P}_r(v: \{\chi_j\}) = \exp\left\{\sum_{r=1}^{\infty} u^r \chi_{r+2}(v'^{r+2})/(r+2)!\right\} \tag{2.7}$$

where $\{\chi_j\}$, $j \in \mathbf{N}$ are the cumulants of a distribution. Also, if $\Sigma \in B_2(V, \mathbf{R})$ is regular $P_r(-\phi_{\mu, \Sigma}: \{\chi_j\})$ is the density of the finite signed measure with characteristic function $\tilde{P}_r(iv: \{\chi_j\}) \exp\{-\frac{1}{2}\Sigma(v, v) + i\langle v, \mu\rangle\}$ obtained by formally substituting the differential operator for $(-iv)$ in $\tilde{P}_r(iv: \{\chi_j\})$ and using this on $\phi_{\mu, \Sigma}$. In particular $P_r(-\phi_{0, \Sigma}: \{\chi_j\})(v)$ is a polynomial in $v \in V$ multiplied by $\phi_{0, \Sigma}(v)$.

The order symbols o and O are unless otherwise stated used in the sense "as $n \to \infty$".

The paper has been written mainly in a coordinate free notation, which has the advantage, that multivariate results may be stated and proved almost as their one-dimensional analogues. The notation does not rely on tensor algebra, but the reader, who wants a better understanding of the multilinear algebra, will find a study of tensor algebra fruitful. Two introductory textbooks are Eisenhart (1960), (a coordinate version), and Greub (1967), (a coordinate free version). The reader, who wishes mainly to get an idea of the results without worrying about the multivariate case, will easily recognize the one-dimensional meaning of the notation.

The *coordinate versions* of the results of Sections 3, 4 and 5 employ the following notation. Indices referring to particular coordinates will be written as superscripts or subscripts (corresponding to covariant and contravariant tensors; see Eisenhart (1960), but this distinction is of no importance here). E.g. F_i^{jk} will refer to the (i, j, k)-coordinate of the array (F_i^{jk}) $i, j, k \in \{1, ..., m\}$, belonging to $(\mathbf{R}^m)^3$. We do not state the range of the indices, since this will be clear from the context. *If an index appears twice in a term summation over its range is understood.* This is known in tensor calculus as the Einstein summation convention.

Let $(E, \mathcal{E})$ be a measurable space, and (P_θ), $\theta \in \Theta \subseteq V$ a family of probability measures on $(E, \mathcal{E})$ *dominated* by a measure μ. V is a finite dimensional Euclidean vector space and Θ is open in V. Define

$$f(x; \theta) = (dP_\theta/d\mu)(x), \quad x \in E, \theta \in \Theta \tag{2.8}$$

as some version of the Radon-Nikodym derivative of P_θ with respect to μ. Throughout the paper the following regularity conditions will be assumed to hold.

Assumptions 2.1. *A version $f(.;.)$ of the Radon-Nikodym derivatives* (2.1) *exists, such that for each fixed $\theta_0 \in \Theta$ the following conditions hold. Define*

$$E_0 = \{x \in E \mid f(x; \theta_0) > 0\}. \tag{2.9}$$

Then for some integer $p \geq 2$,

I. *$f(x; \theta)$ is p times continuously differentiable at θ_0 with respect to θ for all $x \in E_0$*

II. $E\{\|D^j \log f(X;\theta_0)\|^2\} < +\infty, \quad 1 \leqslant j \leqslant p$ (2.10)

III. $\int Df(x;\theta_0)\,d\mu = \int D^2 f(x;\theta_0)\,d\mu = 0$ (2.11)

IV. $V\{D \log f(X;\theta_0)$ *is regular* (2.12)

Here and in the sequel X is a random variable having distribution P_{θ_0}, $E\{\cdot\}$ and $V\{\cdot\}$ denote expectation and variance in this distribution.

Remark 2.2. Condition (2.11) is the identity obtained by differentiating the integral of the density with respect to θ. The assumption that this may be done inside the integral holds, if $\|Df\|$ and $\|D^2 f\|$ are bounded on $E \times U(\theta_0)$ by functions independent of θ and with finite integrals (w.r.t. μ), where $U(\theta_0)$ is a neighbourhood of θ_0.

Neither (2.11) nor (2.12) are necessary assumptions, but they are assumed to avoid technical problems. Notice, that (2.10) makes sense because $D^j \log f(X;\theta_0)$ is defined with probability one.

Define

$$E_j = E\{D^j \log f(X;\theta_0)\} \in B_j(V, \mathbf{R}) \tag{2.13}$$

$$S_j = D^j \log f(X;\theta_0) - E_j \in B_j(V, \mathbf{R}) \tag{2.14}$$

$$\Sigma_j = V\{S_j\} \in B_2(B_j(V, \mathbf{R}), \mathbf{R}). \tag{2.15}$$

By (2.11) we have

$$E_1 = 0,\ \Sigma_1 = -E_2 \in B_2(V, \mathbf{R}) \simeq \mathrm{Hom}\,(V, V). \tag{2.16}$$

3. Main results

In this section we consider a sequence of experiments indexed by $n \in \mathrm{N}$, each setup of the form introduced in Section 2. All the quantities used except the parameter thus depend on n, but for notational simplicity we shall not always write the index n. The index $n \in \mathrm{N}$ may be replaced by any $i \in I$, where I is a set directed to the right, indexing a system of statistical fields with the same parameter space Θ. The purpose of this section is to derive an Edgeworth expansion of the distribution of the maximum likelihood estimator (MLE) of $\theta \in \Theta$, based on the assumption, that the first p derivatives of the logarithm of the likelihood function at θ_0 may be approximated in distribution by an Edgeworth series when θ_0 is the true value of the parameter. The notation used is coordinate-free, but the main results of the paper are summarized in terms of coordinates at the end of each section.

Assumptions 3.1. *Integers $s \geqslant 2$ and $m \in \mathrm{N}$ exist, such that $(S_1, \ldots, S_p)$ has absolute moment of order s, and a sequence of linear mappings $A_n\colon B_1(V, \mathbf{R}) \times \ldots \times B_p(V, \mathbf{R}) \to \mathbf{R}^m$ exists, satisfying*

I. $V\{A_n(S_1, \ldots, S_p)\} = 1_{\mathbf{R}^m}$ (3.1)

II. $P\{(S_1, \ldots, S_p) \in B\} = \int_{A_n(B)} \xi_n(t)\,dt + o(\beta_n)$ (3.2)

uniformly in the system of Borel-subsets of the linear space spanned by $(S_1, \ldots, S_p)$, where

$$\xi_n(t) = \left(\sum_{r=0}^{s-2} P_r(-\phi : \{\chi_\nu\})\right)(t) \tag{3.3}$$

$\chi_\nu = \nu$*'th cumulant of* $A_n(S_1, \ldots, S_p)$, $2 \leqslant \nu \leqslant s$

$$\beta_n = \begin{cases} (\sup\{\|\chi_\nu\|^{1/(\nu-2)} \mid 3 \leqslant \nu \leqslant s\})^{s-2} & \text{if } s \geqslant 3 \\ 1 & \text{if } s = 2 \end{cases} \tag{3.4}$$

Assumptions 3.2. *An $\alpha > 0$ and a sequence (λ_n), $n \in \mathrm{N}$, of positive real numbers exist, such that*

I. $\lambda_n^{p-1} = o(\beta_n)$ (3.5)

II. $\|E_{j+1} \circ (\Sigma_1^{-\frac{1}{2}})^{\prime j+1}\|/j! = O(\lambda_n^{j-1}), \quad 2 \leqslant j \leqslant p-1$ (3.6)

III. $\|\Sigma_j \circ (\Sigma_1^{-\frac{1}{2}})^{\prime 2j}\|^{\frac{1}{2}}/(j-1)! = O(\lambda_n^{j-1}), \quad 2 \leqslant j \leqslant p$ (3.7)

IV. *A sequence (D_n), $n \in \mathrm{N}$ of sets exists, such that $P\{X_n \in D_n\} = 1 - o(\beta_n)$, and uniformly in $x \in D_n$, and $\theta \in \theta_0 + \Sigma_1^{-\frac{1}{2}}(H_n(\alpha))$, $D^p \log f(x;\theta)$ exists, and*

$$\sup\{\|(D^p \log f(x;\theta) - (S_p + E_p)) \circ (\Sigma_1^{-\frac{1}{2}})^{\prime p}\| \mid \theta \in \theta_0 + \Sigma_1^{-\frac{1}{2}}(H_n(\alpha))\} = o(\beta_n) \tag{3.8}$$

where

$$H_n(\alpha) = \{z \in V \mid \|z\| \leqslant \varrho_n(\alpha)\}$$
$$\varrho_n(\alpha) = ((2+\alpha)\log \beta_n^{-1})^{\frac{1}{2}} \tag{3.9}$$

Remark 3.3. Assumptions 3.1 assure that an Edgeworth expansion of the distribution of $(S_1, \ldots, S_p)$ is available, and Assumptions 3.2 that the derivatives of the MLE of θ w.r.t. $(S_1, \ldots, S_p)$ are sufficiently well behaved.

Remark 3.4. Notice, that since Σ_1 is regular, we have

$$\|E_{j+1} \circ (\Sigma_1^{-\frac{1}{2}})^{\prime j+1}\| = \sup\{|E_{j+1}(\Sigma_1^{-\frac{1}{2}}(u)^{\prime j+1})| \mid u \in V\}$$
$$= \sup\{|E_{j+1}(v^{\prime j+1})|/\Sigma_1(v, v)^{(j+1)/2} \mid v \in V\}$$

such that (3.6) and (3.7) are not as hard to prove as it may seem; see the exponential regression example in Section 5.

As in Skovgaard (1981) we define the *formal cumulants* (and formal moments) of polynomials of $(S_1, ..., S_p)$ as the cumulants (and moments) computed in the usual way in terms of the cumulants of $(S_1, ..., S_p)$, except that the cumulants of $(S_1, ..., S_p)$ of order higher than s are defined as zero.

Theorem 3.5. *Suppose Assumptions 2.1, 3.1 and 3.2 hold. Then a sequence $\hat{\theta}_n$ of estimators of θ exists, such that with probability $1-o(\beta_n)$, $\hat{\theta}_n$ is a unique maximum of the likelihood function in the interior of $\theta_0+\Sigma_1^{-\frac{1}{2}}(H_n(\alpha))$, and the following expansion holds*

$$P\{\hat{\theta}_n-\theta_0\in B\}$$

$$=\int_{B-\varkappa_1}\eta_n(t)\,dt+o(\beta_n)\quad \text{uniformly in } B\in\mathcal{B}(V) \tag{3.10}$$

where $B-\varkappa_1=\{t\,|\,t+\varkappa_1\in B\}$ *and*

$$\eta_n(t)=\sum_{r=0}^{q-2}P_r(-\phi_{0,\varkappa_2}:\{\varkappa_\nu\})(t) \tag{3.11}$$

$$q=\max\{p,s\} \tag{3.12}$$

and $\{\varkappa_\nu\}$ are approximate cumulants of the polynomial

$$Y_1+\sum_{j=2}^{p-1}A_j(Y_1,...,Y_j),\quad Y_j=\Sigma_1^{-1}S_j \tag{3.13}$$

where A_j is a homogeneous polynomial of degree j in $(Y_1, ..., Y_j)$ computed as described in Remark 3.6, and $\{\varkappa_\nu\}$, $1\leqslant\nu\leqslant q$ are computed as described in Remark 3.7. A_1 through A_4 are given at the end of this section.

Remark 3.6. Computation of the A's. Consider the Taylor-series expansion of the likelihood equation

$$\Sigma_1^{-1}S_1\approx-\sum_{j=2}^{p}\Sigma_1^{-1}(E_j+S_j)(\hat{\theta}_n-\theta_0)^{j-1}/(j-1)! \tag{3.14}$$

Considering $(S_2, ..., S_p)$ as fixed the derivatives of $(\hat{\theta}_n-\theta_0)$ with respect to S_1 at $S_1=0$ may be expressed in terms of the derivatives of S_1 with respect to $(\hat{\theta}_n-\theta_0)$ at zero. These (former) derivatives are easily derived recursively and it is seen, that they are polynomials in $((1_V-Y_2)^{-1}, Y_3, ..., Y_p)$. Expanding

$$(1_V-Y_2)^{-1}=\sum_{j=0}^{N}(Y_2)^j+o(\|Y_2\|^N)\quad\text{as } \|Y_2\|\to 0 \tag{3.15}$$

we obtain an expansion of $(\hat{\theta}_n-\theta_0)$ as a polynomial in the Y's around $(Y_1, Y_2)=(0, 0)$. In particular the Taylor series expansion of $(\hat{\theta}_n-\theta_0)$ with respect to $(Y_1, ..., Y_p)$ around $(0, ..., 0)$ is obtained as (3.13) by equating $A_j(Y_1, ..., Y_p)$ to the sum of the terms of power j in $(Y_1, ..., Y_p)$. It is easy to see that A_j only depends on $(Y_1, ..., Y_j)$.

Remark 3.7. Computation of the $\varkappa$'s. By the results of Leonov & Shiryaev (1959) and Skovgaard (1981) it follows, that the approximate cumulants $(\varkappa_\nu)$, $1\leqslant\nu\leqslant q$ may be calculated as follows. Recall, that $q=\max\{p,s\}$.

To calculate $\varkappa_\nu$, $1\leqslant\nu\leqslant q$, raise (3.13) to the power ν and consider each term, omitting terms of power greater than $\nu+p-2$ in the Y's, and also of power $\nu+p-2$ if this is odd. For each of the remaining terms compute its mean in terms of the cumulants of the Y's, and omit terms for which

I. The "partition" corresponding to the cumulants is *decomposable*; see Leonov & Shiryaev (1959) or for a short description Brillinger (1975).

II. The number of cumulants entering the term is strictly less than $x-(\nu+q-2)/2$, where x is the degree (in Y) of the term.

$\varkappa_\nu$ is then obtained as the sum of the remaining terms.

Using this method, $\varkappa_\nu$ may be written down almost immediately from (3.13), although the final expression may be rather involved.

Remark 3.8. Notice, that $\hat{\theta}_n$ need not be the maximum likelihood estimator (MLE) of θ; it is only proved that $\hat{\theta}_n$ is a maximum in a neighbourhood of θ_0. To prove that $\hat{\theta}_n$ is the MLE other (nonlocal) techniques must be used, e.g. as in Wald (1949) or Ivanov (1976). If the likelihood equation has a unique solution, then $\hat{\theta}_n$ must obviously coincide with the MLE.

Remark 3.9. The inversion of a power series f, which is locally one-to-one may be obtained recursively by differentiation of f^{-1}, expressing the derivatives in terms of the derivatives of f. An explicit formula in the one-dimensional case is given in Skovgaard (1981). In the multivariate case Bolotov & Yuzhakov (1978) gives an explicit formula in terms of coordinates even for implicit functions, but no coordinate-free version seems to be known.

Corollary 3.10. *Let the assumptions of Theorem 3.5 be fulfilled, and let $g\in C^p(\Theta, W)$ be a fixed function satisfying*

$$Dg(\theta_0)\text{ is non-singular} \tag{3.16}$$

If also $\|\Sigma_1^{-1}\|=o(1)$ then the distribution of $g(\hat{\theta}_n)$ may be expanded in an Edgeworth series of the form (3.10) replacing (3.13) by the stochastic expansion

$$Dg(\theta_0)\,Y_1 + \sum_{j=2}^{p-1} \tilde{A}_j(Y_1, \ldots, Y_j) \tag{3.17}$$

where

$$\tilde{A}_j(Y_1, \ldots, Y_j) = \sum_{\mu \in T(j)} D^{\Sigma\mu_i} g(\theta_0) [Y_1'^{\mu_1}, \ldots, A_j(Y_1, \ldots, Y_j)'^{\mu_j}] \Big/ \prod_{i=1}^{j} \mu_i! \tag{3.18}$$

where $T(j) = \{(\mu_1, \ldots, \mu_j) \in \mathbf{N}_0^j \mid \Sigma i\mu_i = j\}$.

Results in coordinates. Define

$$E^{i_1 \ldots i_m} = E\left\{ \frac{d}{d\theta_{i_1}} \cdots \frac{d}{d\theta_{i_m}} \log f(X;\theta)\Big|_{\theta=\theta_0} \right\} \tag{3.19}$$

$$S^{i_1 \ldots i_m} = \left(\frac{d}{d\theta_{i_1}} \cdots \frac{d}{d\theta_{i_m}} \log f(X;\theta)\Big|_{\theta=\theta_0} \right) - E^{i_1 \ldots i_m} \tag{3.20}$$

where $\theta = (\theta_1, \ldots, \theta_k) \in \mathbf{R}^k$. Also

$$Y_i^{i_2 \ldots i_m} = g_{ij} S^{j i_2 \ldots i_m} \tag{3.21}$$

$$F_i^{i_2 \ldots i_m} = -g_{ij} E^{j i_2 \ldots i_m} \tag{3.22}$$

where (g_{ij}) is the inverse of $(g^{ij}) = (-E^{ij})$, i.e. the inverse Fisher-information.

The first four terms of (3.13)

$$(A_1)_i = Y_i$$

$$(A_2)_i = Y_i^j Y_j - \tfrac{1}{2} F_i^{jk} Y_j Y_k$$

$$(A_3)_i = Y_i^j Y_j^k Y_k - \tfrac{1}{2} Y_i^j F_j^{kl} Y_k Y_l - F_i^{jk} Y_j Y_k^l Y_l + \tfrac{1}{2} Y_i^{jk} Y_j Y_k - \tfrac{1}{6} F_i^{jkl} Y_j Y_k Y_l + \tfrac{1}{2} F_i^{jk} Y_j F_k^{lm} Y_l Y_m$$

$$\begin{aligned}(A_4)_i = {} & Y_i^j Y_j^k Y_k^l Y_l - \tfrac{1}{2} Y_i^j Y_j^k F_k^{lm} Y_l Y_m \\ & - Y_i^j F_j^{kl} Y_k Y_l^m Y_m - \tfrac{1}{2} F_i^{jk} Y_j^l Y_l Y_k^m Y_m \\ & - F_i^{jk} Y_j Y_k^l Y_l^m Y_m + \tfrac{1}{2} Y_i^j Y_j^{kl} Y_k Y_l + Y_i^{jk} Y_j Y_k^l Y_l \\ & - \tfrac{1}{6} Y_i^j F_j^{klm} Y_k Y_l Y_m - \tfrac{1}{2} F_i^{jkl} Y_j Y_k Y_l^m Y_m \\ & + \tfrac{1}{6} Y_i^{jkl} Y_j Y_k Y_l + \tfrac{1}{2} Y_i^j F_j^{kl} Y_k F_l^{mn} Y_m Y_n \\ & + \tfrac{1}{2} F_i^{jk} Y_j^l Y_l F_k^{mn} Y_m Y_n + F_i^{jk} Y_j F_k^{lm} Y_l Y_m^n Y_n \\ & + \tfrac{1}{2} F_i^{jk} Y_j Y_k^l F_l^{mn} Y_m Y_n - \tfrac{1}{2} Y_i^{jk} Y_j F_k^{lm} Y_l Y_m \\ & - \tfrac{1}{2} F_i^{jk} Y_j Y_k^{lm} Y_l Y_m - \tfrac{1}{24} F_i^{jklm} Y_j Y_k Y_l Y_m \\ & + \tfrac{1}{6} F_i^{jk} Y_j F_k^{lmn} Y_l Y_m Y_n + \tfrac{1}{4} F_i^{jkl} Y_j Y_k F_l^{mn} Y_m Y_n \\ & - \tfrac{1}{8} F_i^{jk} Y_j F_k^{lm} Y_l F_m^{no} Y_n Y_o \\ & - \tfrac{1}{8} F_i^{jk} F_j^{lm} Y_l Y_m F_k^{no} Y_n Y_o \end{aligned} \tag{3.23}$$

4. Exponential family models

In this section we consider (for each $n \in \mathbf{N}$) a setup of the form given below. Assume that E is a finite dimensional Euclidean space, $\mathcal{E}$ the Borel σ-field on E and μ a measure on $(E, \mathcal{E})$. Define

$$\psi(\eta) = \log \int \exp\{\langle \eta, x \rangle\}\, d\mu(x), \quad \eta \in H \subseteq E \tag{4.1}$$

where H is the subset of E for which the integral is positive and finite. Define the family (P_η), $\eta \in H$ of probability measures on $(E, \mathcal{E})$ by

$$(dP_\eta/d\mu)(x) = f(x;\eta) = \exp\{\langle \eta, x \rangle - \psi(\eta)\} \tag{4.2}$$

The model we shall consider is given by a differentiable parametrization

$$\eta \in C^p(\Theta, H), \quad \eta = \eta(\theta), \quad \theta \in \Theta \subseteq V \tag{4.3}$$

where V is a finite dimensional Euclidean space independent of n, and usually of lower dimension than E. The cumulants of P_{η_0}, $\eta_0 = \eta(\theta_0)$, are

$$\chi_k = D^k \psi(\eta_0), \quad k \in \mathbf{N} \tag{4.4}$$

Also

$$D \log f(x; \eta(\theta_0)) = \langle x - D\psi(\eta_0), D\eta(\theta_0) \rangle \tag{4.5}$$

and accordingly

$$-E_k = \sum_{\nu \in T'(k)} k!\, \chi_{\Sigma\nu_i} \circ [(D\eta_0)'^{\nu_1}, \ldots, (D^k \eta_0)'^{\nu_k}] \Big/ \prod_{i=1}^{k} \nu_i!\,(i!)^{\nu_i} \tag{4.6}$$

$$S_k = \langle x - \chi_1, D^k \eta_0 \rangle \tag{4.7}$$

$$\Sigma_k = \chi_2 \circ (D^k \eta_0, D^k \eta_0) \tag{4.8}$$

where E_k, S_k and Σ_k are defined in (2.13), (2.14) and (2.15), $D^k\eta_0 = D^k\eta(\theta_0)$ and $T'(k) = T(k) \setminus \{(0, \ldots, 0, 1)$.

Thus the approximate cumulants in Theorem 3.5 may be expressed explicitely in terms of (χ_k), $(D^k\eta_0)$, $k \geq 1$. Some of these cumulants are given below in a coordinate version. The expressions may be somewhat simplified using a coordinate-free notation, but for computations this is not useful. Recall, that for fixed p in (3.13), only the first p cumulants are needed.

Remark 4.1. There are a number of situations, where the expression (3.13) and its cumulants are considerably simpler. These include

(a) *A canonical model*, i.e. η is affine. Then

$$-E_k = \chi_k \circ (D\eta^{\prime k}), \quad k \geq 2$$

$$S_k = 0, \quad \Sigma_k = 0, \quad k \geq 2 \tag{4.9}$$

(b) *An affine mean value structure*, i.e. $(D\psi)\circ\eta$ is affine. Then

$$-E_k = k\chi_2 \circ (D\eta_0, D^{k-1}\eta_0), \quad k \geq 3 \tag{4.10}$$

where E_k is understood to be symmetric.

(c) *The normal case* (with fixed variance), where

$$\chi_k = 0, \quad k \geq 3.$$

The normal regression models will be discussed further in the next section.

If, in particular, both (a) and (b) are fulfilled, then the MLE is an affine function of the minimal sufficient statistic S_1, and the transformation of an Edgeworth expansion of S_1 to an Edgeworth expansion of the MLE is trivially valid.

Results in cordinates. Using the method described in Remark 3.7 it is straight forward to calculate the approximate cumulants $(\varkappa_j)$ of (3.23). Let $(\varkappa_m)_{i_1 \dots i}$ denote the jth cumulant of the mth approximation, i.e. with $q = m+1$ in (3.11). Thus $(\varkappa_1)_i$ and $(\varkappa_1)_{ij}$ denote mean and variance in the first (normal) approximation. With obvious modification of the notation in Section 4 we define

$$[i, j] = (\chi_2)^{\alpha\beta}(D\eta_0)^i_\alpha (D\eta_0)^j_\beta = g^{ij}$$

$$[i, j, k] = \chi_3{}^{\alpha\beta\gamma}(D\eta_0)^i_\alpha (D\eta_0)^j_\beta (D\eta_0)^k_\gamma$$

$$[i, jk] = \chi_2^{\alpha\beta}(D\eta_0)^i_\alpha (D^2\eta_0)^{jk}_\beta \quad \text{etc.}$$

where

$$(D^m\eta_0)^{i_1 \dots i_m}_\alpha = \frac{d}{d\theta_{i_1}} \cdots \frac{d}{d\theta_{i_m}} \eta(\theta)_\alpha \bigg|_{\theta=\theta_0}.$$

Then we have

$m = 1$:

$$(\varkappa_1)_i = 0, \quad (\varkappa_1)_{ij} = g_{ij}$$

$m = 2$:

$$(\varkappa_2)_i = -\tfrac{1}{2} g_{ij}([j, kl] + [j, k, l])\, g_{kl}$$

$$(\varkappa_2)_{ij} = (\varkappa_1)_{ij} = g_{ij}$$

$$(\varkappa_2)_{ijk} = -\text{sym}\,\{g_{il}\, g_{jm}\, g_{kn}(2[l, m, n] + 3[l, mn])\}$$

$m = 3$:

$$(\varkappa_3)_i = (\varkappa_2)_i$$

$$\begin{aligned}
(\varkappa_3)_{ij} = {} & g_{ij} + g_{ik}\, g_{jl}([km, nl] - [kl, mn] \\
& -[k, m, n, l])g_{mn} + \text{sym}\,\{g_{ik}\, g_{jl}\, g_{mn} \\
& (-[k, lmn] - [k, l, mn] - [m, n, kl] \\
& -2[k, m, nl]) + g_{ik}\, g_{jl}\, g_{mn} g_{op}([k,l,m][n,o,p] \\
& +\tfrac{3}{2}[k, m, o][l, n, p] + [k, l, m][n, op] \\
& +[k, lm][n, o, p] + [m, kl][n, o, p] \\
& +2[k, m, o][l, np] + 2[k, m, o][n, lp] \\
& +2[k, mo][n, lp] - [m, ko][n, lp] \\
& +\tfrac{1}{2}[k, mo][l, np] + [m, kl][n, op] \\
& +[k, lm][n, op])\}
\end{aligned}$$

$$(\varkappa_3)_{ijk} = (\varkappa_2)_{ijk}$$

$$\begin{aligned}
(\varkappa_4)_{ijkl} = {} & \text{sym}\,\{g_{i\alpha} g_{j\beta} g_{k\gamma} g_{l\delta}(-3[\alpha, \beta, \gamma, \delta] - 4[\alpha, \beta\gamma\delta] \\
& -12[\alpha, \beta, \gamma\delta] + 12[\alpha, \beta, m]g_{mn}[n, \gamma, \delta] \\
& 12[\alpha, \beta, m]g_{mn}[n, \gamma\delta] \\
& +24[\alpha, \beta, m]g_{mn}[\gamma, \delta n] \\
& +12[\alpha, \beta m]g_{mn}[n, \gamma\delta] \\
& +12[\alpha, \beta m]g_{mn}[\gamma, \delta n])\}.
\end{aligned} \tag{4.11}$$

where sym {...} means the average over all permutations of the indices appearing on the left hand side on the equation. Actually taking this average is not necessary in applications, because the appearance of the cumulants in (3.11) is symmetric in their indices. This fact is a considerable relief in calculations.

The variance term for $m = 3$ may be identified with that given in Efron (1975) in the one-dimensional case and with its multivariate generalization in L. T. Madsen (1979). That our formula seems more complicated is only because of the less directly computable terms appearing in the above mentioned papers. All the terms, except $\varkappa_4$, may be found in Shenton & Bowman (1977). Notice, however that their square brackets have a meaning different from ours.

An interesting feature of the correction terms for $m = 2$ (i.e. the first correction to the normal distribution) is, that since $\gamma_{il}\gamma_{jm}\gamma_{kn}\,[l, m, n]$ is invariant under reparametrizations in the one-dimensional case, and in the multivariate case its range is invariant, then the first correction term of (3.11) cannot be removed by a reparametrization, unless this invariant vanishes, e.g. if the third cumulant of the exponential family is zero.

5. Normal non-linear regression

Consider a sequence $X_1, X_2, \dots$ of independent random vectors, X_i normally distributed on $\mathbf{R}^m$ with mean $\mu_i(\theta)\in \mathbf{R}^m$ and variance $\Sigma=\sigma^2\Sigma_0$, $\sigma^2>0$, $\Sigma_0\in B_2(\mathbf{R}^m,\mathbf{R})$. Σ_0 is supposed to be known, $\theta\in V$ unknown. Whether σ^2 is known or unknown is immaterial, when considering maximum likelihood estimation of θ. We shall consider σ^2 as known for simplicity. With notation as in the previous sections, we have

$$\log f(x;\theta)=\text{const}-\tfrac{1}{2}\sum_{i=1}^{n}\Sigma^{-1}(x_i-\mu_i(\theta),x_i-\mu_i(\theta)) \tag{5.1}$$

from which we derive

$$-E_k=\tfrac{1}{2}\sum_{i=1}^{n}\sum_{j=1}^{k-1}\binom{k}{j}\Sigma^{-1}\circ(D^j\mu_i(\theta_0),D^j u_i(\theta_0)),\quad k\geq 2 \tag{5.2}$$

$$S_k=\sum_{i=1}^{n}\Sigma^{-1}(X_i-\mu_i(\theta_0),D^k\mu_i(\theta_0)),\quad k\geq 1 \tag{5.3}$$

$$\Sigma_k=\sum_{i=1}^{n}\Sigma^{-1}\circ(D^k\mu_i(\theta_0),D^k\mu_i(\theta_0)),\quad k\geq 1 \tag{5.4}$$

Since this class of models is widely used, we shall in somewhat more details investigate under which conditions Assumptions 3.2 are fulfilled. Notice, that Assumptions 3.1 are fulfilled with $\beta_n=0$, because $(S_1,\dots,S_p)$ are exactly normally distributed. Of Assumptions 2.1 only IV needs to be checked.

Lemma 5.1. *Let* (E_k) *and* (S_k) *be given by* (5.2) *and* (5.3). *Then* (3.5) *and* (3.7) *implies* (3.6).

Notice, that since $(S_1,\dots,S_p)$ is exactly normally distributed, the sequence (λ_n) may be chosen as any sequence, which is $o(1)$. The next lemma shows that also (3.8) may be deduced under simple conditions.

Lemma 5.2. *Suppose, that the functions* (μ_i) *are analytic in a neighbourhood of* θ_0, *and that* (3.7) *holds uniformly in* $j\geq 2$, *then Assumptions 3.2 hold with* $\beta_n=\lambda_n^{p-2}$.

Remark 5.3. The conditions of Lemma 5.2 may be stated in the following form. The functions (μ_i) are analytic, and (by Remark 3.4)

$$\left|\sum_{i=1}^{n}\Sigma^{-1}(D^j\mu_i(\theta_0)(v^j)^2)\right|^{\frac{1}{2}}\Bigg/\left(\sum_{i=1}^{n}\Sigma^{-1}(D\mu_i(\theta_0)(v)^2)\right)^{j/2}$$
$$=O(\lambda_n^{j-1})\ \text{uniformly in } v\in V \text{ and } j\geq 2 \tag{5.5}$$

Remark 5.4. Another interesting case, closely connected with the one discussed above, occurs if, in the non-linear regression models described above, we fix n, and consider the limiting behaviour as $\sigma^2\to 0$. It is quite trivial to check, that Assumptions 3.2 are fulfilled with $\lambda=\sigma$, and hence that the conclusion of Theorem 3.5 holds. This proves that the asymptotic results may be applied if the variance is small, *even if the nubber of observations is small.*

An example: exponential regression

Let $X_1,\dots,X_n$ be independent, $x_i\in\mathbf{R}$ normally distributed with mean

$$\mu_i(\theta_1,\theta_2)=\theta_1 e^{\theta_2 t_i},\quad \theta_1>0,\ \theta_2\in\mathbf{R},\ t_i\in\mathbf{R} \tag{5.6}$$

and variance $\sigma^2>0$. The conditions of Lemma 5.2 are verified as follows. First note that the functions (μ_i) and hence the likelihood functions are analytic. Let $\eta=(\eta_1,\eta_2)\in\mathbf{R}^2$, $\|\eta\|=1$. Then for any $\theta=(\theta_1,\theta_2)$

$$\left[\sum_{i=1}^{n}(D^j\mu_i(\theta)(\eta^j))^2/\sigma^2\right]\Bigg/\left[\sum_{i=1}^{n}(D\mu_i(\theta)(\eta))^2/\sigma^2\right]^j$$
$$=\sum_{i=1}^{n}(\eta_1\eta_2^{j-1}t_i^{j-1}e^{\theta_2 t_i}+\eta_2^j\theta_1 t_i^j e^{\theta_2 t_i})^2\Bigg/\left(\left[\sum_{i=1}^{n}(\eta_1 e^{\theta_2 t_i}+\eta_2\theta_1 t_i)^2\right]^j/(\sigma^2)^{j-1}\right)$$
$$\leq(\max\{t_i\mid i=1,\dots,n\})^{2(j-1)}/\Sigma_1(\eta,\eta)^{j-1} \tag{5.7}$$

where $\Sigma_1(\eta,\eta)=\sum_{i=1}^{n}(D\mu_i(\theta)(\eta))^2/\sigma^2$ is the Fisher-information of η. Thus if

$$\lambda_n=(\max\{t_i\mid i=1,\dots,n\})\|\Sigma_1^{-\frac{1}{2}}\|=o(1)$$

then Theorem 3.5 is applicable. E.g. if $\theta_2>0$ and $t_i=i$, then λ_n will decrease exponentially fast. Thus, if one has observations at equidistant points (t's) a very good agreement between the correct distribution and the approximations may be expected with relatively few points, but, of course, this can only be proved by estimating the difference.

Although this example is of practical interest in itself, it is unusually simple. The condition (5.5) is however so simple, that further simplification of importance is hardly obtainable.

Results in coordinates. In the normal regression model as discussed in this section the cumulants (4.11) are still valid, but important simplifications are achieved, because only the square bracket factors of the form $[i_1\dots i_2,j_1\dots j_2]$ are different from zero. Thus $[i,j,k]$, $[i,j,kl]$ etc. vanish. In the second-order expansion, i.e. $m=2$, we get (cf. (4.11))

$$(\varkappa_2)_i=-\tfrac{1}{2}g_{ij}[j,kl]g_{kl}$$

$$(\varkappa_2)_{ij}=g_{ij}$$

$$(\varkappa_2)_{ijk}=-3\,\text{sym}\{g_{il}g_{jm}g_{kn}[l,mn]\}.$$

Here we have $(g_{ij})=\Sigma^{-1}$ (cf. (5.4)) and

$$[j,kl]=\sum_{i=1}^{n}\Sigma^{-1}\left(\frac{d}{d\theta_j}\mu_i(\theta_0),\ \frac{d^2}{d\theta_k\,d\theta_l}\mu_i(\theta_0)\right).$$

The case $m=3$ is obtained from (4.11) quite similarly.

6. Proofs

Proof of Theorem 3.5. The likelihood equation

$$D\log f(x;\theta)=0 \tag{6.1}$$

may be expanded around $\theta=\theta_0$ yielding

$$S_1+\sum_{j=2}^{p}(E_j+S_j)(\theta-\theta_0)^{\prime j-1}/(j-1)!+R_1(\theta-\theta_0)=0 \tag{6.2}$$

where $R_1(\theta-\theta_0)$ is stochastic. Write

$$Z=\Sigma_1^{\frac{1}{2}}(\theta-\theta_0),\quad U_1=\Sigma_1^{-\frac{1}{2}}S_1,\quad U_2=B(S_1,\ldots,S_p)$$

where $B: B_1(V,\mathbf{R})\times\ldots\times B_p(V,\mathbf{R})\to V_2$ is a linear mapping into a Euclidean space V_2, and (U_1,U_2) is a normalization of $(S_1,\ldots,S_p)$, i.e. $\dim V+\dim V_2$ equals the dimension of the support of $(S_1,\ldots,S_p)$ and the variance of (U_1,U_2) is the identity on $V\times V_2$. Define

$$g: V\times V\times V_2\to V$$

$$g(z,u_1,u_2)=u_1+\sum_{j=2}^{p}\Sigma_1^{-\frac{1}{2}}(E_j+S_j(u_1,u_2))\times(\Sigma_1^{-\frac{1}{2}}(z)^{\prime j-1})/(j-1)!+R_2(z) \tag{6.3}$$

where $R_2(z)=\Sigma_1^{-\frac{1}{2}}R_1(\Sigma_1^{-\frac{1}{2}}(z))$ and $(S_1,\ldots,S_p)(u_1,u_2)$ is the solution of $(U_1,U_2)(S_1,\ldots,S_p)=(u_1,u_2)$ belonging to the affine support of $(S_1,\ldots,S_p)$. Thus (6.2) may be written

$$g(Z,U_1,U_2)=0 \tag{6.4}$$

Using Assumptions 3.2 (and 2.1) we obtain

$$g(0,0,0)=0,\quad Dg(0,0,0)(z,u_1,u_2)=u_1-z$$

$$\begin{aligned}\|D^kg(0,0,0)(z,u_1,u_2)^{\prime k}\|&\\ \leqslant\|\Sigma_1^{-\frac{1}{2}}E_{k+1}(\Sigma_1^{-\frac{1}{2}}(z)^{\prime k})\|&\\ +k\|\Sigma_1^{-\frac{1}{2}}S_k(u_1,u_2)(\Sigma_1^{-\frac{1}{2}}(z)^{\prime k-1})\|&\\ \leqslant\|E_{k+1}(\Sigma_1^{-\frac{1}{2}})^{\prime k+1}\|\,\|z\|^k&\\ +k\|S_k(u_1,u_2)\circ(\Sigma_1^{-\frac{1}{2}})^{\prime k}\|\,\|z\|^{k-1}&\\ =O(\lambda_n^{k-1})\ \text{ if }\ \|z\|\leqslant 1,\ \|(u_1,u_2)\|\leqslant 1,\ k\leqslant p&\end{aligned} \tag{6.5}$$

Here we have used the fact that, since the variance of (U_1,U_2) is the identity, then the differential, D_k say, of $S_k(u_1,u_2)\circ(\Sigma_1^{-\frac{1}{2}})^{\prime k}$ with respect to (u_1,u_2) satisfies

$$\begin{aligned}\|D_k\|^2=\|D_k\circ D_k^*\|&=\|V\{S_k(U_1,U_2)\circ(\Sigma_1^{-\frac{1}{2}})^{\prime k}\}\|\\ &=\|\Sigma_k\circ(\Sigma_1^{-\frac{1}{2}})^{\prime 2k}\|\end{aligned} \tag{6.6}$$

Using (6.5) and (3.8) in (6.3) we obtain

$$\begin{aligned}g(z_1,u_1,u_2)-g(z_2,u_1,u_2)&\\ =-(z_1-z_2)+o(\sqrt{\lambda_n})\,\|z_1-z_2\|&\\ \text{if}\quad z_1,z_2\in H_n(\alpha),\ \|(u_1,u_2)\|\leqslant\varrho_n(\alpha),&\\ X\in D_n&\end{aligned} \tag{6.7}$$

because $\varrho_n^m(\alpha)\lambda_n=o(1)$ for any $\alpha>0$, $m>0$. Thus for any fixed (u_1,u_2), $(\|(u_1,u_2)\|\leqslant\varrho_n(\alpha))$ and n sufficiently large there is with probability $1-o(\beta_n)$ at most one solution $z\in H_n(\alpha_1)$ to the likelihood equation, $\alpha_1<\alpha$.

Let $\delta>0$ be fixed. Then if $\|(u_1,u_2)\|\leqslant\varrho_n(\alpha_1)$, $\alpha_1<\alpha$, $\|z-u_1\|<\delta$ and n is sufficiently large we have $z\in H_n(\alpha)$. To prove the existence of a solution $z\in H_n(\alpha)$ of (6.4) we apply *Brauer's fixpoint theorem* to the function

$$\tilde g(y)=g(u_1+y,u_1,u_2)+y,\quad \|y\|<\varepsilon,\ 0<\varepsilon<\delta.$$

By the remark above and (6.7) we have

$$\|y\|<\varepsilon\Rightarrow\|\tilde g(y)\|<\varepsilon$$

because $g(u_1,u_1,u_2)=u_1+o(1)$. Also, by (6.7)

$$\|\tilde g(y_1)-\tilde g(y_2)\|=o(1)\,\|y_1-y_2\|,\quad \|y_1\|,\ \|y_2\|<\varepsilon$$

proving, that $\tilde g$ has a fixpoint in $\{\|y\|<\varepsilon\}$, implying the existence of a solution $z(=y+u_1)\in H_n(\alpha)$ to the likelihood equation when $X\in D_n$ and n is sufficiently large.

By the uniqueness of power series expansions, (3.13) must be the $p-1$ order Taylor series expansion of $\hat\theta_n=\Sigma_1^{-\frac{1}{2}}Z_n$ in terms of $(S_1,\ldots,S_{p-1})$ around zero, where Z_n is the solution of (6.4) and hence $\hat\theta_n$ a solution of the likelihood equation (6.1). Thus, it only remains to be proved, that the derivatives of Z_n with respect to (U_1,U_2) satisfies Assumptions 3.1 of Skovgaard (1981), since the expansion (4.5) of Skovgaard (1981) then implies that $\hat\theta_n$ locally maximizes the likelihood function.

Write $u=(u_1,u_2)$ and let $z=\psi(u)$ be the solution (in $H_n(\alpha)$) of the equation (6.4). Also, if $\omega(u)=(\psi(u),u)$

$$D^k(g\circ\omega)(u_0)=0,\quad k\geqslant 0,\ \|u_0\|\leqslant\varrho_n(\alpha_1) \tag{6.8}$$

and using a general formula (see Federer (1969), 3.1.11)

$$D^k(g\circ\omega)(u_0)(u'^k)$$
$$=\sum_{\mu\in T(k)} k!\, D^{\Sigma\mu_i}g(z_0,u_0)$$
$$[D\omega(u_0)(u)'^{\mu_1},\dots,D^k\omega(u_0)(u'^k)'^{\mu_k}]$$
$$\Big/\prod_{i=1}^{k}\mu_i!\,(i!)^{\mu_i},\quad z_0=\psi(u_0) \tag{6.9}$$

where $T(k)$ is given in Corollary 3.10. From (6.9) and (6.5) we obtain by induction

$$D\psi(0)(u)=u_1$$
$$\|D^k\psi(0)\|\leqslant\sum_{\mu\in T'(k)}k!\,\|D^{\Sigma\mu_i}g(0,0)\|$$
$$\times\prod_{i=1}^{k}(\|D^i\omega(0)\|/i!)^{\mu_i}/\mu_i!$$
$$=\sum_{\mu\in T'(k)}O(\lambda_n^{\Sigma\mu_i-1})\prod_{i=1}^{k}O(\lambda_n^{i-1})$$
$$=0(\lambda_n^{k-1}),\quad 2\leqslant k\leqslant p-1 \tag{6.10}$$

where $T'(k))=T(k)\setminus\{(0,\dots,0,1)\}$.

Using (3.8) and (6.3) it is seen that, if $\|u\|$, $\|z\|<\varrho_n(\alpha)$, then $\|D^kg(z,u)\|=O(\lambda_n^{k-1})$, $k\leqslant p-1$ and $\|D^pg(z,u)\|=o(\beta_n)$ if $X\in D_n$, and as above it follows, that

$$\|D^p\psi(u)\|=o(\beta_n)\quad\text{uniformly in }\{\|u\|\leqslant\varrho_n(\alpha)\} \tag{6.11}$$

By (6.10) and (6.11), Assumptions 3.1 in Skovgaard (1981), and hence our Theorem 3.5, is proved. □

Proof of Corollary 3.10. The formula (3.18) is easily obtained using the formula for derivatives of composite functions, see Federer (1969), 3.1.11. By this formula and the assumption $\|\Sigma_1^{-1}\|=o(1)$, which says that the eigenvalues of the Fisher-information tends uniformly to infinity, it follows, that the assumptions of Theorem 3.2 in Skovgaard (1981) are fulfilled, proving the corollary. □

Proof of Lemma 5.1. We shall prove, that

$$|E_{k+1}(v'^{k+1})|=(\Sigma_1(v,v))^{(k+1)/2}O(\lambda_n^{k-1}),$$
$$2\leqslant k\leqslant p-1,\ v\in V \tag{6.12}$$

By (5.2) and Cauchy-Schwarz inequality we have

$$|E_{k+1}(v'^{k+1})|$$
$$=\tfrac{1}{2}\sum_{j=1}^{k}\binom{k+1}{j}\sum_{i=1}^{n}$$
$$\Sigma^{-1}(D^j\mu_i(\theta_0)(v'^j),D^{k+1-j}\mu_i(\theta_0)(v'^{k+1-j}))$$
$$\leqslant\tfrac{1}{2}\sum_{j=1}^{k}\binom{k+1}{j}\left[\sum_{i=1}^{n}\Sigma_j(v'^{2j})\right]^{\frac{1}{2}}$$
$$\times[\sum_{i=1}\Sigma_{k+1-j}(v'^{2k+2-2j})]^{\frac{1}{2}}$$
$$=\tfrac{1}{2}\sum_{j=1}^{k}\binom{k+1}{j}\Sigma_1(v,v)^{(k+1)/2}O(\lambda_n^{j-1})\,O(\lambda_n^{k-j})$$
$$=(\Sigma_1(v,v))^{(k+1)/2}O(\lambda_n^{k-1}).\quad\square$$

Proof of Lemma 5.2. For sufficiently large n, $\log f(x;\theta)$ will coincide with its Taylor series expansion around $\theta=\theta_0$, when $\|\theta-\theta_0\|$ is less than the radius of convergence. Hence

$$D^p\log f(x;\theta)\circ(\Sigma_1^{-\frac{1}{2}})'^p$$
$$=D^p\log f(x;\theta_0)\circ(\Sigma_1^{-\frac{1}{2}})'^p$$
$$+\sum_{j=p+1}^{\infty}D^j\log f(x;\theta_0)((\Sigma_1^{-\frac{1}{2}})'^p,(\theta-\theta_0)'^{j-p})/(j-p)! \tag{6.13}$$

on the set

$$M_c=\{\theta\in V\mid\|\Sigma_1^{\frac{1}{2}}(\theta-\theta_0)\|^{j-p}\|D^j\log f(x;\theta_0)$$
$$\circ(\Sigma^{-\frac{1}{2}})'^j\|/(j-p)!<c^{j-p}\} \tag{6.14}$$

for any $c\in\,]0,1[$. Rewriting (6.13) we obtain

$$D^p\log f(x;\theta)\circ(\Sigma_1^{-\frac{1}{2}})'^p$$
$$=(E_p+S_p)\circ(\Sigma_1^{-\frac{1}{2}})'^p$$
$$+\sum_{j=p+1}^{\infty}(E_j\circ(\Sigma_1^{-\frac{1}{2}})'^j)(z'^{j-p})/(j-p)!$$
$$+\sum_{j=p+1}^{\infty}(S_j\circ(\Sigma_1^{-\frac{1}{2}})'^j)(z'^{j-p})/(j-p)!,\quad\theta\in M_c \tag{6.15}$$

where $z=\Sigma_1^{\frac{1}{2}}(\theta-\theta_0)$.

By a slight modification of the proof of Lemma 5.1 it follows, that (3.6) holds uniformly in $j\geqslant 2$, hence the first sum in (6.14) is $O(\lambda_n^{p-1})$ if $z\in H_n(\alpha)$.

The next step is to obtain bounds on (S_j), $j\geqslant p$ holding with probability $1-o(\beta_n)=1-o(\lambda_n^{p-2})$. Let d be the dimension of V. Then $D^j\mu(\theta_0)\in B_j(V,\mathbf{R}^{mn})$, $\mu(\theta)=(\mu_1(\theta),\dots,\mu_n(\theta))$, spans a d^j-dimensional subspace, L_j say, of $\mathbf{R}^{mn}$. Let p_j denote the projection on L_j w.r.t. the metric Δ on $\mathbf{R}^{mn}$ induced by the metric Σ^{-1} on each component $\mathbf{R}^m$. Then, using (5.3),

$$S_j=\Delta(p_j(\mathbf{X}-\mu(\theta_0)),D^j\mu(\theta_0)),\quad\mathbf{X}=(X_1,\dots,X_n) \tag{6.16}$$

and

$$\begin{aligned}\|S_j \circ (\Sigma_1^{-\frac{1}{2}})'^j\| &= \|\Delta(p_j(\mathbf{X}-\mu(\theta_0)), D^j\mu(\theta_0)\circ(\Sigma_1^{-\frac{1}{2}})'^j)\| \\ &\leq \Delta((p_j(\mathbf{X}-\mu(\theta_0)'^2)^{\frac{1}{2}} \|\Delta(D^j\mu(\theta_0) \\ &\quad \circ(\Sigma_1^{-\frac{1}{2}})'^j)'^2)\|^{\frac{1}{2}} \\ &= \Delta((p_j(\mathbf{X}-\mu(\theta_0)))'^2)^{\frac{1}{2}} \|\Sigma_j \circ (\Sigma_1^{-\frac{1}{2}})'^{2j}\| \end{aligned} \tag{6.17}$$

The first factor is the Δ-norm of a d^j-dimensional normally distributed random vector with mean zero and variance Δ^{-1} restricted to L_j. Thus by Lemma 4.1 in Skovgaard (1981) we have for any $K_j > 0$

$$\begin{aligned}&P\{\Delta((p_j(\mathbf{X}-\mu(\theta_0)))'^2)^{\frac{1}{2}} > K_j\} \\ &\quad \leq \exp(-\tfrac{1}{2}K_j^2)(K^{d^j-2}/\Gamma(d^j/2) + \sqrt{2}^{d^j-2})\end{aligned} \tag{6.18}$$

Chosing $K_j = K\sqrt{\lambda_n}^{-(j-p)}$, $K > 0$, we obtain

$$\begin{aligned}&P \bigcup_{j=p+1}^{\infty} \{\Delta((p_j(\mathbf{X}-\mu(\theta_0)))'^2) > K_j\} \\ &\quad \geq \sum_{j=p+1}^{\infty} \exp\{-\tfrac{1}{2}K^2 \sqrt{\lambda_n}^{p-j}\}(K_j^{d^j-2}/\Gamma(d^j/2) + \sqrt{2}^{d^j})\end{aligned} \tag{6.19}$$

which decresses towards zero at exponential rate in $\sqrt{\lambda_n}^{-1}$. Combining this with (6.17), we have

$$\|S_j \circ (\Sigma_1^{-\frac{1}{2}})'^j\| = K\sqrt{\lambda_n}^{p-j} O(\lambda_n^{j-1})$$

with probability $1 - o(\lambda_n^{p-2})$, implying that the second sum in (6.15) is $O(\lambda_n^{p-\frac{1}{2}}) = o(\beta_n)$ with probability $1 - o(\beta_n)$, $\beta_n = \lambda_n^{p-2}$, on the set $\theta_0 + \Sigma_1^{-\frac{1}{2}} H_n(\alpha)$. □

Acknowledgement

I wish to thank Steffen L. Lauritzen for useful comments and discussions on the subject. Also, I wish to thank a referee for helpful suggestions.

This work was supported by the Danish Natural Science Research Council..

References

Barndorff-Nielsen, O. & Cox, D. R. (1979). Edgeworth and saddlepoint approximations with statistical applications. *J. R. Statist. Soc. B* **41**, 279–312.

Bhattacharya, R. N. & Ghosh, J. K. (1978). On the validity of the formal Edgeworth expansion. *Ann. Statist.* **6**, 434–451.

Bolotov, V. A. & Yuzhakov, A. P. (1978). A generalization of the inversion formulas of systems of power series in systems of implicit functions. (Russian). *Mat. Zametki* **23**, 47–54.

Brillinger, D. R. (1975). *Time series: data analysis and theory.* Holt, Rinehart and Winston, New York.

Daniels, H. E. (1954). Saddlepoint approximations in statistics. *Ann. Math. Statist.* **25**, 631–650.

Efron, B. (1975). Defining the curvature of a statistical problem (with applications to second order efficiency). *Ann. Statist.* **3**, 1189–1242.

Eisenhart, L. P. (1960). *Riemann geometry.* 2nd ed. Princeton Univ. Press, Princeton.

Federer, H. (1969). *Geometric measure theory.* Springer, New York.

Greub, W. H. (1967). *Multilinear algebra.* Springer, Berlin.

Ivanov, A. V. (1976). An asymptotic expansion for the distribution of the least squares estimator of the non-linear regression parameter. *Theor. Probability Appl.* **21**, 557–570.

Leonov, V. P. & Shiryaev, A. N. (1959). On a method of calculation of semi-invariants. *Theor. Probability Appl.* **4**, 319–329.

Madsen, L. T. (1979). The geometry of statistical models. A generalization of curvature. Research report 79/1, Statistical Research Unit, Copenhagen.

Shenton, L. R. & Bowman, K. O. (1977). *Maximum likelihood estimation in small samples.* Griffin, London.

Skovgaard, I. M. (1981). Transformation of an Edgeworth expansion by a sequence of smooth functions. *Scand. J. Statist.* **8**, 207–217.

Wald, A. (1949). Note on the consistency of the maximum likelihood estimate. *Ann. Math. Stat.* **20**, 595–601.

Ib Skovgaard
Institute of Math. Statist.
Royal Vet. & Agr. Univ.
Thorvaldsensvej 40
DK-1871 Copenhagen V
Denmark

Chapter 2

A second-order investigation of asymptotic ancillarity

Introduction by Peter McCullagh

University of Chicago

2.1 Background

In his 1925 paper, Fisher coined the term ancillary statistic, by which he meant any statistic to be used in conjunction with the estimator for inferential purposes. This is to be contrasted with his 1934 paper on location-scale problems, where the configuration statistic is used as an illustration of the role of an ancillary statistic for purposes of setting fiducial or confidence limits. In that setting, the ancillary statistic, being group-invariant, automatically has a fixed distribution independent of the parameter.

The mathematical difficulties surrounding ancillary statistics are of a fundamental nature. Two statistics or events that are ancillary in a model need not be jointly ancillary, so the set of ancillary events is not a σ-field. It is most unlikely that Fisher paid any attention to σ-fields, but by the mid 1930s, he was well aware of some of the difficulties. In his 1936 tercentenary address to the American Academy of Arts and Sciences, he presented an optimistic assessment of the state of mathematical statistics at the time, and posed the celebrated problem of the Nile as a mathematical challenge stripped of its statistical origins.

The agricultural land of a pre-dynastic Egyptian village is of unequal fertility. Given the height to which the Nile will rise, the fertility of every portion of it is known with exactitude, but the height of the flood affects different parts of the territory unequally. It is required to divide the area, between the several households of the village, so that the yields of the lots assigned to cach shall be in pre-determined proportion, whatever may be the height to which the river rises.

He continued, *If this problem is capable of a general solution, then it is possible in general to recognize something corresponding with the configuration of the sample ... and one of the primary problems of uncertain inference will have reached its complete solution.* This uncharacteristic and over-optimistic tone may be excusable in view of Fisher's audience and his goal of stimulating mathematical work on the topic.

The problem of the Nile is a generalized sandwich problem, one version of which was stated and proved by Stone and Tukey (1942), another more general and more relevant version by Halmos (1948). Although it was clearly Fisher's intention to stimulate work on this topic, it cannot be said that he was successful, because none of these authors shows any awareness of the statistical implications or of Fisher's challenge. Given Halmos's inscrutable title, *On the range of a vector measure*, it is likely that Fisher never became aware of the paper or its implications for inference.

The role of ancillary statistics for inferential purposes was later examined by Cox (1958), who pointed out that unconditional Neyman-Pearson sample-space calculations could be misleading if used for inferential purposes related to parametric inference. Maximization of power for significance tests may lead to procedures that are misleading. As Fisher (1956) had pointed out, there are usually recognizable subsets for which the error rates are consistently higher than the quoted significance level, and others for which the error rate is consistently lower. Given the data at hand, the relevance of the long-run frequency for inferential purposes is not apparent.

Efron (1975) stimulated a long line of work on connections between the geometry of a statistical model and its role in frequency-based inferential procedures. Efron and Hinkley (1978) may be seen as a partial, or asymptotic, answer to some of the questions originally encountered by Fisher. Together, these papers stimulated at least two decades of work on asymptotic distribution theory, and its use for conditional inference, the present paper being a highlight of that era.

Ib's paper focuses on statistics generated by the local behaviour of the log likelihood function in the vicinity of its maximum. He points out that standard first-order asymptotics is inadequate to discriminate between conditional and unconditional likelihood-based procedures, so a second-order theory is needed. In this respect, the paper follows on naturally from Efron–Hinkley plus related developments by Cox (1980) and Amari (1982).

On the mathematical front, Ib uses a multivariate Edgeworth expansion, which is justified on the grounds that the log likelihood derivative for the full sample is a sum of independent contributions. The asymptotic joint distribution is naturally expressed in terms of multivariate cumulants, which are the Hermite coefficients in the Edgeworth expansion. All of this is carefully done in an elegant way, and the conclusions mesh well with previously known one-dimensional results related to local curvature of the model manifold.

Although a full appreciation of the technique demands a certain mathematical fortitude to cope with multi-dimensional arrays, multi-linear forms and tensor contractions, the conclusions reached are succinct and easily comprehended. First, observed Fisher information is more relevant than expected information for inferential purposes. Second, the information lost in the reduction to the maximum-likelihood estimate, which is closely related to Le Cam's deficiency, is recovered by conditioning on the approximate ancillary. Finally, the second-order analysis is sufficient to discriminate between three first-order equivalent test statistics, showing that the likelihood-ratio statistic and the conditionally-normalized Wald statistic, using observed Fisher information, are second-order independent of the ancillary, whereas the conventional Wald statistic is not.

The Annals of Statistics
1985, Vol. 13, No. 2, 534–551

A SECOND-ORDER INVESTIGATION OF ASYMPTOTIC ANCILLARITY

BY IB M. SKOVGAARD

Royal Veterinary and Agricultural University, Copenhagen

The paper deals with approximate ancillarity as discussed by Efron and Hinkley (1978). In the multivariate i.i.d. case we derive the second-order Edgeworth expansion of the MLE given a normalized version of the second derivative of the log-likelihood at its maximum. The expansion agrees with the one derived by Amari (1982a) for curved exponential families, but holds for any family satisfying the regularity conditions given in the paper. It is shown that the Fisher information lost by reducing the data to the MLE is recovered by the conditioning, and it is sketched how the loss of information relates to the deficiency as defined by LeCam. Finally, we investigate some properties of three test statistics, proving a conjecture by Efron and Hinkley (1978) concerning the conditional null-distribution of the likelihood ratio test statistic, and establishing a kind of superiority of the observed Fisher information over the expected one as estimate of the inverse variance of the MLE.

1. Introduction. The purpose of this paper is to investigate some properties related to the conditioning on asymptotic ancillaries as proposed by Efron and Hinkley (1978). Since exact properties are hard to derive in general, the investigation is carried out in terms of second-order asymptotic distributions, i.e., including the $n^{-1/2}$ terms in the asymptotic expansions. It turns out that *first-order asymptotics fail to discriminate between the conditional approach and the usual (marginal) approach.* Emphasis will be on the results, since the techniques used to prove these are largely well-known, but in Section 7 we shall sketch the ideas of the proofs.

Since the arguments for conditioning on (approximately) ancillary statistics are outlined in Efron and Hinkley (1978), we shall not discuss the issue at length, but merely give an example, essentially based on Pierce (1975), illustrating the advantages of this approach.

EXAMPLE 1.1 Let $(\bar{X}, \bar{Y})$ be the average of n independent two-dimensional normal variables, each with the identity matrix as covariance and with mean $\mu(\beta) \in \mathbb{R}^2$, where β is a real parameter, and μ is some smooth function. For each β, let L_β denote the line through $\mu(\beta)$ orthogonal to the tangent at $\mu(\beta)$. If $\hat{\beta}$ is the maximum likelihood estimator of β, then the observation $(\bar{x}, \bar{y})$ must be on the line $L_{\hat{\beta}}$; see Efron (1978) for further geometrical details. If n is large, we may for inferential purposes approximate $\mu(\beta)$ locally by a segment of a circle (see Figure 1). Let P denote the center of this circle; then the lines L_β will for β near to $\hat{\beta}$ approximately go through P. Now, if we want a confidence interval for β, a

Received June 1981; revised September 1984.
AMS 1980 *subject classifications.* Primary 62E20, secondary 62F12.
Key words and phrases. Ancillarity, deficiency, Edgeworth expansions, loss of information, maximum likelihood, observed Fisher information, second-order asymptotics, Wald's test.

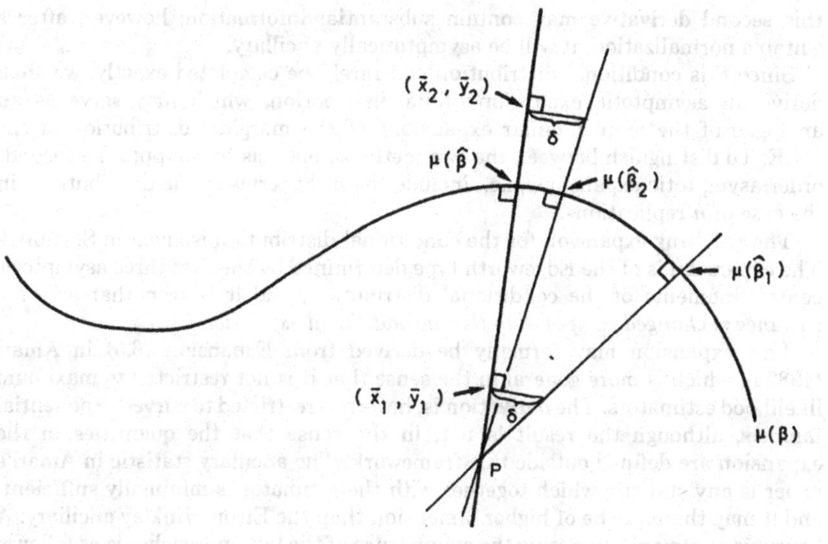

FIG. 1. *Accuracy of the maximum likelihood estimate. The sensitivity of the estimate due to a displacement δ of the observation depends on the distance of $(\bar{x}, \bar{y})$ to P.*

common method will be to "center" this interval at $\hat{\beta}$ and let the length be approximately proportional to the standard deviation of $\hat{\beta}$, disregarding the position of $(\bar{x}, \bar{y})$ on the line $L_{\hat{\beta}}$. However, if the observation is $(\bar{x}_1, \bar{y}_1)$, a displacement of this by an amount δ orthogonal to $L_{\hat{\beta}}$ would change the estimate from $\hat{\beta}$ to $\hat{\beta}_1$, whereas, if the observation is $(\bar{x}_2, \bar{y}_2)$, a similar displacement would only change the estimate to $\hat{\beta}_2$. This suggests that the "accuracy" of the estimate is somehow increasing with the distance of $(\bar{x}, \bar{y})$ from the center P. It may be noted that confidence intervals constructed using the likelihood ratio test would certainly reflect this fact. This may be seen by noticing that if the observation is near the center P, the distance to the curve $\mu(\beta)$ is almost constant in the neighbourhood of $\mu(\hat{\beta})$. In more general examples, similar considerations hold, but the geometrical picture is not equally obvious. □

The example shows that the estimator may not be sufficiently informative. A natural way to try to improve it is to look for an ancillary statistic and replace the marginal distribution of the estimator by its conditional distribution given the ancillary statistic.

Several suggestions of ancillaries capturing some of this additional information have been put forward (see e.g. Barndorff-Nielsen, 1980), but except for one they are related to exponential models or other specific classes of models, e.g., translation models. The remaining one is essentially the second derivative of the log-likelihood function at its maximum. This idea goes back to Fisher, but was suggested by Efron and Hinkley (1978) in more explicit form. A problem is that this second derivative may contain substantial information; however, after a suitable normalization, it will be asymptotically ancillary.

Since this conditional distribution can rarely be calculated exactly, we shall derive an asymptotic expansion of its distribution, which may serve as an analogue of the more familiar expansions of the marginal distribution of the MLE. To distinguish between the two methods, one has to compute the second-order asymptotic expansions, i.e., include the $n^{-1/2}$ terms of the distributions in the case of n replications.

The resulting expansion for the conditional distribution is given in Section 4. The expansion is of the Edgeworth type determined by the first three asymptotic central moments of the conditional distribution, and it is seen that *only the variance is changed compared to the unconditional expansion.*

This expansion may formally be derived from Expansion (6.5) in Amari (1982a), which is more general in the sense that it is not restricted to maximum likelihood estimators. The derivation is, however, restricted to curved exponential families, although the result is not, in the sense that the quantities in the expansion are defined outside this framework. The ancillary statistic in Amari's paper is any statistic which together with the estimator is minimally sufficient, and it may therefore be of higher dimension than the Efron-Hinkley ancillary. A heuristic argument suggesting the equivalence of the two approaches is as follows: The maximum likelihood estimator together with the second derivative of the log-likelihood function at its maximum is a sufficient statistic to second order (see Section 5). Therefore the model can be approximated by a curved exponential family generated by these two statistics, and within this family, the ancillaries of Amari's paper will coincide (locally) with the standardized versions of the second derivatives of the log-likelihood, i.e., the Efron-Hinkley ancillaries.

Knowledge of this expansion makes the conditional approach feasible, but provides no justification for the method. An investigation in this direction is given in Section 5 in terms of (loss of) *Fisher information.* It is shown that the average Fisher information contained in the conditional distribution differs from that of the whole set of data only by an amount which tends to zero as the number of replications increases, whereas the corresponding deficit for the unconditional distribution converges to a fixed quantity depending on the curvature of the model (see Efron, 1975).

A more operationally meaningful way of defining the loss of information involved in a data reduction is in terms of the *deficiency* (LeCam, 1964), which roughly speaking measures how well any test based on the full data set can be mimicked using only the reduced data. The deficiencies for the reductions of the data to the MLE and to the MLE supplemented by the ancillary statistic are also investigated in Section 5, where it is shown as a general result that *the deficiency is bounded by an amount proportional to the square root of the relative loss of Fisher information.* Thus the results concerning loss of Fisher information may be interpreted as providing bounds for the deficiencies for the various reductions.

The variance of the second-order conditional distribution as calculated in Section 4 is, when evaluated at the MLE, equal to the observed Fisher information. Hence, a natural consequence would be to replace the (expected) Fisher information in the Wald test statistic by the observed Fisher information (see Efron and Hinkley, 1978). A comparison of the behaviors of these two test statistics compared to the likelihood ratio test statistic is carried out in Section 6 in two ways. First it is shown that the null-distribution given the asymptotic ancillary statistic converges more rapidly (namely as $O(n^{-1})$) towards its chi-squared limit if the observed Fisher information is used than if the expected one is used, when the convergence rate is $O(n^{-1/2})$. This was conjectured by Efron and Hinkley (1978), and has been shown by Peers (1978) for the one-parameter case. Next, it is shown that the use of the observed information is also superior in the sense that it provides a better stochastic approximation to the likelihood ratio test than the usual Wald test, when null-distributions are considered.

Proofs have been carried through for the case of n replications under the assumptions given in Section 7, which also contains sketches of the proofs, but we shall not give these in detail since they are based mainly on well-known techniques. Some basic ideas of the proofs are, however, given along with the results.

To introduce the results, some examples are given in Section 3; in particular, Example 3.2 should give an impression of the kind and amount of computations required for applications.

2. Preliminaries. Let $X_1, \cdots, X_n$ be independent identically distributed random variables on some measurable space, and suppose that the distribution of X_i is a member of a family $\{P_\beta, \beta \in B \subseteq \mathbb{R}^p\}$, where B is some subset of $\mathbb{R}^p$. We assume that the family is dominated by some measure μ on $\mathbb{R}^p$, and let $f(x, \beta)$ denote the densities. We also assume that the conditions of Section 7 are fulfilled, these essentially being various kinds of smoothness conditions.

For each $\beta \in \mathrm{int}(B)$ define

$$D_j(\beta) = (1/n) \textstyle\sum_{i=1}^n D^j \log f(X_i, \beta), \quad E_j(\beta) = E_\beta\{D_j(\beta)\}$$

where $D^j \log f(X_i, \beta)$ denotes the j-sided array of jth derivatives with respect to β. Also, we define the joint cumulants of these derivatives by

$$\chi_{j\cdots k}(\beta) = \mathrm{cum}_\beta\{D^j \log f(X_i, \beta), \cdots, D^k \log f(X_i, \beta)\}$$

which is of dimension $p^j \cdots p^k = p^{j+\cdots+k}$. In particular the (expected) *Fisher information* per observation is

$$I(\beta) = -E_2(\beta) = \chi_{11}(\beta),$$

whereas the *observed information* per observation is

$$\hat{J} = J(\hat{\beta}) = -D_2(\hat{\beta}).$$

For convenience we shall use the following notational conventions. If the argument β is omitted, a fixed point β_0 (the "true" parameter) is understood, while a circumflex indicates evaluation at the maximum likelihood estimate $\hat{\beta}$, e.g. $I = I(\beta_0)$ and $\hat{I} = I(\hat{\beta})$. Dependence on n is usually not explicitly indicated.

For multiplying vectors, matrices and arrays we shall sometimes use coordinates to clarify definitions and results; otherwise we use some notation that is most easily explained by a few examples given below. When $v = (v_k)$, $k = 1, \cdots, r$ is a vector, $M = (m_{jk})$ is a $q \times r$ matrix, and $A = (a_{ijk})$ is a $p \times q \times r$ array, then:

$$Mv = \textstyle\sum_k m_{jk} v_k \qquad \text{[vector]}$$

$$M(v^2) = \textstyle\sum_{j,k} m_{jk} v_j v_k, \quad \text{if} \quad q = p \qquad \text{[scalar]}$$

$$Av = (\textstyle\sum_k a_{ijk} v_k) \qquad \text{[matrix]}$$

$$A(v^3) = \textstyle\sum_{ijk} a_{ijk} v_i v_j v_k \qquad \text{[scalar]}.$$

Finally we use $\langle\ ,\ \rangle$ to denote the inner product, e.g.

$$\langle v, w \rangle = \textstyle\sum_k v_k w_k, \quad \langle M, N \rangle = \sum_{j,k} m_{jk} n_{jk},$$

where w is an r-vector and N is a $q \times r$-matrix (as is M). The coordinates are collected in an obvious manner, e.g.

$$(\chi_{22})_{ij,kl} = \mathrm{Cov}((d^2/d\beta_i d\beta_j)\log f(X_1, \beta), (d^2/d\beta_k d\beta_l)\log f(X_1, \beta)).$$

The *Efron-Hinkley ancillary statistic* A is defined as a standardized version of the observed information, i.e.

$$A = \sqrt{n}\, \hat{F}^{-1/2}(\hat{J} - \hat{I}) \in \mathbb{R}^d, \tag{2.1}$$

where $F(\beta)$, defined as

$$F(\beta)_{ij,kl} = (\chi_{22})_{ij,kl} - \textstyle\sum_{\alpha,\gamma} (\chi_{21})_{ij,\alpha} (I^{-1})_{\alpha\gamma} (\chi_{12})_{\gamma,kl} \tag{2.2}$$

is the asymptotic variance of $\sqrt{n}\,(\hat{J} - \hat{I})$. Here d is the rank of F, and $F^{-1/2}$ is defined as any smooth "square-root" of F, i.e.

$$(F^{-1/2})' F^{-1/2} = 1_d (\text{the identity matrix on } \mathbb{R}^d),$$

which is the asymptotic variance of A. It may be noted that F is the residual variance of D_2 after regression on D_1.

In the one-dimensional case $F = (\gamma I)^2$, where γ is the statistical curvature defined in Efron (1975). In the multidimensional case F is the square of the second fundamental form in a differential geometry corresponding to that used by Efron (1975); see Reeds (1975) and Madsen (1979, page 24).

3. Examples.

EXAMPLE 3.1. *Non-linear normal regression.* Let $X_1, \cdots, X_n$ be independent normal random variables with variance σ^2 and expectation vector $\mu(\beta)$, where $\beta \in \mathbb{R}^p$ and μ: $\mathbb{R}^p \to \mathbb{R}^n$ is a known function. The variance σ^2 is considered known, since this is notationally more convenient and has no influence on the estimate of β and its distribution. These models do not fit into the i.i.d. framework, but the asymptotics $\sigma^2 \to 0$ corresponds to independent replications of the entire experiment (since in that case $\sigma^2/n \to 0$ as $n \to \infty$), and is furthermore reasonable in cases where σ is small.

The information matrix is

$$I = \textstyle\sum_{i=1}^{n} (D\mu_i)'(D\mu_i)/\sigma^2, \quad \mu_i(\beta) = E_\beta X_i \tag{3.1}$$

and the difference $\hat{J} - \hat{I}$ between the observed and expected Fisher information at $\hat{\beta}$ is

$$\Delta = \hat{J} - \hat{I} = -\textstyle\sum_{i=1}^{n} (X_i - \hat{\mu}_i)D^2\hat{\mu}_i/\sigma^2, \tag{3.2}$$

where $D^2\hat{\mu}_i$ is the matrix of second derivatives of μ_i at $\hat{\beta}$. The ancillary statistic A is a normalized version of the statistic Δ, but it need not be calculated in applications.

The asymptotic unconditional distribution of $\sqrt{n}(\hat{\beta} - \beta)$ is the well-known normal approximation with variance matrix I^{-1}, and the second-order expansion of this distribution is given by an Edgeworth expansion with the same variance, but with a bias and third cumulant of order $O(n^{-1/2})$, which are easily calculated, see, e.g., Skovgaard (1980b).

The conditional second-order distribution is obtained from the unconditional one merely by replacing the inverse variance matrix I by $I + \Delta$ (see (4.12)), or equivalently we may write the variance matrix itself as

$$\mathrm{Var}(\hat{\beta} \mid A) \approx I^{-1}\{1_p + (\textstyle\sum_{i=1}^{m} (x_i - \hat{\mu}_i)D^2\hat{\mu}_i)I^{-1}\}. \tag{3.3}$$

It should be noted that this depends on the true parameter value β_0 through I^{-1}, and that if the common practice of replacing I by $\hat{I}$ is adopted, the distributional approximation will in general only be of first order. However, if $\hat{J}^{-1}$ is used as an approximation to the variance (or equivalently I replaced by $\hat{I}$ in (3.3)), this will be superior to the usual approximation $\hat{I}^{-1}$ for testing an hypothesis about β, in two specific ways stated in Theorem 6.1 and Theorem 6.2.

If the random variables $X_1, \cdots, X_n$ are correlated with covariance matrix $\sigma^2\Sigma$, where Σ is known, the same results hold, except that the sums in (3.1), (3.2) and (3.3) are replaced by inner products with respect to Σ^{-1}. □

EXAMPLE 3.2. To be more specific we shall consider an example of nonlinear normal regression. Consider a *logistic growth function* of the form $e^{\alpha(t-\gamma)}/\{1 + e^{\alpha(t-\gamma)}\}$, where t is time (the independent variable), α and γ are unknown parameters and the growth is supposed to be scaled, such that the limiting "size" is 1. We shall consider this model in logarithmic scale, i.e. we assume that $X_1, \cdots, X_n$ are independent normal variables with variance σ^2, and with expectations

$$EX_i = \mu_i(\alpha, \gamma) = \alpha(t_i - \gamma) - \log\{1 + e^{\alpha(t_i-\gamma)}\}, \tag{3.4}$$

where $t_1, \cdots, t_n$ are known time points.

The information matrix is

$$I = \textstyle\sum_{i=1}^{n} (D\mu_i)'D\mu_i/\sigma^2 = \sum_{i=1}^{n} M_i/(d_i^2\sigma^2) \tag{3.5}$$

where $d_i = 1 + e^{\alpha(t_i-\gamma)}$ and

$$M_i = \begin{pmatrix} (t_i - \gamma)^2 & -\alpha(t_i - \gamma) \\ -\alpha(t_i - \gamma) & \alpha^2 \end{pmatrix}$$

and the difference $\Delta = \hat{J} - \hat{I}$ equals (cf. (3.2))

(3.6) $$\Delta = \sum_{i=1}^{n} (x_i - \hat{\mu}_i) e^{\hat{\alpha}(t_i - \hat{\gamma})} \hat{M}_i / (\hat{d}_i^2 \sigma^2).$$

To derive the conditional distribution of $(\hat{\alpha} - \alpha, \hat{\gamma} - \gamma)$ one needs to compute the bias and the third cumulant of the second-order approximation to this vector. The general formulae are well-known, and in this example the bias is

(3.7) $$\frac{1}{2} \hat{I}^{-1} \sum_{i=1}^{n} \begin{pmatrix} t_i - \hat{\gamma} \\ -\hat{\alpha} \end{pmatrix} e^{\hat{\alpha}(t_i - \hat{\gamma})} \mathrm{trace}\{\hat{M}_i \hat{I}^{-1}\} / d_i^3,$$

and the third cumulant is of similar complexity, but more cumbersome to write, because it is a $2 \times 2 \times 2$ array. The conditional distribution of $(\hat{\alpha}, \hat{\gamma})$ is now given by (4.11) in terms of these cumulants. □

EXAMPLE 3.3. This example is the one from the end of the paper by Hinkley (1980). Let (Y_i, Z_i) be i.i.d. bivariate normal variables with Z_i distributed as $N(\theta_1, 1)$ and $Y_i = \theta_2 Z_i + \varepsilon_i$, where ε_i is $N(0, 1)$. By simple computations we get

$$\hat{\theta}_1 = \bar{Z} = \sum Z_i / n, \qquad \hat{\theta}_2 = \sum Z_i Y_i / \sum Z_i^2$$

$$I(\theta) = \mathrm{diag}(1, 1 + \theta_1^2), \quad \hat{J} = \mathrm{diag}(1, \sum Z_i^2 / n).$$

Since $\hat{J} - \hat{I} = \mathrm{diag}(0, \sum (Z_i - \bar{Z})^2 / n - 1)$ has one-dimensional support, we only compute the corresponding element of F, i.e. $F_{2222} = 2$, and define (see (2.1))

$$A = (\sum (Z_i - \bar{Z})^2 - n) \sqrt{2n}.$$

A is seen to be exactly ancillary and $(\hat{\theta}, A)$ is sufficient. $\hat{\theta}_1$ is independent of A, and since the conditional distribution of $\hat{\theta}_2$ given $Z_1, \cdots, Z_n$ is $N(\theta_2, (\sqrt{2n}A + n(1 + \hat{\theta}_1^2))^{-1})$, it follows, that to second order the conditional distribution of $(\hat{\theta}_1, \hat{\theta}_2)$ given $A = a$ is normal with mean zero and

$$V\{(\hat{\theta}_1, \hat{\theta}_2)\} \sim \mathrm{diag}(n^{-1}, (\sqrt{2n}a + n(1 + \theta_1^2))^{-1}) = n^{-1}(\hat{J} + I - \hat{I})^{-1}$$

in agreement with (4.12). Also, if L is the likelihood ratio statistic, $W = (\hat{\theta} - \theta)' \hat{I} (\hat{\theta} - \theta)$ is the Wald test statistic, and $W_c = (\hat{\theta} - \theta) \hat{J} (\hat{\theta} - \theta)$ is the modified Wald test statistic obtained by using the observed information J instead of I, then, as noted by Hinkley, $L = W_c = n(\bar{Z} - \theta_1)^2 + (\sum Z_i \varepsilon_i)^2 / \sum Z_i^2$ is exactly distributed as X_2^2, whereas W deviates from this by an amount of order $n^{-1/2}$ (cf. Theorem 6.1 and 6.2).

4. Expansion of the conditional distribution. In this section we shall expand the conditional distribution of $Z = \sqrt{n}(\hat{\beta} - \beta_0)$ given A under the distribution P_{β_0}. It is not hard to prove that to first order Z and A are asymptotically independent. Thus to obtain any interesting results, we must carry the expansion to second order, i.e., include the $n^{-1/2}$ terms. The first step is to expand the joint distribution of (Z, A). This is done in the following three steps:

(i) the second-order (stochastic) Taylor-series expansion of (Z, A) in terms of the derivatives of the log-likelihood at β_0 is computed;

(ii) the first three joint cumulants of these approximating polynomials are computed—these will be functions of the E's and χ's;

(iii) by insertion of these cumulants into the general formula for the Edgeworth approximation, the joint distribution is obtained. Since the expansion obtained in this way is the basis of all our results, we shall state it in detail in Theorem 4.1 below.

We shall refer to the cumulants of the approximating distribution as κ_z resp. κ_a for the first cumulants (the means) of Z resp. A, κ_{zzz} the third cumulant of Z, κ_{zza} the mixed third cumulant of Z, Z, A, etc. The cumulants needed in the expansion of the conditional distribution are given by

$$\langle \kappa_z, Iz \rangle = -\tfrac{1}{2}\langle I^{-1}, \chi_{111}(z) + \chi_{21}(z)\rangle/\sqrt{n} \tag{4.1}$$

$$\kappa_{zzz}((Iz)^3) = -(2\chi_{111}(z^3) + 3\chi_{12}(z^3))/\sqrt{n} \tag{4.2}$$

$$\kappa_{zza}(Iz, Iz, a) = -\langle F(z^2), (F^{-1/2})'(a)\rangle/\sqrt{n} \tag{4.3}$$

$$\kappa_{zaa} = 0 \tag{4.4}$$

for all $z \in \mathbb{R}^p$, $a \in \mathbb{R}^d$.

THEOREM 4.1. *Under Conditions* 7.1 *we have the following local expansions for any* $c > 0$:

$$\sup\{|g_n(z, a) - \gamma_n(z, a)|;\ \|z, a\|^2 \le c \log n\} = O(n^{-1}) \tag{4.5}$$

$$\sup\{|h_n(a) - \xi_n(a)|;\ \|a\|^2 \le c \log n\} = O(n^{-1}), \tag{4.6}$$

where $\|\cdot\|$ *denotes the Euclidean norm,* g_n *and* h_n *are the densities of* (Z, A) *and* A, *and*

$$\begin{aligned}\gamma_n(z, a) &= (2\pi)^{-(p+d)/2}(\det I)^{1/2}\{\exp -\tfrac{1}{2}(I(z^2) + \langle a, a\rangle)\}\\ &\quad\times (1 + \langle \kappa_z, Iz\rangle + \langle \kappa_a, a\rangle + \tfrac{1}{6}\kappa_{zzz}((Iz)^3) + \tfrac{1}{6}\kappa_{aaa}(a^3)\\ &\quad + \tfrac{1}{2}\kappa_{zza}(Iz, Iz, a) - \tfrac{1}{2}\langle I, \kappa_{zzz}(Iz)\rangle - \tfrac{1}{2}\langle I, \kappa_{zza}(a)\rangle\\ &\quad - \tfrac{1}{2}\operatorname{trace}\{\kappa_{aaa}(a)\}),\end{aligned} \tag{4.7}$$

$$\begin{aligned}\zeta_n(a) &= (2\pi)^{-d/2}\exp\{-\tfrac{1}{2}\langle a, a\rangle\}\\ &\quad\cdot(1 + \langle \kappa_a, a\rangle + \tfrac{1}{6}\kappa_{aaa}(a^3) - \tfrac{1}{2}\operatorname{trace}\{\kappa_{aaa}(a)\}),\end{aligned} \tag{4.8}$$

are the Edgeworth approximations to the two densities, in which d *is the dimension of* A.

To clarify the meaning of the notation, we shall give formula (4.7) for the case where Z and A (and hence all the κ's) are one-dimensional. We then have

$$\begin{aligned}\gamma_n(z, a) &= (2\pi)^{-(p+d)/2}I^{-1/2}\{\exp -\tfrac{1}{2}(Iz^2 + a^2)\}\\ &\quad\times (1 + \kappa_z Iz + \kappa_a a + \tfrac{1}{6}\kappa_{zzz}I^3z^3 + \tfrac{1}{6}\kappa_{aaa}a^3\\ &\quad + \tfrac{1}{2}\kappa_{zza}I^2z^2a - \tfrac{1}{2}\kappa_{zzz}I^2z - \tfrac{1}{2}I\kappa_{zza}a - \tfrac{1}{2}\kappa_{aaa}a),\end{aligned} \tag{4.9}$$

which is, in fact, not much different from (4.7).

It is seen from (4.8) combined with the fact that the first and third cumulant (κ_a and κ_{aaa}) of A are both $O(n^{-1/2})$, that A is not, in general, second-order ancillary in the sense that the second-order distribution is independent of β_0; but it is locally second-order ancillary in the sense of Cox (1980). This means that in any set of the form $\{\beta;\ \|\beta - \beta_0\| \leq c/\sqrt{n}\}$ with $c > 0$ fixed, the distribution of A is constant except for terms of order $O(n^{-1})$. This is the property that turns out to be important to avoid loss of information (see Section 5).

The expansion of the conditional distribution of Z given A may now be obtained by dividing $\gamma_n(z, a)$ by $\zeta_n(a)$, although the proof requires further expansion than given in Theorem 4.1.

THEOREM 4.2. *Under Conditions 7.1 we have the following expansion of the conditional distribution of* $Z = \sqrt{n}\,(\hat{\beta} - \beta_0)$ *given* $A = a$,

$$P\{Z \in B \mid A = a\} = \int_B \eta_n(z \mid a)\, dz + O(n^{-1}) \tag{4.10}$$

uniformly over all Borel sets $B \in \mathbb{R}^p$ *and* $\|a\|^2 \leq (2 + \alpha) \log n$, *for some* $\alpha > 0$, *where*

$$\begin{aligned}\eta_n(z \mid a) = {} & (2\pi)^{-p/2}\{\det(I + F^{1/2}(a))\}^{1/2}\exp\{-\tfrac{1}{2}(I + F^{1/2}(a))(z^2)\} \\ & \times \{1 + \langle \kappa_z, Iz\rangle + \tfrac{1}{6}\kappa_{zzz}((Iz)^3) - \tfrac{1}{2}\langle I, \kappa_{zzz}(Iz)\rangle\}\end{aligned} \tag{4.11}$$

is the second-order expansion of the condition density.

REMARK. It is important to note that the event $\{\|A\|^2 \leq (2 + \alpha)\log n\}$ has probability $1 - O(n^{-1})$, so that Theorem 4.2 together with (4.6) implies, that

$$P\{Z \in B\} = \int_{\|a\|^2 \geq (2+\alpha)\log n} \zeta_n(a) \int_B \eta_n(z \mid a)\, dz\, da + O(n^{-1}).$$

A local expansion of the conditional density of Z given A holding uniformly only on a bounded set, would not suffice to prove this, and in this sense the result would be incomplete.

There are some points worth noting about the moments of η_n. The first and third moment are (to second order) independent of a, and the same as in the unconditional second-order expansion, whereas the variance depends on a. The theorem says nothing about the conditional moments of the exact distribution, but if these are to be used as descriptive quantities of the distribution, then rather than expanding these, it is the moments of the approximating distribution that are relevant.

To second order we have

$$\tilde{V}(Z \mid A = a)^{-1} = I + n^{-1/2}F^{1/2}(a) \sim \hat{J} + I - \hat{I}, \tag{4.12}$$

where $\tilde{V}$ is the variance of the approximate distribution. Thus it is seen that if the common practice of inserting the estimate $\hat{\beta}$ for the unknown parameter β_0 is used in Formula (4.12), one arrives at the observed Fisher information $\hat{J}$ as an estimate of the inverse variance in the approximate conditional distribution, as noted by Efron and Hinkley (1978) and Amari (1982a), Formula (6.12). If, however, this approximation is used, the distributional approximation is no longer of second order, and it is questionable whether anything has been gained compared to the usual unconditional first-order approximation.

In the special case when the derivative of $I(\beta)$ at β_0 vanishes, as is the case of the translation models considered in Efron and Hinkley (1978), then the approximation

$$\tilde{V}(Z \mid A = a) \sim \hat{J}^{-1}$$

will, however, lead to a second-order approximation to the conditional distribution. In contrast the bias and the third cumulant of Z may in general be replaced by estimates obtained through evaluation at $\hat{\beta}$ without changing the order of magnitude of the distributional approximation in (4.10).

5. Recovery of information. Fisher's main reason for considering ancillaries and, more specifically, conditional distributions given ancillaries was that by the reduction to a single statistic, such as the MLE, one might lose a certain amount of (Fisher) information, which might be "recovered" by a conditional approach.

The total amount of Fisher information in the experiment is $nI(\beta_0) = \inf(X)$, say, where $X = (X_1, \cdots, X_n)$. In general we let $\inf(T)$ denote the Fisher information (at β_0) contained in an experiment where only T is observed. Also, we shall consider the information $\inf(T \mid A = a)$ in the experiment, where $A(X) = a$ is fixed and T is observed, and its expected value is $\inf_A(T) = E\{\inf(T \mid A)\}$. The well-known identity $\inf(T) = \inf(X) - E\{\mathrm{Var}\{D \log f(X; \beta_0) \mid T\}\}$, see e.g. Fisher (1925), is useful in computing $\inf(T)$. It is well-known, see Fisher (1925), that $\inf(X) - \inf(\hat{\beta})$ tends to a finite limit as $n \to \infty$, which Efron (1975) identified as $\gamma^2 I$ in the one-dimensional case, where γ is the statistical curvature of the model at β_0. The following theorem shows that this information lost by the reduction of X to $\hat{\beta}$ is indeed recovered by conditioning by A as defined in (2.1).

THEOREM 5.1. *Under Conditions* 7.1 *we have*

$$\inf(X) - \inf(\hat{\beta}) = F(\cdot, I^{-1}, \cdot) + O(n^{-1}) \tag{5.1}$$

$$\inf(X) - \inf(\hat{\beta}, A) = O(n^{-1}) \tag{5.2}$$

$$\inf(A) = O(n^{-1}) \tag{5.3}$$

$$\inf_A(\hat{\beta}) = \inf(X) - O(n^{-1}) \tag{5.4}$$

where $F(\cdot, I^{-1}, \cdot)$ *is the matrix with entry* (i, j) *given by* $\sum_{k,l} F_{ik,lj}(I^{-1})_{kl}$

Note that (5.4) follows from (5.2) and (5.3), since $\inf_A(\hat{\beta}) = \inf(\hat{\beta}, A) - \inf(A)$. Formal proofs of (5.1) and (5.2) go back to Fisher (1925), whereas Rao (1961) gave a strict proof of (5.1) in the multinomial case; see Efron (1975), Section 9, for further discussion and references. Strict proofs may be given under weaker assumptions than those of Section 7, but we shall not elaborate on this point.

If one does not believe, as Fisher seemed to, that the (Fisher) information is an absolute measure of information, then it would be natural to look for other interpretations or implications of Theorem 5.1 and similar results; see LeCam (1975). A reasonable possibility would be to measure the information lost in the reduction from $X=(X_1,\cdots,X_n)$ to $T=T_n(X)$ by the *deficiency* of the experiment $(Q_\beta, \beta\in B)$ with respect to the experiment $(P_\beta, \beta\in B)$, where Q_β is the distribution of T, as defined by LeCam (1964). The deficiency is defined as the "distance" between the original distribution and the best "reconstruction" of it based on T by a randomization which is independent of the parameter. As the distance is used the maximal difference in probability over all sets, and the supremum over all parameters is the deficiency. This is an intuitively appealing measure, because it tells something about the probabilistic performance of the two experiments, namely how well *any* test based on X can be approximated by use of T only.

In agreement with LeCam (1956) and Michel (1978) we shall use a slight modification of the deficiency by restricting attention to compact sets of parameter values. Define

$$
\begin{aligned}
\delta_K(T, X) &= \inf_\Pi \sup_{\beta\in K} \tfrac{1}{2}\, \| P_\beta - \Pi Q_\beta \| \\
&= \inf_\Pi \sup_{\beta\in K} \sup_A \{ | P_\beta(A) - (\Pi Q_\beta)(A) | \}, \quad K\subseteq \mathbb{R}^P \text{ compact}
\end{aligned}
\tag{5.5}
$$

where Π varies over the class of Markov-kernels and A over all measurable sets. Except for minor technical differences concerning the class of kernels Π this is the deficiency of $(Q_\beta, \beta \in K)$ with respect to $(P_\beta, \beta\in K)$. Attention is restricted to compact sets $K\subseteq B$, since uniform approximation over B can hardly be obtained in general. Notice that $\delta_K(T, X) = 0$ if T is sufficient.

Let us now assume, that $\hat\beta$ is a function of T, although another first-order efficient estimator might do as well as $\hat\beta$, and let us define $\Pi = P^t_{\hat\beta}$, i.e. the (ΠQ_β)-conditional distribution of X given $T=t$ is $P^t_{\hat\beta}$, where P^t_β is the P_β-conditional distribution of X given $T=t$. We shall give a formal proof that $\delta_K(T, X)$ *is asymptotically bounded by the maximum over K of the square root of the relative loss of Fisher information.* More precisely

$$
\| P_\beta - \Pi Q_\beta \| \le \sqrt{p}(\text{trace } R_\beta(T))^{1/2}(1+o(1)) \tag{5.6}
$$

where p is the dimension of β and $R_\beta(T) = \inf(X)^{-1}(\inf(X) - \inf_\beta(T))$ is the relative loss of Fisher information.

Let $f^t(x;\beta)$ denote the density of P^t_β with respect to μ. The proof of (5.6) then goes as follows

$$
\begin{aligned}
\| P_\beta - \Pi Q_\beta \| &= \int\!\!\int | f^t(x;\beta) - f^t(x;\hat\beta) |\, d\mu(x)\, dQ_\beta(t) \\
&\sim \int\!\!\int | (D_\beta \log f^t(x;\beta))(\hat\beta - \beta) |\, dP_\beta(x)\, dQ_\beta(t) \\
&\le \int\!\!\int \| I(\beta)^{-1/2}(D_\beta \log f^t(x;\beta)) \| \\
&\qquad \cdot \| I(\beta)^{1/2}(\hat\beta - \beta) \|\, dP_\beta(x)\, dQ_\beta(t)
\end{aligned}
$$

$$\leq (E_\beta\{(nI(\beta))(\hat{\beta} - \beta)^2\})^{1/2}$$
$$\cdot (E_\beta\{\langle (nI(\beta))^{-1}, \inf_\beta(X \mid T)\rangle\})^{1/2}$$
$$\leq \sqrt{p}(E_\beta\{\text{trace}((ni(\beta))^{-1}\inf_\beta(X \mid T))\})^{1/2}$$
$$= \sqrt{p}(\text{trace}\{\inf_\beta(X)^{-1}(\inf_\beta(X) - \inf_\beta(T))\})^{1/2}$$

where the second inequality follows from Hölders inequality.

Using this result together with Theorem 5.1 we see, that $\delta_K(\hat{\beta}, X) = O(n^{-1/2})$ and $\delta_K((\hat{\beta}, A), X) = O(n^{-1})$, which may be viewed as a special case of the result in Michel (1978), where the statistics of the form $T = (\hat{\beta}, \hat{D}_2, \cdots, \hat{D}_m)$ are considered. We also see that in the case $T = \hat{\beta}$, we have

$$n^{1/2}\| P_\beta - \Pi Q_\beta \| \leq \sqrt{p}(\langle I^{-1}, F(\cdot, I^{-1}, \cdot)\rangle)^{1/2} = \sqrt{p}(\textstyle\sum_{i,j,k,l} F_{ik,lj}(I^{-1})_{ij}(I^{-1})_{kl})^{1/2}$$

which reduces to the statistical curvature $|\gamma(\beta)|$ in absolute value in the case $p = 1$, whereas in the multivariate case this quantity equals $\sqrt{p\ \text{trace}(IE)}$, where E is the $p \times p$ matrix termed the "Efron excess" by Reeds (1975).

6. Comparison of test statistics. Consider a hypothesis of the form H_0: $H\beta = h_0$, where H is a $q \times p$ matrix of rank $q \leq p$ and $h_0 \in V_0$ a known point. The most interesting example of this kind is testing that a coordinate of β takes a fixed value, but any "smooth hypothesis" may be written in this way, if necessary by a reparametrization. Let $\tilde{\beta}$ be the maximum likelihood estimate under H_0, and let H' be the transpose of H. We shall consider the following three test statistics of the hypothesis H_0:

$$L = 2\textstyle\sum_{i=1}^n (\log f(X_i, \hat{\beta}) - \log f(X_i, \tilde{\beta}))$$
$$W = (H\hat{\beta} - h_0)'(H'\hat{I}^{-1}H)^{-1}(H\hat{\beta} - h_0)$$
$$W_c = (H\hat{\beta} - h_0)'(H'\hat{J}^{-1}H)^{-1}(H\hat{\beta} - h_0).$$

Here L is the likelihood ratio test statistic, and W and W_c are quadratic test statistics in $(H\hat{\beta} - h_0)$ normalized by different estimates of its variance. In particular, W is the Wald test statistic and W_c a modified Wald test with $\hat{J}^{-1}$ as variance estimates of $\hat{\beta}$ instead of $\hat{I}^{-1}$. The index c means "conditional", although $\hat{J}^{-1}$ is not in general the conditional variance of $\sqrt{n}(\hat{\beta} - \beta_0)$ given A. The following theorem confirms a conjecture by Efron and Hinkley (1978), that even conditionally W_c follows a chi-squared distribution with error term of order $O(n^{-1})$.

THEOREM 6.1. *Under Conditions* 7.1 *and the assumption that* 7.1 (vi) *holds for the restricted model* H_0, *we have the following expansions under* H_0, *i.e. if* $H\beta_0 = h_0$,

$$P_{\beta_0}\{L \leq t \mid A = a\} = \chi^2_{p-q}(t) + O(n^{-1}) \tag{6.1}$$

$$P_{\beta_0}\{W_c \leq t \mid A = a\} = \chi^2_{p-q}(t) + O(n^{-1}) \tag{6.2}$$

$$(6.3) \qquad P_{\beta_0}\{W \le t \mid A = a\} = \chi^2_{p-q}(t) + O(n^{-1/2})$$

uniformly in $t \ge 0$ *for all a in* $\{\| a \|^2 \le (2+\alpha)\log n\}$, *where* χ^2_{p-q} *is the chi-squared distribution function with* $p-q$ *degrees of freedom.*

The statement concerning W is in a sense negative and stated for comparison only. The important point is that the error is not in general $O(n^{-1})$. Note that marginally all three test statistics are asymptotically chi-squared distributed with error $O(n^{-1})$; see Chandra and Ghosh (1979).

Although this result indicates that L and W_c behave more like conditional tests than W does, it says nothing about the (marginal) properties of the tests. A possibility would be to compare the (asymptotic) powers of the tests, but a uniform superiority of any of these could hardly be expected. If one takes the standpoint in accordance with Example 1.1 that L is theoretically preferable to W and W_c, then one could compare W and W_c by their performance relative to L. This leads to the following result.

THEOREM 6.2. *Under the conditions of Theorem 6.1* W_c *is stochastically closer to* L *than* W *is, in the sense that for any continuous function* h: $\mathbb{R} \to [0, \infty)$, $h(0) = 0$, $h(x_2) > h(x_1)$ *if* $0 < x_1 < x_2$ *or* $x_2 < x_1 < 0$, *we have*

$$(6.4) \qquad P_{\beta_0}\{h(\sqrt{n}\,(W_c - L)) < h(\sqrt{n}\,(W - L))\} = \delta(h) + o(1)$$

with $\delta(h) \ge 1/2$, *and* $\delta(h) = 1/2$ *if and only if* $F = 0$, *and hence* $W - W_c = O(n^{-1})$ *with probability* $1 - O(n^{-1})$.

Note that the function h is included to show that the result holds in "any scale", rather than, e.g., in the absolute values $|W_c - L|$ and $|W - L|$.

Both of the theorems suggest that W_c should be preferred to W, whereas it is hard to see any reason for preferring W to W_c in general. Moreover in connection with numerical maximization of the log-likelihood, W_c is easily computed because $-\hat{J}$ is just the matrix of second derivatives at the maximum. The results are, however, only asymptotic, and in particular cases W may well be preferable.

7. Conditions and proofs.

CONDITIONS 7.1. Let $\beta_0 \in \text{int}(B)$ be a fixed parameter value, then

(i) If $x \in \{x; f(x;\beta_0) > 0\}$, then $f(x;\beta)$ is 7 times continuously differentiable w.r.t. β in a neighbourhood of β_0.

(ii) $I(\beta_0)$ is nonsingular and 5 times continously differentiable in a neighbourhood of β_0.

(iii) $E_{\beta_0}\{\| D^j \log f(X;\beta_0) \|^7\} < \infty$, $1 \le j \le 7$.

(iv) $\exists \delta_0 > 0$:

$$E_{\beta_0}\{(\sup\{\| D^7 \log f(X;\beta) \|; \|\beta - \beta_0\| \le \delta_0\})^7\} < \infty.$$

(v) For $n = 1$ the characteristic function of $U(D_1, \cdots, D_7)$ belongs to L_m for some $m \in \mathbb{N}$, where U is an affine function mapping the affine support of $(D_1, \cdots, D_7)$ bijectively onto a real space, such that $\mathrm{Var}_{\beta_0}\{U\}$ equals the identity and $E_{\beta_0}\{U\} = 0$.

(vi) For sufficiently large n the MLE $\hat{\beta}_n$ of β exists with P_{β_0} – probability one, and for all $c > 0$

$$P_{\beta_0}\{\| \sqrt{n}(\hat{\beta}_n - \beta_0) \|^2 > c \log n\} = o(n^{-5/2}).$$

(vii) Expectations with respect to P_{β_0} of all linear and bilinear functions of $D \log f(X; \beta_0)$, $D^2 \log f(X; \beta_0)$ and $D^3 \log f(X; \beta_0)$ may be differentiated by differentiation under the integral sign.

We have not tried to minimize the assumptions of each theorem; instead, since the purpose of this section is to outline the techniques, they are a compromise between the demand that they should be easily verifiable, and the desire to avoid too great technicalities. In particular in (vi), probability one could be replaced by probability $1 - o(n^{-5/2})$. It may seem somewhat odd that 5 times differentiability is considered in (ii), and that 7 derivatives of $\log f$ are considered in (iv). These high numbers, compared to the theorems in which only second-order expansions are considered, are first of all used to derive the higher order expansions of $(\hat{\beta}, A)$ and A needed to control the error term of the expansion of the conditional distribution. In the sequel we shall refer to the assumptions as (i)–(vii), and it should be clear from the proofs what the purpose of each assumption is. Before going on to these we shall state a lemma of some independent interest.

LEMMA 7.2. *Let P be a probability measure and Q a finite signed measure both dominated by a measure μ on some measurable space (E, S). Let $f = dP/d\mu$ and $g = dQ/d\mu$ denote the densities. If $Q(E) = 1$ and a set $A \in S$ exists, such that for some $\varepsilon_1 \geq 0$, $\varepsilon_2 \geq 0$*

(a) $$\sup\{| f(x) - g(x) |; x \in A\} \leq \varepsilon_1$$

(b) $$\int_{A^c} | g(x) | \, d\mu(x) \leq \varepsilon_2$$

then

(7.1) $$\sup\{| P(B) - Q(B) |; B \in S\} \leq 2(\varepsilon_1 \mu(A) + \varepsilon_2).$$

PROOF.

$$\begin{aligned} | P(B) - Q(B) | &\leq | P(B \cap A) - Q(B \cap A) | + | P(B \cap A^c) - Q(B \cap A^c) | \\ &\leq \varepsilon_1 \mu(A) + 1 - P(A) + \varepsilon_2 \leq 2(\varepsilon_1 \mu(A) + \varepsilon_2). \quad \square \end{aligned}$$

We shall now proceed to comment on the proofs, avoiding details that may in essence be found elsewhere.

Expansion of the distribution of $(D_1, \cdots, D_7)$. By the conditions (iii) and (v) we may apply Theorem 19.2 of Bhattacharya and Rao (1976) to obtain an asymptotic expansion as $n \to \infty$ in powers of $n^{-1/2}$ of the density of $n^{1/2}U(D_1, \cdots, D_7)$, the error term being $o(n^{-5/2})$ uniformly over the whole set.

PROOF OF THEOREM 4.1. We shall use Theorem 3.2 of Skovgaard (1980a) to transform the local expansion of U to a local expansion of (Z, A). This theorem is stated in terms of distributions, but since it is proved by the use of local expansions, it may be applied here in modified form. The technique was first used by Bhattacharya and Ghosh (1978) to derive an expansion of the distribution of Z under similar, but more general, assumptions. In Theorem 4.1 only the second-order expansions are stated, but to prove Theorem 4.2 we need to establish the validity of a local Edgeworth expansion with error term $O(n^{-2-\delta})$ for some $\delta > 0$. To do this a Taylor-series expansion of the form

$$Z \sim A_1(D_1) + n^{-1/2}A_2(D_1, D_2) + \cdots + n^{-5/2}A_6(D_1, \cdots, D_6) + o(n^{-5/2})$$

uniformly in $\| U(D_1, \cdots, D_7) \|^2 \leq c \log n$, is required. This is constructed as in Bhattacharya and Ghosh (1978) using conditions (i), (iv) and (vi). A similar expansion is needed for A, and this is obtained by expanding around $\hat{\beta} = \beta_0$ using the expansion of Z and conditions (i), (ii) and (iv). The expansion of A is only needed up to an error of order $O(n^{-2-\delta})$. On transforming the expansion of U, the validity of local Edgeworth expansions of (Z, A) and A including the n^{-2} terms is established, the errors being $O(n^{-2-\delta})$. Condition (vii) is needed to compute the second-order expansions, whereas we need not actually compute the higher-order expansions.

There is a slight technical problem in computing the differential $DF^{-1/2}(\hat{\beta} - \beta_0)$ of $F^{-1/2}$ in the direction $\hat{\beta} - \beta_0$. Since

$$(F^{-1/2})'F^{-1/2} = F^{-1}$$

and

$$DF^{-1}(\hat{\beta} - \beta_0) = -F^{-1}(DF(\hat{\beta} - \beta_0))F^{-1}$$

we obtain by the product rule

$$(DF^{-1/2}(\hat{\beta} - \beta_0))'F^{-1/2} + (F^{-1/2})'DF^{-1/2}(\hat{\beta} - \beta_0) = -F^{-1}(DF(\hat{\beta} - \beta_0))F^{-1},$$

which turns out to be all that is needed. Note that the right-hand side is independent of which "square root" of F is used. Based on the Taylor-series expansions, the computations of the κ's and the second-order expansions are straightforward; see, e.g., Skovgaard (1980b).

PROOF OF THEOREM 4.2. The method used to prove this is essentially the one given in Michel (1980). (4.11) is obtained by dividing (4.7) by (4.8); the problem is to prove the validity. To do this we need the expansions of $g_n(z, a)$ and $h_n(a)$ with error terms $O(n^{-2-\delta})$ as constructed above. The ratio of these will, on expanding in powers in $n^{-1/2}$ and keeping only the first- and second-order terms, give the same result as the ratio of the second-order expansions. The point is now that if α in Theorem 4.2 is sufficiently small, then the relative error of the higher-order expansion of $h_n(a)$ within the set $\| a \|^2 \leq (2+\alpha)\log n$ is $O(n^{-1-\epsilon})$ for some $\varepsilon > 0$. On this set also the error of the higher order expansion of $g_n(z, a)$ is $O(n^{-1-\epsilon})$, when divided by $h_n(a)$. The theorem then follows from Lemma 7.2.

PROOF OF THEOREM 5.1. The main computations leading to (5.1) and (5.2) are quite similar to those given by Fisher (1925) (or Amari, 1982b). For example, to prove (5.2), an expansion of the likelihood equation gives

$$0 \sim D_1 + D_2(\hat{\beta} - \beta_0) + \tfrac{1}{2} D_3(\hat{\beta} - \beta_0)^2 + O(n^{-1/2}),$$

which in turn leads to an expansion of the form

$$D_1 \sim I(\hat{\beta} - \beta_0) + F^{1/2}A(\hat{\beta} - \beta_0) + Q(\hat{\beta} - \beta_0)^2 + O(n^{-1/2}),$$

where $Q(\hat{\beta} - \beta_0)^2$ is some quadratic form in $\hat{\beta} - \beta_0$. This shows intuitively that the conditional variance of D_1 given $(\hat{\beta}, A)$ is $O(n^{-1})$, although there are still some technical problems left. These problems are essentially overcome by showing that in the calculations of variances of D_1, one may neglect a region of the form $\| D_1 \|^2 > cn \log n$, where c is some constant. In this way the problems with integration of the error term may be avoided. It should also be noted that the results of Section 4 do not suffice to prove this theorem, but the (higher order) expansions used to prove Theorem 4.2 may again be used to establish the results.

Expansions of L, W and W_c. In the proofs of Theorem 6.1 and Theorem 6.2 we shall confine ourselves to the case of a simple hypotheses, i.e. H_0: $\beta = \beta_0$, since the ideas of the proofs are the same in the more complicated setting. Note that we then have $W = \hat{I}((\hat{\beta} - \beta_0)^2)$ and $W_c = \hat{J}((\hat{\beta} - \beta_0)^2)$. The Taylor-series expansions to second order of L, W and W_c around $Z = 0$ can be expressed as

$$L \sim (I + n^{-1/2}F^{1/2}(A))(Z^2) + n^{-1/2}(\chi_{12}(Z^3) + \tfrac{2}{3}\chi_{111}(Z^3))$$

$$W \sim I(Z^2) + n^{1/2}(2\chi_{12}(Z^3) + \chi_{111}(Z^3))$$

$$W_c \sim (I + n^{-1/2}F^{1/2}(A))(Z^2) + n^{-1/2}(2\chi_{12}(Z^3) + \chi_{111}(Z^3)),$$

the error being $O(n^{-1})p(A, Z)$ with probability $1 - O(n^{-1})$ uniformly on each set of the form $\| (Z, A) \|^2 \leq c \log n$, where p is a polynomial independent of n. These expansions are the key to the proofs of the two theorems of Section 6. Notice that the quadratic terms in Z are the squared length of Z as measured by the inverse *conditional variance* (cf. (4.12)) in L and W_c, whereas the *unconditional variance* is used in W.

PROOF OF THEOREM 6.1. Using the expansions above, (6.1) and (6.2) follows from Theorem 1 of Chandra and Ghosh (1979); see their Remark 2.2. Their condition (2.2) is not exactly fulfilled, because it only holds in sets of "size" $O(\log n)$ instead of $O(\sqrt{n})$, but it makes no essential difference in the proof. (6.3) is obvious, and it is seen that since $n^{-1/2}F^{1/2}(A)$ is in general *not* $O(n^{-1})$, neither is the error in (6.3).

PROOF OF THEOREM 6.2. Consider the differences

$$D = W - L \sim n^{-1/2}(-F^{1/2}(A)(Z^2) + \chi_{12}(Z^3) + \tfrac{1}{3}\chi_{111}(Z^3))$$

$$D_c = W_c - L \sim n^{-1/2}(\chi_{12}(Z^3) + \tfrac{1}{3}\chi_{111}(Z^3))$$

both being of order $O(n^{-1/2})$. To a first approximation, Z and $F^{1/2}(A)$ are independent, normally distributed with means zero and variances $\mathrm{Var}\{Z\} \sim I^{-1}$, $\mathrm{Var}\{F^{1/2}(A)\} \sim F$. Thus, to order $n^{-1/2}$, the conditional distribution of $\sqrt{n}D$ given Z is normal with mean $\sqrt{n}D_c$ and variance $F(Z^4)$, while D_c is a function of Z. In this approximate distribution it is seen that the probability of $h(\sqrt{n}D)$ being greater than $h(\sqrt{n}D_c)$ is a least ½, since the probability of the event that this occurs with D and D_c of the same sign equals ½. Since the other part of the event $h(\sqrt{n}D) > h(\sqrt{n}D)$ has probability zero if and only if F is zero, and hence $W = W_c + O(n^{-1})$, the theorem follows.

REFERENCES

AMARI, S. (1982a). Geometrical theory of asymptotic ancillarity and conditional inference. *Biometrika* **69** 1–18.

AMARI, S. (1982b). Differential geometry of curved exponential families—curvature and information loss. *Ann. Statist.* **10** 357–385.

BARNDORFF-NIELSEN, O. (1980). Conditionality resolutions. *Biometrika* **67** 293–310.

BHATTACHARYA, R. N. and RAO, R. R. (1976). *Normal Approximation and Asymptotic Expansions.* Wiley, New York.

BHATTACHARYA, R. N. and GHOSH, J. K. (1978). On the validity of the formal Edgeworth expansion. *Ann. Statist.* **6** 434–451.

CHANDRA, T. K. and GHOSH, J. K. (1979). Valid asymptotic expansions for the likelihood ratio statistic and other perturbed chi-squared variables. *Sankhya A* **41** 22–47.

COX, D. R. (1975). Discussion to Efron (1975). *Ann. Statist.* **3** 1211.

COX, D. R. (1980). Local ancillarity. *Biometrika* **67** 279–286.

EFRON, B. (1975). Defining the curvature of a statistical problem. *Ann. Statist.* **3** 1189–1242.

EFRON, B. (1978). The geometry of exponential families. *Ann. Statist.* **6** 362–376.

EFRON, B. and HINKLEY, D. V. (1978). Assessing the accuracy of the maximum likelihood estimator: Observed versus expected Fisher information. *Biometrika* **65** 457–482.

FISHER, R. A. (1925). Theory of statistical estimation. *Proc. Cambridge Philos. Soc.* **122** 700–725.

HINKLEY, D. V. (1980). Likelihood as approximate pivotal distribution. *Biometrika* **67** 287–292.

LECAM, L. (1956). On the asymptotic theory of estimation and testing hypotheses. *Proc. Third Berkeley Symp. Math. Statist. Probab.* **1** 129–156.

LECAM, L. (1964). Sufficiency and approximate sufficiency. *Ann. Math. Statist.* **35** 1419–1455.

LECAM, L. (1975). Discussion to Efron (1975). *Ann. Statist.* **3** 1223–1224.

MADSEN, L. T. (1979). The geometry of statistical models. A generalization of curvature. Research report 79/1, Statistical Research Unit, Copenhagen.

MICHEL, R. (1978). Higher order asymptotic sufficiency. *Sankhya A* **40** 76–84.

MICHEL, R. (1979). Asymptotic expansions for conditional distributions. *J. Multivariate Anal.* **9** 393–400.

PEERS, H. W. (1978). Second-order sufficiency and statistical invariants. *Biometrika* **65** 489–496.

PIERCE, D. A. (1975). Discussion to Efron (1975). *Ann. Statist.* **3** 1219–1221.

RAO, C. R. (1961). Asymptotic efficiency and limiting information. *Proc. Fourth Berkeley Symp. Math. Statist. Probab.* **1** 531–545.

REEDS, J. (1975). Discussion to Efron (1975). *Ann. Statist.* **3** 1234–1238.

SKOVGAARD, I. M. (1980a). Transformation of an Edgeworth expansion by a sequence of smooth functions. *Scand. J. Statist.* **8** 207–217.

SKOVGAARD, I. M. (1980b). Edgeworth expansions of the distributions of maximum likelihood estimators in the general (non i.i.d.) case. *Scand. J. Statist.* **8** 227–236.

ROYAL VETERINARY AND AGRICULTURAL UNIVERSITY
DEPARTMENT OF MATHEMATICS AND STATISTICS
THORVALDSENSVEJ 40
DK 1871 COPENHAGEN V
DENMARK

Chapter 3

On multivariate Edgeworth expansions

Introduction by John Robinson
University of Sydney

This work follows a tradition of rigorous examination of limit theory extending back more than a half century among Scandinavian probabilists. It has a degree of subtlety mixed with complexity which challenge the reader. The work is technically difficult so this comment really consists of a summary with a few observations. The first sentence of Ib's summary is "Conditions are given for the validity of multivariate Edgeworth expansions for a sequence of random variables". It is these conditions, required to give existence of densities and rates of convergence to zero for errors of approximation of these densities by the Edgeworth expansions, that are central to this work. They permit consideration of special cases to be reduced to proving that these conditions hold.

Two seminal books written a decade before this paper described the current knowledge on asymptotic approximations. The first, appearing in Russian in 1972, but published in English as Petrov (1975), gave an account of limit theorems, Edgeworth approximations and large deviation results for sums of independent variables. The second, Bhattacharya and Rao (1976), gave an account of "Normal Approximations and Asymptotic Expansions" for sums of independent random vectors. I believe that these provide a background for Ib's work, in particular, in Skovgaard (1981a,b), where the methods were extended so they could be applied to problems associated with maximum likelihood estimates. These studies, in turn, led to this paper which gives a general approach.

This work looks at Edgeworth expansions and the error rates of these approximations in a new light and uses methods developed in Ib's earlier two papers to generalize and simplify results such as those in Bhattacharya and Rao (1976). First, moments and cumulants are introduced via multilinear symmetric forms and the Hermite polynomials of the Edgeworth expansion are obtained using multivariate differentials. These two devices, both used in Ib's two earlier works, allow considerable improvement in notation and permit the later examination of cases where there are different rates of convergence for different directions. The paper considers a general sequence of random vectors, S_n, in a finite dimensional real vector space V_n with probability distribution P_n. Here S_n is not necessarily assumed to be a sum of independent random vectors as in the classical case. A further notational device is introduced by considering the inner product for V_n based on Σ_n^{-1} and the corresponding norm $||.||_n$, where Σ_n is the covariance matrix of S_n, assumed nonsingular. The characteristic function of S_n is denoted by $\xi_n(t)$ for $t \in V_n^*$, the dual of V_n, which is given the inner product based on Σ_n and the corresponding norm $||t||_n$. Canonical Lebesgue measures, λ_n, corresponding to these inner products assign unit mass to the unit cube. Formal Edgeworth expansions are derived from expansions of the logarithms of the characteristic functions of the sequence of general random vectors followed by an inversion of a truncated series.

The major result in Theorem 3.1 on local approximations considers four conditions under which the density f_n of P_n with respect to λ_n, exists and an Edgeworth approximation, $f_{s,n}$, of order s, holds uniformly with absolute error $O(\epsilon_n)$, where ϵ_n is a positive sequence converging to zero. It is assumed that a sequence of positive numbers $a_n(t)$ exists for each $t \in V_n^*$ depending on t only through its direction. I will not set these conditions out in full but they are central to the results so I will attempt to describe them.

Condition (I), relating $a_n(t)$ to the required errors, is $a_n(t)^{-(s-1)} = O(\epsilon_n)$. In the classical case of sums of independent random vectors $a_n(t) = \sqrt{n}$ does not depend on t and $\epsilon_n = n^{-(s-1)/2}$. The condition (II) corresponds to bounding the standardized cumulant forms by $1/a_n(t)$. (An error in placement of brackets in this condition is corrected in Skovgaard (1989a).) Condition (III) is essentially a smoothness condition, giving a bound for the tail integral of the absolute value of the characteristic function outside $||t||_n > \delta a_n(t)$ as $O(\epsilon_n)$. Condition (IV) puts bounds on standardized derivatives of the characteristic function. These conditions reflect those required in the classical Edgeworth expansions for sums of independent random variables and demonstrate, in the relative simplicity of the proof under them, the basic requirements to obtain appropriate rates of convergence.

Essentially the same conditions are used to obtain results on Edgeworth approximations for probabilities of Borel sets. A corollary gives uniform rates of convergence for these whenever the local approximation for densities is valid. To improve these results to cover cases where densities do not exist requires an adjustment of condition (III). The modification required is given as a variation of a Cramér type condition in Albers *et al.* (1976), given in (III''_α) and Ib gives a related slightly weaker condition (III'_α). This enables proof of Theorem 3.4 giving rates of convergence for Edgeworth approximations of probabilities of a wide class of Borel sets having a smoothness condition on their boundaries.

Next the case of sums of independent random vectors, $S_n = X_1 + \cdots + X_n$, is considered. Lemma 4.1 gives two conditions on the set of characteristic functions of $X_1, \cdots, X_n$; the first implies (III'_α), and the second, together with the first, implies (III). Skovgaard (1989a) gives a minor correction to (4.3) used in the first condition, but the proof as it stands is actually written in terms of the adjusted condition. Now Corollary 4.2 gives a simple proof of the rate of convergence for the special case of $S_n = a_1Y_1 + \cdots + a_nY_n$ where $Y_1, \cdots, Y_n$ are independent and identically distributed and $a_1, \cdots, a_n$ are constants, under mild conditions.

Example 4.1 is used to demonstrate that the methods proposed in this work are needed to give results under weaker conditions when the rates of convergence in different directions differ. The example, first considered in Skovgaard (1981a), concerns rates of convergence for the score functions associated with a sequence of independent random variables $Y_1, \cdots, Y_n$, where $Y_j/(\alpha + \beta u_j)$ are Gamma distributed with index ω_j and $u_1, \cdots, u_n$ and $\omega_1, \cdots, \omega_n$ are known positive constants. The score statistic for likelihood estimation of (α, β), is $S_n = \sum_{j=1}^n X_j$, where $X_j = \{Y_j - \omega_j/(\alpha + \beta u_j)\}(1, u_j)/(\alpha + \beta u_j)^2$. Ib notes that the single condition for the validity of the Edgeworth expansion of S_n is that both eigenvalues of the variance matrix of the scores tend to infinity and gives necessary and sufficient conditions for this as $\sum \omega_j(u_j - \bar{u})^2 \to \infty$ and $\sum \omega/u_j^2 \to \infty$, where $\bar{u} = \sum \omega_j u_j / \sum \omega_j$. He then points out that with the directional method employed here there is no necessity to impose restrictions on the relative rates of convergence of these eigenvectors as was required in Skovgaard (1981a). Finally he gives an extra condition for the validity of Edgeworth expansions of $(\hat{\alpha}, \hat{\beta})$ derived from the expansions for S_n.

§5 illustrates the application of the results to sums of dependent random vectors by considering sums of a vector valued moving average process, pointing out that for this example the mixing conditions of Götze and Hipp (1983) are stronger than required for this example. The proof proceeds in a straightforward, but far from simple, manner, showing that each of the conditions of Theorem 3.1 hold for this example.

The final section gives an admirably brief treatment for a sequence of lattice random variables in p-dimensional space maintaining the generality of Chapter 5 of Bhattacharya and Rao (1976). A lattice in a p-dimensional vector space, V, is defined as

$$L = \{v_0 + Z_1 v_1 + \cdots , Z_p v_p : \; Z_j \in \mathbb{Z}, j = 1, \cdots , p\},$$

where $v_0 \in V$ and $v_1, \cdots , v_p$ are linearly independent vectors in V. Lattice random variables S are then defined to take values on the lattice with probability 1 and to have nonsingular variance. If no sublattice contains the support of S then we refer to a minimal lattice for S. Next the region F^* in V^*, required for the treatment of characteristic functions, is taken as

$$F^* = \{t_1\eta_1 + \cdots + t_p\eta_p : \; |t_j| < \pi, j = 1, \cdots , p\},$$

where $\eta_1, \cdots , \eta_p \in V^*$ is a dual basis to $v_1, \cdots , v_p$. For a sequence of lattice random variables the point probability is given and a lattice norm is defined. Then a theorem analogous to Theorem 3.1 is stated for sequences of lattice random variables with four conditions closely related to those of Theorem 3.1 and a fourth condition bounding the ratio of the norm $||t||_n$ to the lattice norm.

This work contains methods with far reaching consequences. It has been cited only 26 times but has provided an approach for a number of extremely difficult problems. I will cite only four of these where the results of this work have been decisive in obtaining proofs, namely, Lieberman *et al.* (2003), Andrews *et al.* (2006), He and Severini (2007) and Kline and Santos (2012). In my view the work will continue to find applications in non-standard problems where its generality and depth will be crucial.

International Statistical Review (1986), **54**, 2, pp. 169–186. Printed in Great Britain

On Multivariate Edgeworth Expansions

Ib M. Skovgaard

Department of Mathematics, The Royal Veterinary and Agricultural University, Thorvaldsensvej 40, DK-1871 Frederiksberg C, Denmark

Summary

Conditions are given for the validity of multivariate Edgeworth expansions for a sequence of random variables. It is shown that by the use of a directional approach in the expansions of characteristic functions it is possible to sharpen the results to cover cases where the rate of convergence is different for different linear functions of the variables. Furthermore, this approach simplifies the proof by essentially reducing the problem to dimension one. Uniform expansions are established for densities, for point probabilities, for probabilities of Borel sets, and for probabilities of sets within certain uniformity classes. The conditions simplify if the variables have finite Laplace transforms in a neighbourhood of zero, and also if the statistics are sums of independent variables. The conditions are proved for sums of an infinite moving average process under weak conditions on the coefficients.

Key words: Directional derivatives; Edgeworth expansion; Independent variables; Lattice variables; Local expansion; Moving average; Random vectors; Symmetric forms.

1 Introduction

The classical theory of Edgeworth expansions deals with expansions of distributions of sums of independent and identically distributed random variables. An excellent account of this theory, together with generalizations to nonidentically distributed variables and related topics, may be found in the monograph by Bhattacharya & Rao (1976). In a more recent work Götze & Hipp (1978) improved the results for expansions of moments of smooth functions, and they also gave an important contribution to the theory of Edgeworth expansions for dependent variables (Götze & Hipp, 1983). For inferential purposes, the starting point is typically a 'classical' Edgeworth expansion, e.g. for the score statistic, and the problem is then to transform this expansion to an expansion for the estimator or test statistic. Major contributions to this aspect are Pfanzagl (1973) dealing with one-dimensional minimum contrast estimators, Bhattacharya & Ghosh (1978) with a general theorem on transformation of Edgeworth expansions in the multivariate independently and identically distributed case and an identification with the expansion obtained by the delta method, Chibisov (1979) with an accurate treatment of moment conditions for such a transformation, and Chandra & Ghosh (1979) with a transformation result for asymptotically quadratic statistics (chi-squared expansions). A more direct approach was taken by Albers, Bickel & van Zwet (1976) to the problem of expanding distributions for distribution free tests. This paper gives a good account of the problem of establishing Edgeworth expansions for the distributions of discrete statistics, e.g. like rank statistics. A quite general framework for expansions of distributions was given by Götze (1985), suitable also, for example, for expansions of von Mises statistics. The Edgeworth expansions also form the basis of the numerically quite accurate saddlepoint expansions, an account of which was given by Barndorff-Nielsen & Cox (1979), where also the mixed Edgeworth-saddlepoint expansions and approximations to conditional distributions are discussed. The conditions under which Edgeworth expansions of densities may be divided to yield expansions of conditional distributions were given by Michel (1979). For expansions of conditional distributions in the lattice case, see Hipp (1984).

The scope of the present paper lies within the classical approach, and the purpose is to sharpen the results of Bhattacharya & Rao (1976), mainly for the case of multivariate random variables. Much is achieved merely by replacing the coordinate based approach given, for example, by Bhattacharya & Rao (1976), with a coordinate free approach in which characteristic functions are expanded in one direction at a time. This is most naturally done in connection with a change of notation from a coordinate based notation to an invariant notation. By the 'directional' approach it becomes possible to handle cases in which the variance tends to infinity at different rates in different directions. Such cases would usually not be covered by the classical results as may be seen from Remark 3.3 and Example 4.1. Since the characteristic function in any particular direction is the characteristic function of a one-dimensional statistic, the invariant method is just as easy for the multivariate as for the univariate case, except, possibly, for unfamiliarity with notation. It is, however, possible to expand only the characteristic function, not its derivatives, by considering one direction of a time; therefore it is important to show, as in Corollary 3.3, that this is sufficient to obtain an expansion of the multivariate distribution. Apart from these considerations there is a distinct advantage in picking out the assumptions required to prove the expansion from the theorems of Bhattacharya & Rao (1976), but at an earlier stage instead of carrying through the proof for sums of independent variables. This was done in much the same way by Durbin (1980) and Skovgaard (1981a); Durbin, however, at the same time proved uniformity with respect to a parameter. In any case this makes the theorem (corresponding to Theorem 3.1 in the present paper) applicable also to dependent variables.

The present paper is formulated primarily in terms of uniform expansions for densities and consequently for the measure of all Borel sets (see Corollary 3.3), but it is a minor technical matter, through smoothing techniques, to relax the conditions to give uniform expansions for classes of sets with uniformly negligible boundary, amongst other things for all convex sets.

In § 2 we give the basic construction of the Edgeworth expansions in terms of Edgeworth approximations, of a certain order, to a (fixed) probability distribution. The main results of the behaviour of these approximations as asymptotic expansions for sequences of distributions are given in § 3. For the case of lattice random variables the analogous results are given in § 6. Sections 4 and 5 contain some specializations of the general results; namely to sums of independent random variables and to sums of moving average processes, respectively. These two sections may be read independently of each other.

Readers who are not familiar with the coordinate free notation may have to consult Appendices 1 and 2 on multivariate moments and cumulants and on multivariate differentials, respectively.

2 Edgeworth approximations

Let P be a probability measure on a finite-dimensional real vector space V of dimension $p \in \mathbb{N}$, and let S denote a random variable on V with distribution P. We assume that S has finite cumulants of order $s \geqslant 2$ and we let $\chi_1, \ldots, \chi_s$ denote the first s cumulant forms, i.e.

$$\chi_k(t_1, \ldots, t_k) = \text{cum}(\langle t_1, S\rangle, \ldots, \langle t_k, S\rangle)$$

is the joint cumulant of the real random variables $\langle t_1, S\rangle, \ldots, \langle t_k, S\rangle$, where $k = 1, \ldots, s$ and $t_i \in V^*$ is a linear mapping of V into $\mathbb{R}$, that is V^* is the dual of V. We assume that $\chi_1 = 0$, or equivalently we choose χ_1 as the origin, and also that the variance $\Sigma = \chi_2$ is nonsingular. Let

$$\xi(t) = E\{e^{i\langle t, S\rangle}\}, \quad t \in V^*,$$

denote the characteristic function of P and consider the expansion

$$\log \xi(t) = \tfrac{1}{2}i^2\chi_2(t^2) + \cdots + \frac{1}{s!}i^s\chi_s(t^s) + o(\|t\|^s) \tag{2.1}$$

as $t \to 0$. The Edgeworth approximations are motivated by the case when P is a normalized sum of n independent replications of some random variable. In that case $\xi(t)$ will be of the form $\xi_1(t/\sqrt{n})^n$, where ξ_1 is the characteristic function of one of the components of the sum. Therefore, consider the function

$$\tau^{-2}\log \xi(\tau t) = \tfrac{1}{2}i^2\chi_2(t^2) + \tau\frac{1}{3!}i^3\chi_3(t^3) + \ldots + \tau^{s-2}\frac{1}{s!}i^s\chi_s(t^s) + o(\tau^{s-2}\|t\|^s) \tag{2.2}$$

as $\tau t \to 0$, where τ is a positive real variable playing the role of $n^{-\frac{1}{2}}$ in the case of independent replications. Taking the exponential of both sides of (2.2) and expanding around $\tau = 0$ we get the following expansion for each fixed t:

$$\xi(\tau t)^{\tau^{-2}} = \exp\{-\tfrac{1}{2}\Sigma(t^2)\}\left[1 + \sum_{k=1}^{s-2}\tau^k\tilde{P}_k(it:\{\chi_j\})\right] + o(\tau^{s-2}) \tag{2.3}$$

as $\tau \to 0$, where $\tilde{P}_k(it:\{\chi_j\})$ is the kth (multivariate) Cramér–Edgeworth polynomial in (it) with coefficients depending on $\chi_3, \ldots, \chi_{k+2}$. By definition $\tilde{P}_k$ is the kth derivative at $\tau = 0$ of the exponential of the right-hand side of (2.2) divided by $k!$. An explicit formula is given by Bhattacharya & Rao (1976, § 7). Putting $\tau = 1$ in (2.3) we obtain the approximation

$$\xi(t) \simeq \xi_s(t:\{\chi_j\}) = \exp\{-\tfrac{1}{2}\Sigma(t^2)\}\left[1 + \sum_{k=1}^{s-2}\tilde{P}_k(it:\{\chi_j\})\right] \tag{2.4}$$

to the characteristic function of S. Notice that if S is, in fact, a normalized sum of independently and identically distributed random variables, then we get the same approximation whether we use (2.4) or we let $\tau = n^{-\frac{1}{2}}$ and replace the χ's in (2.3) by the cumulants of one of the random variables in the sum. This is easily seen from (2.2).

Through inversion of the characteristic function $\xi_s(t:\{\chi_j\})$ we obtain a signed measure P_s, say, which is called the Edgeworth approximation of order $s-1$ to P. Thus, the first-order Edgeworth approximation is the normal approximation. The densities $f_s(x:\{\chi_j\})$, say, are the well-known linear combinations of Hermite polynomials multiplied by the normal density. They are given by the inversion formula

$$f_s(x:\{\chi_j\}) = (2\pi)^{-p/2}\int \xi_s(t:\{\chi_j\})e^{-i\langle t, x\rangle}\,d\lambda(t), \tag{2.5}$$

with respect to the Lebesgue measure 'corresponding' to the one on V^* used in the integral. In this paper we choose to let f_s denote the density with respect to the Lebesgue measure generated by the variance Σ; see Appendix 1. This merely means that the usual normalization factor $(\det\Sigma)^{-\frac{1}{2}}$ is included in the Lebesgue measure on V, while units on V^* are defined by the inner product Σ.

With $\Delta = \Sigma^{-1}$ and $\phi_\Delta(x) = (2\pi)^{-p/2}\exp\{-\frac{1}{2}\Delta(x^2)\}$ being the normal density with variance Σ, the second-order approximation to the density is

$$f_3(x:\{\chi_j\}) = \phi_\Delta(x)[1 + \tfrac{1}{6}\chi_3(\Delta(x)^3) - \tfrac{1}{2}\text{trace}\{\Delta\chi_3(\Delta(x))\}],$$

where the notation is transparent from the following coordinate version

$$\phi_\Delta(x)[1 + \tfrac{1}{6}\chi_{ijk}\sigma^{i\alpha}\sigma^{j\beta}\sigma^{k\gamma}x_\alpha x_\beta x_\gamma - \tfrac{1}{2}\sigma^{ij}\chi_{ijk}\sigma^{k\alpha}x_\alpha],$$

where we have used the summation convention that summation is understood over all indices that appear twice in a term, and where (σ^{ij}) and (χ_{ijk}) are coordinate representations of Σ^{-1} and χ_3, respectively. With obvious extensions of this notation, the next correction term is

$$\begin{aligned}\phi_\Delta(x)[&\tfrac{1}{24}\chi_{ijkl}\{\sigma^{i\alpha}\sigma^{j\beta}\sigma^{k\gamma}\sigma^{l\delta}x_\alpha x_\beta x_\gamma x_\delta - 6\sigma^{i\alpha}\sigma^{j\beta}\sigma^{kl}x_\alpha x_\beta + 3\sigma^{ij}\sigma^{kl}\}\\
&+ \tfrac{1}{72}\chi_{ijk}\chi_{lmn}\{\sigma^{i\alpha}\sigma^{j\beta}\sigma^{k\gamma}\sigma^{l\delta}\sigma^{m\varepsilon}\sigma^{n\tau}x_\alpha x_\beta x_\gamma x_\delta x_\varepsilon x_\tau\\
&- (6\sigma^{i\alpha}\sigma^{j\beta}\sigma^{k\gamma}\sigma^{l\delta}\sigma^{mn} + 9\sigma^{i\alpha}\sigma^{j\beta}\sigma^{l\gamma}\sigma^{m\delta}\sigma^{kn})x_\alpha x_\beta x_\gamma x_\delta\\
&+ (9\sigma^{i\alpha}\sigma^{l\beta}\sigma^{jk}\sigma^{mn} + 18\sigma^{i\alpha}\sigma^{j\beta}\sigma^{kl}\sigma^{mn} + 18\sigma^{i\alpha}\sigma^{l\beta}\sigma^{jm}\sigma^{kn})x_\alpha x_\beta\\
&- (9\sigma^{ij}\sigma^{kl}\sigma^{mn} + 6\sigma^{il}\sigma^{jm}\sigma^{kn})\}].\end{aligned}$$

In the calculation of this and higher-order terms it is a help to use the following explicit formula for the Hermite polynomials, derived from work of Grad (1949). The kth Hermite polynomial corresponding to the metric Δ on V is a polynomial in $x \in V$ defined by

$$H_{k,\Delta}(x) = (-1)^k\phi_\Delta(x)^{-1}D^k\phi_\Delta(x),$$

for $x \in V$. Thus $H_{k,\Delta}(x)$ is a k-linear symmetric mapping of V into $\mathbb{R}$; see Appendices 1 and 2. The diagonal values are consequently defined by

$$(H_{k,\Delta}(x))(y^k) = (-1)^k\phi_\Delta(x)^{-1}\frac{d^k}{dh^k}\phi_\Delta(x+hy)\bigg|_{h=0}$$

for $y \in V$, and the density $(H_{k,\Delta}(x))(y^k)\phi_\Delta(x)$ has characteristic function $(i\langle t, y\rangle)^k\exp\{-\frac{1}{2}\Sigma(t^2)\}$. The explicit formula is

$$(H_{k,\Delta}(x))(y^k) = \sum_{m=0}^{[k/2]}\frac{(-1)^m k!}{(k-2m)!\,m!\,2^m}\Delta(y,x)^{k-2m}\Delta(y^2)^m \tag{2.6}$$

for the diagonal values. Because $H_{k,\Delta}(x)$ is symmetric, the nondiagonal values, $(H_{k,\Delta}(x))(y_1, \ldots, y_k)$, are obtained from expression (2.6) by insertion of the y_i's in the places of the y's, averaged over all permutations of the y_i's.

In applications the cumulants (χ_j) used to construct the approximation may not be the cumulants of P itself, but some approximate cumulants. We shall, however, not notationally distinguish such cases, although such 'cumulants' may not be the cumulants of any probability measure.

3 Edgeworth expansions

In § 2 the Edgeworth approximations were defined for a fixed measure, but there were no results on the errors of these approximations. In this section we shall establish conditions under which the Edgeworth approximations to a sequence of distributions are, in fact, asymptotic expansions; i.e. have error terms that tend to zero. Under independent replications, the Edgeworth series will generally be an asymptotic expansion in powers of $n^{-\frac{1}{2}}$, and the error term to the Edgeworth approximation of order $s-1$ will be

$O(n^{-(s-1)/2})$ as $n \to \infty$. Here we shall work more generally with an error term of order $O(\varepsilon_n)$, where ε_n will then, typically, be a power of $n^{-\frac{1}{2}}$ in applications.

We consider a sequence (S_n) of random variables with distributions (P_n), on vector spaces (V_n) of dimensions $p_n \leqslant p$, for $n \in \mathbb{N}$, and we want to derive conditions for the Edgeworth approximations to form an asymptotic expansion with uniform error term $O(\varepsilon_n)$.

Denote the cumulants of S_n by $\chi_{1,n}, \chi_{2,n}, \ldots$ and the characteristic function of S_n by $\xi_n : V^* \to \mathbb{C}$. As in the previous section we translate the distributions such that $\chi_{1,n} = 0$ and we assume that the variances $\Sigma_n = \chi_{2,n}$ are nonsingular. These assumptions cause no loss of generality, since we might otherwise define the vector space V_n accordingly. We equip V_n with the inner product Σ_n^{-1} and the corresponding norm, denoted $\|\,.\,\|_n$, while we use the dual inner product Σ_n on V_n^*. This means that we may write $\|t\|_n$ instead of $(\Sigma_n(t^2))^{\frac{1}{2}}$, etc., for $t \in V_n^*$, and in this way the standardization of the variances is avoided. Corresponding to these inner products are canonical Lebesgue measures on V_n and V_n^*, both denoted λ_n and assigning unit mass to the unit cube.

An important function for the determination of the rate of convergence is the directional analogue of β_n of Bhattacharya & Rao (1976, formula (9.19)),

$$\rho_{s,n}(t) = \sup\left\{\frac{1}{k!}|\chi_k(t^k)| \Big/ (\|t\|_n^k)^{1/(k-2)};\ 3 \leqslant k \leqslant s\right\}, \tag{3.1}$$

which depends on $t \in V_n^*$ only through its direction. For the purpose of bounding the error term of certain Taylor series expansions, we also need the quantities

$$M_{s,n}(t) = \sup\left\{\left|\frac{d^{s+1}}{dh^{s+1}} \log \xi_n(ht)\right| \Big/ (\|t\|_n^{s+1}(s+1)!);\ 0 < h < 1\right\}, \tag{3.2}$$

which depend on the magnitude of t as well as its direction.

THEOREM 3.1. *Given the sequence (S_n) of random variables in (V_n) and a sequence (ε_n) of positive numbers converging to zero, suppose that for each $t \in V_n^*$ a positive number $a_n(t)$ exists, depending on t only through its direction $t/\|t\|_n$, such that for some fixed $s \geqslant 2$ the following conditions hold*:

(I) *$a_n(t)^{-(s-1)} = O(\varepsilon_n)$ as $n \to \infty$ uniformly in $t \in V_n^*$;* (3.3)

(II) *if $s \geqslant 3$, the following inequality holds for n sufficiently large, for all $t \in V_n^*$,*

$$\rho_{s,n}(t) \leqslant a_n(t)^{-1}; \tag{3.4}$$

(III) *for all $\delta > 0$, as $n \to \infty$,*

$$\int_{\|t\|_n > \delta a_n(t)} |\xi_n(t)|\, d\lambda_n(t) = O(\varepsilon_n); \tag{3.5}$$

(IV) *for some $\delta > 0$, as $n \to \infty$,*

$$M_{s,n}(\delta a_n(t)t/\|t\|_n) = O(a_n(t)^{-(s-1)}) \tag{3.6}$$

uniformly in $t \in V_n^*$.

Then, for sufficiently large n, the density f_n of P_n with respect to λ_n exists, and, as $n \to \infty$,

$$|f_n(x) - f_{s,n}(x)| = O(\varepsilon_n) \tag{3.7}$$

uniformly in $x \in V_n$, where $f_{s,n}$ is the density of the Edgeworth approximation of order $(s-1)$ with respect to λ_n.

Proof. From the inversion formula we have

$$|f_n(x) - f_{s,n}(x)| \leq (2\pi)^{-p_n} \int |\xi_n(t) - \xi_{s,n}(t)|\, d\lambda_n(t) \leq (2\pi)^{-p_n}(I_1 + I_2 + I_3),$$

say, where p_n is the dimension of V_n and

$$I_1 = \int_{\|t\|_n < \delta a_n(t)} |\xi_n(t) - \xi_{s,n}(t)|\, d\lambda_n(t), \quad I_2 = \int_{\|t\|_n > \delta a_n(t)} |\xi_n(t)|\, d\lambda_n(t),$$

$$I_3 = \int_{\|t\|_n > \delta a_n(t)} |\xi_{s,n}(t)|\, d\lambda_n(t),$$

where δ is chosen to satisfy (IV). The integral I_2 is $O(\varepsilon_n)$ by assumption, and so is I_3 by standard arguments because $\exp\{-\frac{1}{2}a_n(t)^2\delta^2\}a_n(t)^m$ is $O(\varepsilon_n)$ for any power $m \in \mathbb{N}$. The estimation of I_1 is taken care of through a Taylor series expansion of the log-characteristic function around zero. For this purpose fix t in $\{\|t\|_n < \delta a_n(t)\}$ and, for δ sufficiently small, bound the function $|\xi_n(ut) - \xi_{s,n}(ut)|$, for $0 < u < 1$, as Bhattacharya & Rao (1976, Th. 9.9), by $\exp\{-\frac{1}{4}\|ut\|_n^2\}(M_{s,n}(ut) + \rho_{s,n}(t)^{s-1})$ multiplied by a polynomial in $\|ut\|_n$. On integration this yields a bound on I_1 which is

$$O(\sup\{M(\delta a_n(t)t); \|t\|_n = 1\} + \rho_{s,n}(t)^{s-1}) = O(\varepsilon_n),$$

which proves the theorem. Notice that the analogue of Theorem 9.9 of Bhattacharya & Rao (1976) is much easier to prove here, because we do not have to expand the derivatives of the characteristic functions; i.e. we need only the case $|\alpha| = 0$ in their formulation. □

Remark 3.1. Quite generally, if $s \geq 3$, $a_n(t)$ should be chosen as $\rho_{s,n}(t)^{-1}$, but exceptionally, e.g. if $\rho_{s,n}(t) = 0$, it may be necessary to choose $a_n(t)$ otherwise to prove (IV). Notice that (I) and (II) implies that $\rho_{s,n}(t)^{s-1}$ is $O(\varepsilon_n)$, which will usually define the sequence (ε_n). In the case of identical replications, $\varepsilon_n = n^{-(s-1)/2}$, $\rho_{s,n}(t) = cn^{-\frac{1}{2}}$ and $a_n(t) = n^{\frac{1}{2}}/c$, for some $c > 0$.

Remark 3.2. From inspection of the expansion it is seen that its terms will be bounded by powers of $\rho_{s,n} = \sup\{\rho_{s,n}(t); t \in V^*\}$, but in general the expansion will not, strictly speaking, be a power series expansion, because the terms depend on n in a more complicated way (through the cumulants).

Remark 3.3. The main difference between the present approach and the one of Bhattacharya & Rao (1976, §§ 9, 19) first of all has to do with the allowance for different rates of convergence in different directions. It may appear that if the variance tends to infinity at different rates in different directions, then a standardization would take care of this, and hence the well-known results would be applicable. This is, however, not generally so, because the rates of convergence, e.g. for the Taylor series expansions of the characteristic function, in different directions, as determined by $\rho_{s,n}$, will be different even after a standardization, and hence the cut-off points $\delta a_n(t)$, defining the integrals I_1, I_2 and I_3, have to be chosen separately for each direction. In this way the Taylor series approximation, used to bound I_1, is extended to a maximal region. The expansion in Example 4.1 illustrates this point.

Remark 3.4. It is easy to check that an Edgeworth expansion with error $O(\varepsilon_n)$ is also obtained if all or some of the cumulants are replaced by approximate cumulants $(\tilde{\chi}_{k,n})$, say, satisfying

$$|\tilde{\chi}_{k,n}(t^k) - \chi_{k,n}(t^k)| = O(\varepsilon_n)\, \|t\|_n^k.$$

Also, various asymptotically equivalent expansions may be used if the cumulants (including the first two) are themselves given as expansions. For example, obvious modifications of the expansions are obtained if a mean, tending to zero sufficiently quickly, is included in (2.2) instead of in the translation of the distribution. In such cases first and second order Hermite polynomials may be added to the approximations to the density, while variance terms will result in second order Hermite polynomials.

The hardest conditions to verify in Theorem 3.1 are usually (III) and (IV). For the 'intermediate' derivatives in (IV), great simplification can be obtained if the characteristic function is analytic. That will be so if S_n has finite cumulant generating function in some neighbourhood of the origin, in which case all moments of S_n are finite, and the quantity

$$\rho_n(t) = \sup \{\rho_{s,n}(t) \mid s \geqslant 3\} \tag{3.8}$$

is finite. Furthermore $\xi_n(t)$ will be equal to the (infinite) sum of its Taylor-series around zero whenever $\|t\|_n < \rho_n(t)^{-1}$, hence we deduce the following lemma.

Lemma 3.2. *The quantity $\rho_n(t)$, defined in (3.8), is finite for all $t \in V_n^*$ if and only if the cumulant generating function is finite for t in some neighbourhood of zero in V_n^*, and in that case $\rho_n(t)$ is bounded in t, and, for $t \in V_n^*$ satisfying $\|t\|_n < \rho_n(t)^{-1}$, we have*

$$M_{s,n}(t) \leqslant \rho_n(t)^{s-1}/(1-\rho_n(t)\|t\|_n) \tag{3.9}$$

for all $s \geqslant 3$.

Proof. A trivial consequence of well-known results. □

The conclusion is that, if $\rho_n(t)$ is finite and (I) and (II) of Theorem 3.1 hold with $\rho_{s,n}(t)$ replaced by $\rho_n(t) = a_n(t)^{-1}$, then (IV) automatically holds, and the problem is reduced to the verification of (III). If this can be verified for all s, then the validity of the Edgeworth expansion to any order has been proved and the terms will be bounded in order of magnitude by powers of $\rho_n = \sup \{\rho_n(t); t \in V_n^*\}$.

To obtain an expansion of the distribution as a set function, the expansion (3.7) has to be integrated. By use of Lemma 7.2 of Skovgaard (1985) a uniform expansion may be established as follows.

Corollary 3.3. *If the uniform local expansion* (3.7) *is valid, then the expansion* $(P_{s,n})$ *of* (P_n) *satisfies*

$$|P_n(B_n) - P_{s,n}(B_n)| = O(\varepsilon_n(\log \varepsilon_n^{-1})^m), \tag{3.10}$$

as $n \to \infty$, uniformly in all Borel sets B_n in V_n, for some $m > 0$.

Proof. Define the set $A_\delta = \{x \in V; \|x\|_n^2 < (2+\delta)\log \varepsilon_n^{-1}\}$. Then outside A_δ the total variation of $P_{s,n}$ is $O(\varepsilon_n)$ if $\delta > 0$. Inside A_δ the total variation of $P_n - P_{s,n}$ is $O(\varepsilon_n)$ multiplied by the volume of A_δ. Finally, the mass of P_n outside A_δ is one minus its mass inside A_δ, which may again be bounded by $1 - P_{s,n}(A_\delta) + |P_n - P_{s,n}|\,(A_\delta)$. □

The logarithmic factor in the order of magnitude of the error term is of no importance because the error will still be smaller in order of magnitude than the bound for the last term of the expansion.

The condition (III) is hard to simplify in general, but for a sum of independent variables the integrand is often exponentially bounded because ξ_n is a product of n characteristic functions. Combined with a condition of integrability a simplification may be obtained for such cases, as we shall see in § 4. Without the integrability condition but with the integrand itself tending sufficiently rapidly to zero, a local expansion may be proved for a smoothed version of the density, and hence for sets with a sufficiently well-behaved boundary, e.g. convex sets. This is the standard technique for strongly nonlattice distributions (satisfying Cramér's condition); see, for example, Bhattacharya & Rao (1976, § 20). We shall give a version in terms of certain uniformity classes of sets as defined by Bhattacharya & Rao (1976, formula (20.47)), and with a modification of Cramér's condition used, for example, by Albers et al. (1976), and formulated generally by Graves (1983).

THEOREM 3.4. *Let the conditions of Theorem* 3.1 *be fulfilled, except that* (III) *is replaced by the following with* $0<\alpha\leq 1$:

(III$'_\alpha$) *for all* $\delta>0$ *there exists a* $k_1>2$ *and a* $k_2>0$ *such that*

$$\int |\xi_n(t)|\, d\lambda_n(t) = O(\varepsilon_n(\log \varepsilon_n^{-1})^{k_2}), \tag{3.11}$$

as $n\to\infty$, *where the integral is over the range* $\delta a_n(t) < \|t\|_n < \varepsilon^{-1/\alpha}(\log \varepsilon_n^{-1})^{k_1}$ *Then, as* $n\to\infty$,

$$|P_n(B_n) - P_{s,n}(B_n)| = O(\varepsilon_n(\log \varepsilon_n^{-1})^m) \tag{3.12}$$

for some $m>0$, *uniformly over all sets* B_n *in the class of Borel sets satisfying*

$$\Phi_n((\partial B_n)^d) \leq c d^\alpha, \tag{3.13}$$

for all $d>0$, *where* $c>0$ *is any fixed constant,* Φ_n *is the standard normal distribution on* V_n *and* $(\partial B_n)^d$ *is the d-boundary of* B_n, *that is, the set of points at distance less than d from the boundary of* B_n. *In particular, all convex sets belong to such a class with* $\alpha=1$.

Proof. Let $d_n=\varepsilon_n^{1/\alpha}$ and U_n a random variable on V_n independent of S_n, satisfying $P\{\|U_n\|<d_n\}=1$ and with a characteristic function K_n satisfying $|K_n(t)|\leq c_1 \exp\{-\|d_n t\|_n^{\frac{1}{2}}\}$ for some $c_1>0$ (Bhattacharya & Rao, 1976, Corol. 10.4). Then

$$\begin{aligned}\int_{\|t\|_n>\delta a_n(t)} |\xi_n(t)|\,|K_n(t)|\, d\lambda_n(t) &= O(\varepsilon_n(\log \varepsilon_n^{-1})^{k_2}) + \int \exp\{-\|d_n t\|_n^{\frac{1}{2}}\}\, d\lambda_n(t)\\ &= O(\varepsilon_n(\log \varepsilon_n^{-1})^{k}),\end{aligned}$$

as $n\to\infty$ for some $k>0$, where the integral on the right-hand side is over the range $\|t\|_n>d_n^{-1}(\log \varepsilon_n^{-1})^{k_1}$. Hence, the conditions of Theorem 3.1 may be proved (through some standard arguments) and a local expansion with error $O(\varepsilon_n(\log \varepsilon_n^{-1})^k)$ established for the distribution of S_n+U_n, and the result follows from the following simple argument,

$$\begin{aligned}|P_n(B_n) - P\{S_n+U_n\in B_n\}| &\leq P\{S_n+U_n\in(\partial B_n)^{d_n}\}\\ &= O(d_n^\alpha) + O(\varepsilon_n(\log \varepsilon_n^{-1})^{k_3}) = O(\varepsilon_n(\log \varepsilon_n^{-1})^{k_3}),\end{aligned}$$

as $n\to\infty$, for some $k_3>0$. Through considerations of the orders of magnitudes of the cumulants, it is also seen that $P_{s,n}$ agrees with the expansion for S_n+U_n to the order of magnitude considered. □

Remark 3.5. The condition (III$'_\alpha$) is, of course, implied by:

$$\text{(III}''_\alpha\text{)}\ \sup\{|\xi_n(t)|;\ \delta a_n(t) < \|t\|_n < \varepsilon_n^{-1/\alpha}(\log \varepsilon_n^{-1})^{k_1}\} = O(\varepsilon_n^{p/\alpha+1}(\log \varepsilon_n^{-1})^m), \tag{3.14}$$

as $n\to\infty$, for some $m>0$, where $k_1>2$ is as in (III$'_\alpha$). This version resembles more the standard Cramér condition. By limiting the supremum to an upward bounded set, as in (III$''_\alpha$), expansions may be proved for discrete variables on irregular sets or on sufficiently dense lattices.

4 Sums of independent variables

In this section let $S_n = X_1 + \ldots + X_n \in V$, where $X_1, \ldots, X_n$ are independent, X_j with (finite) cumulants $\kappa_{1,j}, \ldots, \kappa_{s,j}$ and characteristic function ψ_j. Thus, the kth cumulant of S_n is $\chi_{k,n} = \kappa_{k,1} + \ldots + \kappa_{k,n}$ and the characteristic function of S_n is given by

$$\xi_n(t) = \prod_{j=1}^{n} \psi_j(t), \tag{4.1}$$

for $t \in V^*$, where V^* is the dual of V. Concerning the verification of the condition (III) of Theorem 3.1, we now have the following lemma which is a directional version of Lemma 5.5 of Skovgaard (1981a). For convenience we assume that fixed norms, denoted $\|.\|_0$, and corresponding Lebesgue measure λ_0 have been chosen on V and V^*.

LEMMA 4.1. *Let the sequences $(a_n(t))$ and (ε_n) of Theorem 3.1 be given. Then the condition* (i) *below implies that* (III$'_\alpha$) *in Theorem 3.4 holds for any α, while* (i) *and* (ii) *together imply that* (III) *in Theorem 3.1 holds.*

(i) *There exists a bounded set $K \subset V^*$ such that*

$$\gamma_n(t) = \inf\left\{\sum_{j=1}^{n} (1 - |\psi_j(ht)|^2);\, h > 0,\, ht \notin K\right\}, \tag{4.2}$$

which is a function of the direction of t, satisfies

$$\gamma_n(t)^{-1} = o((1 + \|t\|_n^2/a_n(t)^2)/\log(\|t\|_n/\varepsilon_n)), \tag{4.3}$$

as $n \to \infty$, uniformly in all unit vectors $t \in V^*$, $\|t\|_0 = 1$, where $\|t\|_n^2 = \Sigma_n(t^2)$.

(ii) *There is a finite subset $J \subset \mathbb{N}$, such that*

$$\int \prod_{j \in J} |\psi_j(t)|\, d\lambda_0(t) < \infty. \tag{4.4}$$

Proof. By repeated use of the general inequality for characteristic functions,

$$1 - |\psi_j(2t)|^2 \leqslant 4(1 - |\psi_j(t)|^2),$$

see, for example, Petrov (1975, Ch. 1, Lemma 1), and the estimate

$$\prod_{j \notin J} |\psi_j(t)| \leqslant \exp\left\{-\frac{1}{2}\sum_{j \notin J} (1 - |\psi_j(t)|^2)\right\},$$

we shall obtain a bound for the integral of $|\xi_n(t)|$. Let $K_n(t) = \sup\{h \geqslant 0;\, ht/\|t\|_n \in K\}$, and let $m_n(t)$ be the smallest nonnegative integer such that

$$2^{m_n(t)}\|t\|_n \geqslant K_n(t),$$

then $2^{m_n(t)} \leqslant \max\{1, 2K_n(t)/(\delta a_n(t))\}$ for all t satisfying $\|t\|_n > \delta a_n(t)$. For all such t we have

$$\sum_{j=1}^{n} (1 - |\psi_j(t)|^2) \geqslant 4^{-m_n(t)}\gamma_n(t) \geqslant \gamma_n(t) \min\{1, \delta^2 a_n(t)^2/(4K_n(t)^2)\}$$

and hence

$$\prod_{j \notin J} |\psi_j(t)| \leqslant \exp\{\tfrac{1}{2}|J|\} \exp\{-\tfrac{1}{2}c\gamma_n(t)/(1 + K_n(t)^2/a_n(t)^2)\}$$

for some constant $c > 0$, depending on δ and the set K, where $|J|$ is the number of elements in J. By assumption (i), since $K_n(t)$ is proportional to $\|t\|_n/\|t\|_0$, the right-hand side is of the form $\exp\{-\frac{1}{2}b_n \log[\|t\|_n/(\|t\|_0\varepsilon_n)]\}$, where $b_n \to \infty$ as $n \to \infty$ uniformly in t. This tends to zero quicker than any power of $(\varepsilon_n \|t\|_0/\|t\|_n)$, thus proving (III''_α) for any α. Furthermore, if (ii) holds,

$$\int_{\|t\|_n > \delta a_n(t)} \prod_{j=1}^{n} |\psi_j(t)|\, d\lambda_n(t) \le \int \Big(\prod_{j\in J} |\psi_j(t)|\Big) \|t\|_n^{-p}\, d\lambda_n(t) O(\varepsilon_n^k) = O(\varepsilon_n^k),$$

as $n \to \infty$, for any $k > 0$, where p is the dimension of V^*. This proves the lemma. □

Remark 4.1. Apart from the question of integrability, the standard condition would require $|\psi_j(t)|$ to be uniformly bounded away from 1 for $t \notin K$, and thus require $\gamma_n(t)$ to grow at the rate of n. In (4.3), we would usually take $a_n(t)$ to be proportional to $\|t\|_n$, and ε_n would be a (negative) power of $\|t\|_n$. Then (4.3) requires $\gamma_n(t)$ to grow quicker than $\log \|t\|_n$, which is comparable to $\frac{1}{2}\log n$ if the variance grows proportionally to n. Condition (4.3) thus allows the 'amount of discreteness' in the independent variables to increase with n.

Remark 4.2. Notice that there is no normalization of the sum S_n. Instead, the density is defined with respect to the Lebesgue measure defined by the variance of S_n. Besides the simplicity of this approach, the arbitrary choice of a square root of the variance matrix is avoided.

Remark 4.3. The generalization of the result to triangular arrays $S_n = X_{n,1} + \ldots + X_{n,n} \in V_n$ with $p_n = \dim V_n$ bounded, is trivial. We would have to replace $\|t\|_n$ in (4.3) by $K_n(t)$ as defined in the proof instead of introducing a fixed norm, and require that $K_n(t)/\varepsilon_n \to \infty$. In (4.4) the integral should be bounded by $K_n(t)$ raised to some power. It might, however, require some ingenuity to choose the sequence of sets K_n in (4.2) such that the conditions were fulfilled.

A simple case of special interest is when S_n is a weighted sum of identically distributed independent random vectors $Y_1, \ldots, Y_n \in V$. Thus let $a_1, a_2, \ldots$ be real numbers and define

$$S_n = a_1 Y_1 + \ldots + a_n Y_n \in V. \tag{4.5}$$

Then, either by use of Lemma 4.1 or directly from the results in § 3, we obtain the following result.

COROLLARY 4.2. *Let S_n be given by* (4.5) *and assume that*:

(i) *the distribution of Y_i has finite cumulant generating function in a neighbourhood of the origin, and nonsingular variance*;

(ii) *the characteristic function ξ of Y_i satisfies Cramér's condition, for some $K > 0$,*

$$\sup\{|\xi(t)|;\ \|t\|_0 > K\} < 1;$$

(iii) *as $n \to \infty$,*

$$m_n^2 = \sup\Big\{a_k^2 \Big/ \sum_{i=1}^{n} a_i^2;\ k = 1, \ldots, n\Big\} \to 0. \tag{4.6}$$

Then conditions (I), (II) *and* (IV) *of Theorem* 3.1 *hold for any s with $\rho_{s,n}(t) = a_n(t)^{-1} = O(m_n)$, and condition (III'_α) and hence the conclusion of Theorem* 3.4 *also hold. If also some power of ξ is integrable then all the conditions of Theorem* 3.1 *hold.*

The proof is trivial.

If instead of (4.5) we define S_n as $\langle t_1, Y_1\rangle + \ldots + \langle t_n, Y_n\rangle$, where $t_1, t_2, \ldots \in V^*$, then we need only change $a_k^2/\Sigma a_i^2$ in (4.6) to $\Lambda(t_k^2)/\Sigma\Lambda(t_i^2)$ in the corollary, where $\Lambda = \text{Var } Y_i$.

Example 4.1. Let $Y_1, \ldots, Y_n$ be independent real random variables, where $Y_j/(\alpha + \beta u_j)$ is Gamma distributed with known index $\omega_j > 0$. The 'covariates' $u_1, \ldots, u_n$ are known and positive while $\alpha > 0$ and $\beta > 0$ are unknown parameters. The score statistic is the (two-dimensional) random vector

$$S_n = \sum_{j=1}^{n} (Y_j - \omega_j/\theta_j)\theta_j^2(1, u_j) = \sum_{j=1}^{n} X_j,$$

say, where $\theta_j = (\alpha + \beta u_j)^{-1}$. The kth cumulant of S_n is given by

$$\chi_{k,n}(t^k) = (k-1)! \sum_{j=1}^{n} \omega_j(t_1 + t_2 u_j)^k/(\alpha + \beta u_j)^k. \tag{4.7}$$

where $t = (t_1, t_2)$. In particular $\Sigma_n = \chi_{2,n}$ is the Fisher information on (α, β), which we assume to be nonsingular. To obtain a bound on $\rho_{s,n}(t)$ in (3.1), we estimate

$$k!^{-1}\,|\chi_k(t^k)|/\|t\|_n^k \leqslant k^{-1}\,\|t\|_n^{-(k-2)} \sup_j\,(|t_1 + t_2 u_j|/(\alpha + \beta u_j))^{k-2}$$

implying that $\rho_{s,n}(t)$ for all $s \geqslant 3$ is bounded by $\rho_n(t)$,

$$\begin{aligned}\rho_n(t) &\leqslant \|t\|_n^{-1} \sup_j\,(|t_1 + t_2 u_j|/(\alpha + \beta u_j)) \\ &\leqslant \max\{\alpha^{-1}, \beta^{-1}\}/\inf\{\|t\|_n, t_1^2 + t_2^2 = 1\}.\end{aligned} \tag{4.8}$$

If both of the eigenvalues of the information tend to infinity as $n \to \infty$, then $\rho_n(t)$ tends to zero at the inverse rate of the square root of the 'slowest' eigenvalue. With $a_n(t) = \rho_n(t)^{-1}$ and $\varepsilon_n = \rho_n^{s-1}$,

$$\rho_n = \sup_t \rho_n(t),$$

it is seen that the conditions (I) and (II) of Theorem 3.1 hold for any s, while (IV) follows by use of Lemma 3.2, because we have established the bound (4.8). It remains to prove (III) by use of Lemma 4.1. The characteristic function of $\langle t, X_j\rangle$ is

$$\begin{aligned}\psi_j(ht) &= Ee^{ih\langle t, X_j\rangle} \\ &= (1 - ih(t_1 + t_2 u_j)/(\alpha + \beta u_j))^{-\omega_j} \exp\{-i\omega_j h(t_1 + t_2 u_j)/(\alpha + \beta u_j)\},\end{aligned}$$

and

$$|\psi_j(ht)|^2 = (1 + h^2(t_1 + t_2 u_j)^2/(\alpha + \beta u_j)^2)^{-\omega_j}$$

is strictly decreasing in $h > 0$. If we choose the set K in (4.3) as a sufficiently small neighbourhood of zero, then $\gamma_n(t)$ is asymptotically proportional to $\|t\|_n$; for details of this argument see Skovgaard (1981a, § 6). Thus (4.3) holds and as the set J in (4.4) we may take any two indices with different values of u_j. This proves the validity of the Edgeworth expansion for S_n to any order. It is constructed from the cumulants in (4.7) by use of the general formula. The single condition that is sufficient to prove the validity is that the information tends to infinity in all directions; that is, that both of the eigenvalues of the variance of S_n tend to infinity. This is so if and only if $\Sigma\omega_j/u_j^2 \to \infty$ and $\Sigma\omega_j(u_j - \bar{u})^2 \to \infty$ where $\bar{u} = \Sigma\omega_j u_j/\Sigma\omega_j$.

It may be noted that, without the present purely directional approach, it would be necessary to impose restrictions on the rate of the two eigenvalues compared to each other. Thus, for the similar example of Skovgaard (1981a, § 6), it was necessary to demand the slowest eigenvalue to grow at least at the rate of $\log n$. The reason for this difference is explained in Remark 3.3.

From a statistical point of view the proof of validity of Edgeworth expansions for the score function is the basis for expanding distributions of estimators and test statistics (Bhattacharya & Ghosh, 1978; Chandra & Ghosh, 1979). In the present example it is fairly easy to show, essentially by the method of Cramér (1946, § 33), that, if the information tends to infinity, the (local) maximum likelihood estimator is asymptotically normally distributed. It seems reasonable to conjecture that this condition is also necessary. The extension to higher-order expansions, by the method of Skovgaard (1981b), is a more complicated matter in this case, but follows the same kind of estimations as above. A sufficient condition for an Edgeworth expansion of order $s-1$ to be valid is that the variance of any polynomial of the form

$$\sum_{j=1}^{n} (y_j - \omega_j/\theta_j)(-\theta_j)^{-s}(a_0 + a_1u_j + \ldots + a_s u_j^s)$$

tends to infinity. This will be so if the u_j's do not converge too rapidly towards s limit points. This condition ensures the validity of a joint Edgeworth expansion of the first $s-1$ derivatives of the log-likelihood function, and these may in turn be transformed to an Edgeworth expansion of the maximum likelihood estimator $(\hat{\alpha}, \hat{\beta})$. It is quite plausible that this further condition is not necessary, but other methods would have to be used to establish the expansion for $(\hat{\alpha}, \hat{\beta})$ if it does not hold.

5 Sums of a certain type of dependent variables

In this section we shall use the results of § 3 to expand the sum of an infinite moving average process. Edgeworth expansions for dependent variables have been proved valid to a considerable generality by Götze & Hipp (1983), but, for the specific class considered here, their mixing conditions are stronger than needed for our purpose.

Let $\{\varepsilon_m, m \in \mathbb{Z}\}$ be independent and identically distributed random variables in a vector space V and define

$$X_j = \sum_{m=-\infty}^{\infty} b_m \varepsilon_{j-m}, \tag{5.1}$$

where the coefficients $\{b_m\}$ satisfy the conditions

$$\sum_{m=-\infty}^{\infty} |mb_m| < \infty, \quad \gamma = \sum_{-\infty}^{\infty} b_m \neq 0. \tag{5.2}$$

We want to expand the distribution of the sum

$$S_n = \sum_{j=1}^{n} X_j = \sum_{m=-\infty}^{\infty} \gamma_{m,n}\varepsilon_m, \tag{5.3}$$

say, where

$$\gamma_{m,n} = \sum_{j=1-m}^{n-m} b_j. \tag{5.4}$$

THEOREM 5.1. *Assume that the distribution of ε_j has finite moments of order $s+1$ and that some power of its characteristic function ψ is integrable. Then the conditions of Theorem 3.1 for $S_n = \Sigma X_j$, where X_j is given by (5.1) and (5.2), are fulfilled with*

$\varepsilon_n = n^{-(s-1)/2}$ and consequently the Edgeworth expansion of order $s-1$ will be valid for the density of S_n, and the terms will be bounded in order of magnitude by powers of $n^{-\frac{1}{2}}$.

Proof. Let $\kappa_1, \ldots, \kappa_s$ denote the first s cumulants of ε_j and assume for convenience that $\kappa_1 = 0$. The kth cumulant of S_n, for $2 \leqslant k \leqslant s$, is

$$\chi_{k,n} = \sum_{m=-\infty}^{\infty} \gamma_{m,n}^k \kappa_k = n\gamma^k \kappa_k + O(1), \tag{5.5}$$

as $n \to \infty$, where $\gamma = \Sigma b_m$. The last equality is obtained by straightforward calculations in which the first condition in (5.2) is used. Thus, in Theorem 3.1, $\rho_{s,n}(t)$ is of order $n^{-\frac{1}{2}}$. We take $a_n(t) = \rho_{s,n}(t)^{-1}$ and $\varepsilon_n = n^{-(s-1)/2}$. The characteristic function of S_n is

$$\xi_n(t) = \prod_{m=1}^{n} \psi(\gamma_{m,n} t), \tag{5.6}$$

which tends to zero at exponential rate when t is bounded away from zero. This follows from the integrability condition on ψ together with the fact that $\gamma_{m,n} \to \gamma$ for all m between δn and $(1-\delta)n$, say, where $0 < \delta < \frac{1}{2}$. The integrability of a fixed number of factors in the product (5.6) follows by Hölder's inequality because a power of each factor is integrable. If these, finitely many, factors are chosen in the interval $\delta n < m < (1-\delta)n$ (as above), then the integral of this finite product with respect to the Lebesgue measure, generated by the variance of S_n, is $O(n^{p/2})$, where $p = \dim V$. Thus, the 'tail integral' in (3.5) tends to zero at exponential rate in n. It remains to prove (3.6). Since $\gamma_{m,n}$ is bounded by $\Sigma\,|b_m| < \infty$, this may be done in terms of the bound of $\psi^{(s+1)}(t)$ in a compact neighbourhood of zero. This bound is finite because ε_j has finite moment of order $s+1$, implying that $\psi^{(s+1)}(t)$ is continuous. □

Remark 5.1. If, instead of the integrability condition on ψ in Theorem 5.1, we require that the Cramér condition

$$\sup\{|\psi(t)|;\ \|t\| > C\} < 1 \tag{5.7}$$

holds for some (and hence all) $C > 0$, then the conditions of Theorem 3.4 may be proved valid instead, such that the version (3.12) of the expansion is valid.

Remark 5.2. It is quite trivial to generalize the result to the (more natural) 'genuine' multivariate moving average process of the form

$$X_j = \sum_{m=-\infty}^{\infty} B_m(\varepsilon_{j-m}), \tag{5.8}$$

where $B_m : V \to W$, say, are linear mappings satisfying

$$\sum_{m=-\infty}^{\infty} |m|\,\|B_m\| < \infty, \quad \text{rank}\left(\sum_{m=-\infty}^{\infty} B_m\right) = \dim W \tag{5.9}$$

instead of (5.2), while the conditions on (ε_j) remain unchanged.

The results of this section do not require the 'directional' multivariate approach and could be proved by use of the theorem of Durbin (1980, § 5) or Skovgaard (1981a, § 5).

6 The lattice case

Suitably interpreted, results quite analogous to those of §§ 3 and 4 hold for the case of lattice random variables. The main changes are that densities are replaced by point probabilities and that integration, as in Corollary 3.3, should be replaced by summation. We shall state the analogue of Theorem 3.1 below, and comment upon the changes of Corollary 3.3 and Lemma 4.1. A basic description of lattice random vectors is given by Bhattacharya & Rao (1976, § 21), from where much of our notation and basic concepts are taken.

A lattice in a p-dimensional vector space V is a set of the form

$$L = \{v_0 + Z_1 v_1 + \ldots + Z_p v_p;\ Z_j \in \mathbb{Z},\ j = 1, \ldots, p\}, \tag{6.1}$$

where $v_0 \in V$ and $v_1, \ldots, v_p$ are linearly independent vectors in V. A random variable S in V is a lattice random variable on L if it takes values in L with probability one and the variance Σ of S is nonsingular. If there is no sublattice containing the support of S, then L is called a minimal lattice for S. Next, let $\eta_1, \ldots, \eta_p \in V^*$ be a dual basis to $v_1, \ldots, v_p$; that is $\langle \eta_j, v_j \rangle = \delta_{ij}$, where δ_{ij} is the Kronecker-delta. Define the set F^* in V^* associated with L by

$$F^* = \{t_1\eta_1 + \ldots + t_p\eta_p;\ |t_j| < \pi,\ j = 1, \ldots, p\}. \tag{6.2}$$

Let $\xi : V^* \to \mathbb{C}$ denote the characteristic function of S distributed on the lattice L, and let λ denote any Lebesgue measure on V^*, then the inversion formula takes the following form:

$$P\{S = x\} = \lambda(F^*)^{-1} \int_{F^*} \xi(t) e^{-i\langle t, x\rangle}\, d\lambda(t), \tag{6.3}$$

for $x \in L$, where $\lambda(F^*)$ is the Lebesgue measure of F^*.

Consider a sequence (S_n) of lattice random variables in vector spaces V_n, and let L_n be the minimal lattice for S_n. The Edgeworth approximation of order $s - 1$ to the point probabilities for S_n is

$$P_{s,n}(x) = (2\pi)^{p_n} f_{s,n}(x)/\lambda_n(F_n^*), \tag{6.4}$$

where $p_n = \dim V_n$, $f_{s,n}$ is the Edgeworth approximation of order $s - 1$ to the (continuous) density and λ_n is the Lebesgue measure generated by the variance Σ_n of S_n. For such an approximation to the point probabilities to be good, the lattice has to be sufficiently dense compared to the variance Σ_n. Thus, besides the norm $\|t_n\|$ of $t \in V^*$ defined as $\Sigma_n(t^2)^{\frac{1}{2}}$, we need a lattice norm of $t = t_1\eta_1 + \ldots + t_p\eta_p$ defined as

$$d_n(t) = (t_1^2 + \ldots + t_p^2)^{\frac{1}{2}}, \tag{6.5}$$

for $t \in V^*$, where we have omitted the index n on $(\eta_1, \ldots, \eta_p)$ from (6.2). This is the norm obtained by defining $\eta_1, \ldots, \eta_p$ as an orthonormal basis in V^*, and it measures t in 'lattice units'. The analogue of Theorem 3.1 is now as follows.

THEOREM 6.1. *Given the sequence of lattice random variables (S_n) on $(L_n \subset V_n)$ and a sequence (ε_n) of positive numbers converging to zero, suppose that for each $t \in V_n^*$, a positive number $a_n(t)$ exists, depending on t only through its direction $t/\|t\|_n$, such that for some fixed $s \geq 2$ the following conditions hold:*

(I) *$a_n(t)^{-(s-1)} = O(\varepsilon_n)$ as $n \to \infty$ uniformly in $t \in V_n^*$;* (6.6)

(II) *if $s \geq 3$, the following inequality holds for n sufficiently large.*

$$\rho_{s,n}(t) \leq a_n(t)^{-1} \tag{6.7}$$

for all $t \in V_n^*$;

(III) *for all* $\delta > 0$,

$$\int_{F_n^* \cap \{\|t\|_n > \delta a_n(t)\}} |\xi_n(t)| \, d\lambda_n(t) = O(\varepsilon_n) \tag{6.8}$$

as $n \to \infty$;

(IV) *for some* $\delta > 0$,

$$M_{s,n}(\delta a_n(t)t/\|t\|_n) = O(a_n(t)^{-(s-1)}), \tag{6.9}$$

as $n \to \infty$, *uniformly in* $t \in V_n^*$;

(V) *a constant* $\alpha > 0$ *exists, such that for n sufficiently large*

$$\|t\|_n^2/d_n(t)^2 \geq (2+\alpha)(\log \varepsilon_n^{-1})/\pi^2, \tag{6.10}$$

for $t \in V_n^*$.

Then

$$|P_n\{S_n = x\} - P_{s,n}(x)| = O(\varepsilon_n)\lambda_n(F_n^*)^{-1}, \tag{6.11}$$

as $n \to \infty$, *uniformly in* $x \in L_n$.

Proof. Quite similar to the proof of Theorem 3.1, except for the estimation

$$\int_{V_n^* \setminus F_n} |\xi_{s,n}(t:\{\chi_j\})| \, d\lambda_n(t) = O(\varepsilon_n),$$

which is proved by use of (6.10) which shows that the factor $\exp\{-\frac{1}{2}\|t\|_n^2\}$ is $O(\varepsilon_n^{1+\alpha/2})$ outside F_n^*, because $d_n(t)^2 \geq \pi^2$ outside F_n^*. □

Remark 6.1. Notice that the infimum of the ratio $\|t\|_n^2/d_n(t)^2$ in (6.10) is the smallest eigenvalue of Σ_n with respect to the inner product generated by the lattice L_n. Thus (6.10) is the condition that ensures the variance to grow sufficiently quickly compared to the distance between lattice points. In the case of sums of independently and identically distributed random vectors, $d_n(t)^2$ will be independent of n, while $\|t\|_n^2$ will be proportional to n, and the condition (V) causes no problem.

Remark 6.2. The bound in (6.11) has to be $O(\varepsilon_n)\lambda_n(F_n^*)^{-1}$ rather than just $O(\varepsilon_n)$ to be useful, because $\lambda_n(F_n^*)^{-1}$ is approximately proportional to the point probabilities near the 'centre' of the distribution.

As in Corollary 3.3 it may now be proved that the approximation may be summed over any subset of L_n, resulting in a uniform bound equal to the bound in (6.11) multiplied by $O((\log \varepsilon_n^{-1})^{p_n/2})$. The problem in the lattice case is that the summation of the Edgeworth densities is difficult, while integration with respect to the Lebesgue measure can be done explicitly. A thorough account of summation results for the lattice case may be found in § 23 of Bhattacharya & Rao (1976).

For sums of independent lattice random variables (on the same lattice), an analogue of Lemma 4.1, providing conditions for (6.8) to hold, may easily be proved. Here, the definition (4.3) of $\gamma_n(t)$ should be restricted to the infimum within $ht \in F_n^*$, then (4.4) remains unchanged, while condition (i) disappears because there is no problem with integrability in the lattice case.

Acknowledgements

I wish to thank Spencer B. Graves for many useful comments and suggestions at an earlier stage of the work, and Jens L. Jensen for valuable comments to the manuscript.

Appendix 1: Moments, cumulants and symmetric forms

Let $X \in V$ be a random variable in a finite-dimensional vector space V with $\dim V = p$, say. Consider the dual space V^* of V, that is the p-dimensional vector space of linear mappings of V into $\mathbb{R}$. If $t \in V^*$ and $x \in V$, we write $\langle t, x\rangle$ for the inner product of t and x, or the value of the mapping t at x. Thus, $\langle t, X\rangle$ is a real random variable, in coordinates it would be of the form $\Sigma t_i X_i$. The kth moment of X is defined as the mapping μ_k of $(V^*)^k$ into $\mathbb{R}$ given by

$$\mu_k(t_1, \ldots, t_k) = E\{\langle t_1, X\rangle \ldots \langle t_k, X\rangle\}, \tag{A.1}$$

where $t_1, \ldots, t_k$ are vectors in V^*. Thus, for suitably chosen t's, μ_k takes the values of the mixed kth moments of the coordinates of X. It is seen that μ_k is linear in each $t_j \in V^*$ and symmetric in $t_1, \ldots, t_k$. Thus, μ_k is a k-linear symmetric form on V^*. Symmetric forms are determined by their values on the 'diagonal', i.e. the values $\mu_k(t, \ldots, t)$, $t \in V^*$, which is the kth moment of the real random variable $\langle t, X\rangle$. In coordinates μ_k would be a (symmetric) k-sided array of mixed moments, $(\mu_{i_1 \ldots i_k})$, say, and, if $t_j = (t_{j1}, \ldots, t_{jp}) \in V^*$, the coordinate analogue of (A.1) would be

$$\sum_{i_1} \cdots \sum_{i_k} \mu_{i_1 \ldots i_k} t_{1 i_1} \ldots t_{k i_k}. \tag{A.2}$$

Similarly, the kth cumulant χ_k may be defined as the k-linear symmetric form of V^* into $\mathbb{R}$ given by

$$\chi_k(t_1 \ldots, t_k) = \text{cum}\,(\langle t_1, X\rangle, \ldots, \langle t_k, X\rangle), \tag{A.3}$$

where $t_1, \ldots, t_k \in V^*$ and cum(. . .) denotes the joint cumulant of the arguments. In particular the values on the 'diagonal' are

$$\chi_k(t, \ldots, t) = \text{cum}_k(\langle t, X\rangle), \tag{A.4}$$

i.e. the kth cumulant of the real random variable $\langle t, X\rangle$. To emphasize the difference from the common definitions, we shall sometimes refer to μ_k and χ_k as the kth moment form and the kth cumulant form, respectively.

For a k-linear form in general we shall use the following abbreviation for the diagonal values

$$A_k(t, \ldots, t) = A_k(t^k), \tag{A.5}$$

for $t \in V^*$, if A_k is a k-linear form on V^*, for example $A_k = \mu_k$. A linear combination of terms of the form (A.5) with $k = 0, \ldots, K$ is called a polynomial of V^* into $\mathbb{R}$, of degree K if A_K is not identically zero. Of particular interest are the bilinear forms, specifically the variance χ_2. If this is positive-definite, i.e.

$$\chi_2(t^2) > 0, \tag{A.6}$$

for all $t \in V^*$, then it is an inner product on V^* and turns V^* into a Euclidean space. Thus, the inner product of t_1 and t_2 in V^* is $\chi_2(t_1, t_2)$, which in matrix notation would be $t_1'\chi_2 t_2$, and the corresponding norm of $t \in V^*$ is $\|t\| = \chi_2(t^2)^{\frac{1}{2}}$. This norm is used frequently in the present paper. The inverse of this inner product is written χ_2^{-1} and is an inner product on V. This inverse is the usual inverse in matrix notation, a strict definition in the present notation is based on an identification of χ_2 with a linear mapping from V^* to V.

Corresponding to an inner product on V or V^*, as χ_2^{-1} and χ_2 above, is a unique (canonical) Lebesgue measure on V, respectively V^*, which assigns unit mass to the unit cube. This defines the measure because all the Lebesgue measures are proportional.

Notice that the mean of X, as defined in (A.1), is a linear mapping of V^* into $\mathbb{R}$, which may be identified with a (unique) point in V.

Appendix 2: Multivariate differentials

This section contains a (brief) definition of differentiable vector functions and their differentials. An introduction to multivariate differentiability is given by Apostol (1981, Ch. 12).

Let $f: V \to W$ be a function from one finite dimensional real vector space to another.

Definition A.1. A function $f: V \to W$ is said to be s times differentiable at $x_0 \in V$ if, for $k = 1, \ldots, s$, there exist k-linear mappings, denoted $D^k f(x_0)$, of V^k into W such that

$$f(x) = f(x_0) + (D^1 f(x_0))(x - x_0) + \ldots + \frac{1}{s!}(D^s f(x_0))((x - x_0)^s) + o(\|x - x\|^s) \quad \text{(A.7)}$$

as $x \to x_0$. The k-linear mapping $D^k f(x_0): V^k \to W$ is called the kth differential of f at x_0.

Definition A.2. A function $f: V \to W$ is said to be s times continuously differentiable at x_0, if f is s times differentiable in a neighbourhood of x_0 and the mapping $x \to D^s f(x)$ is continuous at x_0.

If f is k times differentiable, then the kth differential is given in terms of the real variables $h_1, \ldots, h_k$ by

$$(D^k f(x_0))(x_1, \ldots, x_k) = (d^k / dh_1 \ldots dh_k) f(x_0 + h_1 x_1 + \ldots + h_k x_k), \quad \text{(A.8)}$$

evaluated at $h_1 = \ldots = h_p = 0$, where $x_1, \ldots, x_k \in V$. In particular the diagonal values are

$$(D^k f(x_0))(x^k) = (d^k / dh^k) f(x_0 + hx) \quad \text{(A.9)}$$

evaluated at $h = 0$.

If f is s times continuously differentiable at $x_0 \in V$, then the k-linear differential $D^k f(x_0)$ is symmetric in its k arguments for $k = 1, \ldots, s$, and is therefore determined by its values on the diagonal, given in (A.9).

In coordinates the kth differential of a mapping of V into $\mathbb{R}$ at a point $x_0 \in V$, is the k-sided array of kth mixed partial derivatives. If the image is a vector space W, then a further index will denote the coordinate function in W.

References

Albers, W., Bickel, P.J. & van Zwet, W.R. (1976). Asymptotic expansions for the power of distribution free tests in the one-sample problem. *Ann. Statist.* **4,** 108–156.

Apostol, T. (1981). *Mathematical Analysis,* 2nd ed. Reading, Mass: Addison-Wesley.

Barndorff-Nielsen, O. E. & Cox, D.R. (1979). Edgeworth and saddle-point approximations with statistical applications. *J. R. Statist. Soc.* B **41,** 279–312.

Bhattacharya, R.N. & Ghosh, J.K. (1978). On the validity of the formal Edgeworth expansion. *Ann. Statist.* **6,** 434–451.

Bhattacharya, R.N. & Rao, R.R. (1976). *Normal Approximation and Asymptotic Expansions.* New York: Wiley.

Chandra, T.K. & Ghosh, J.K. (1979). Valid asymptotic expansions for the likelihood ratio statistic and other perturbed chi-square variables. *Sankhyā* A **41,** 22–47.

Chibisov, D.M. (1979). Asymptotic expansion for the distribution of a statistic admitting a stochastic expansion. Preprints in Statistics, No. 47. University of Cologne, Cologne.

Cramér, H. (1946). *Mathematical Methods of Statistics.* Princeton University Press.

Durbin, J. (1980). Approximations for densities of sufficient estimators. *Biometrika* **67,** 311–333.

Federer, H. (1969). *Geometric Measure Theory.* New York: Springer.

Grad, H. (1949). Note on the N-dimensional Hermite polynomials. *Comm. Pure Appl. Math.* **3,** 325–330.
Graves, S.B. (1983). Edgeworth expansions for discrete sums and logistic regression. Ph.D. Thesis. Univ. of Wisconsin, Madison.
Götze, F. (1985). Asymptotic expansions in functional limit theorems. *J. Mult. Anal.* **16,** 1–20.
Götze, F. & Hipp, C. (1978). Asymptotic expansions in the central limit theorem under moment conditions. *Z. Wahr. verw. Geb.* **42,** 67–87.
Götze, F. & Hipp, C. (1983). Asymptotic expansions for sums of weakly dependent random variables. *Z. Wahr. verw. Geb.* **64,** 211–239.
Hipp, C. (1984). Asymptotic expansions for conditional distributions: the lattice case. *J. Prob. Math. Statist.* **4,** 207–219.
Michel, R. (1979). Asymptotic expansions for conditional distributions. *J. Mult. Anal.* **9,** 393–400.
Petrov, V.V. (1975). *Sums of Independent Random Variables.* Berlin: Springer.
Pfanzagl, J. (1973). Asymptotic expansions related to minimum contrast estimators. *Ann. Statist.* **1,** 993–1026.
Skovgaard, I.M. (1981a). Transformation of an Edgeworth expansion by a sequence of smooth functions. *Scand. J. Statist.* **8,** 207–217.
Skovgaard, I.M. (1981b). Edgeworth expansions of the distributions of maximum likelihood estimators in the general (non i.i.d.) case. *Scand. J. Statist.* **8,** 227–236.
Skovgaard, I.M. (1985). A second order investigation of asymptotic ancillarity. *Ann. Statist.* **13,** 534–551.

Résumé

Nous présentons des conditions suffisantes pour la validité du développement d'Edgeworth d'une suite de variables aléatoires vectorielles. Une approche directionelle du développement des fonctions caractéristiques permet l'inclusion du cas où la vitesse de convergence dépend de la direction. Par cette approche, la démonstration est simplifiée parce qu'elle est ramenée au cas unidimensionnel. Développements uniformes sont déduites pour la fonction de la densité de probabilité, pour les probabilités au cas discontinu, et pour les probabilités d'ensembles boréliens. Les conditions se simplifient quand la fonction génératrice des moments existe dans un voisinage de l'origine, et également quand les variables aléatoires sont des sommes de variables indépendentes. Les conditions sont démontrées pour des processus d'ARMA, sous des conditions faibles sur les coefficients.

[*Received May* 1985, *revised October* 1985]

Chapter 4

Saddlepoint expansions for conditional distributions

Introduction by Jens Ledet Jensen
Aarhus University

To put this paper into perspective I will start with the work of Fredrik Esscher. Esscher (1932) was known among the insurance mathematics community, but not among statisticians, and Skovgaard's paper seems to be the first statistical paper to point to Esscher. Esscher's fundamental idea was to relate a tail probability $P(X > x)$ to an integral with respect to an exponential tilted distribution, that is, a member of the natural exponential family generated by X. The wording here is not fully established, and today in the finance community one often refers to the *Esscher transform.* The formula of Esscher states that

$$P(X > x) = \varphi(\theta)e^{-\theta x} \int_0^\infty e^{-\theta\sigma(\theta)y} \bar{F}_\theta(dy), \tag{4.1}$$

where $\varphi(\theta) = Ee^{\theta X}$ is the Laplace transform and, with $\kappa(\theta) = \log \varphi(\theta)$, θ is determined by $\kappa'(\theta) = x$, $\sigma(\theta)^2 = \kappa''(\theta)$ and $\bar{F}$ is the distribution of $(X - x)/\sigma(\theta)$ under the measure P_θ with $dP_\theta/dP = e^{\theta x}/\varphi(\theta)$. Esscher considered replacing $\bar{F}_\theta$ by a standard normal distribution or by the first two terms of an Edgeworth expansion. Esscher stated his formula in the context of an insurance model, but in Esscher (1963) the formula was given in a general form. In an acompanying paper Bohman (1963) stated the formula through an inversion integral:

$$P(X > x) = \frac{1}{2\pi} \int_{-\infty}^\infty \frac{\exp(-itx)}{1 + it/\{\theta\sigma(\theta)\}} \frac{\varphi\{\theta + it/\sigma(\theta)\}}{\varphi(\theta)} dt. \tag{4.2}$$

Bohman used this representation to discuss cases where the accuracy of the approximation improves as x tends to infinity. The (saddlepoint) approximation of Esscher is obtained on replacing $\exp(-itx)\varphi\{\theta + it/\sigma(\theta)\}/\varphi(\theta)$

by an expansion, the first term being the characteristic function of a standard normal distribution. Formulation (4.2) is slightly more convenient than (4.1) when considering the error term. Robinson (1982), not using Esscher's work, used (4.1) for a situation where the characteristic function does not have sufficient integrability properties to use (4.2).

Bahadur and Ranga Rao (1960) used (4.2), but, contrary to Esscher and Bohman, expanded $[1+it/\{\theta\sigma(\theta)\}]^{-1}$. In this way the resulting approximation cannot be used for θ tending to zero, corresponding to x tending to the mean of X. Lugannani and Rice (1980) made a different expansion of (4.2) giving rise to what today is referred to as the Lugannani–Rice formula. The underlying idea is to tranform $\kappa(T) - \kappa(\theta) - (T-\theta)\kappa'(\theta)$ into a quadratic function in a clever way so that the resulting approximation can be used for θ small as well. The expansion part then becomes the classical Laplace method of expanding an integral with a large parameter. The method is described in more detail in Daniels (1987). One reason for the popularity of the Lugannani–Rice formula is that the terms entering the expression have statistical interpretations.

The Lugannani–Rice formula can (of course) also be rederived from the saddlepoint approximation considered by Esscher. More importantly, though, is that the formula can be obtained based on an integration of the saddlepoint approximation to the density of X using a uniform asymptotic expansion framework described in Temme (1982). Consider expansion of

$$\int_\eta^\infty \frac{\sqrt{\alpha}}{\sqrt{2\pi}} \exp\left(-\frac{\alpha}{2}t^2\right) q(t)dt,$$

for α large. When $\eta = 0$ it is the parabolic form t^2 around zero that defines the situation. However, when $\eta > 0$ it is the linear approximation $t^2 \approx \eta^2 + 2\eta(t-\eta)$ that is important. Now, the simple trick is to subtract the main term, corresponding to replacing $q(t)$ by $q(0)$, and then perform integration by parts:

$$\int_\eta^\infty \frac{\sqrt{\alpha}}{\sqrt{2\pi}} \exp\left(-\frac{\alpha}{2}t^2\right) q(t)dt = q(0)\{1-\Phi(\sqrt{\alpha}\eta)\}$$
$$+ \frac{1}{\sqrt{2\pi\alpha}} \exp\left(-\frac{\alpha}{2}\eta\right) \frac{q(\eta)-q(0)}{\eta} + \frac{1}{\alpha}\int_\eta^\infty \frac{\sqrt{\alpha}}{\sqrt{2\pi}} \exp\left(-\frac{\alpha}{2}t^2\right) q_1(t)dt,$$

with $q_1(t) = \frac{d}{dt}[\{q(t)-q(0)\}/t]$ and Φ being the standard normal distribution function. The last integral has been multiplied by $1/\alpha$, but otherwise has the same structure as the original integral. Iterating the process we therefore obtain an expansion in powers of $1/\alpha$.

As mentioned, Ib seems to be the first to point to Esscher's work in a statistics paper. Ib's approach to the conditional distribution is not via Esscher's method, but rather along the lines of Lugannani and Rice. He considers the two-dimensional inversion integral for the one-dimensional tail probability, handles the integration with respect to one variable by a saddlepoint approximation, and the remaining integration along the lines of Lugannani and Rice (1980). Finally, the tail probability is divided by the saddlepoint approximation to the marginal density of a variable. Ib also briefly discusses the possibility of starting from the saddlepoint approximation to the two-dimensional density and then performing an integration, corresponding to the discussion above for the one-dimensional case. Finally, he includes a treatment of the lattice case motivated by Daniels (1987).

In this 1987 paper Ib illustrates the approximation using the hypergeometric distribution. More importantly, perhaps, Ib's approximation has been found very useful in statistical applications, as witnessed by the many citations of his paper. Ib himself has also contributed to the use of conditional test, as seen for example in the paper *Saddlepoint expansions for directional test* included in this volume.

More details on the relation between the different saddlepoint approximations can be found in my own book (Jensen, 1995) together with a discussion as in Bohman (1963) on the properties of the saddlepoint approximation in the extreme tail with $x \to \infty$.

J. Appl. Prob. **24**, 875–887 (1987)
Printed in Israel

SADDLEPOINT EXPANSIONS FOR CONDITIONAL DISTRIBUTIONS

IB M. SKOVGAARD,* *Royal Veterinary and Agricultural University, Copenhagen*

Abstract

A saddlepoint expansion is given for conditional probabilities of the form $P\{\bar{Y} \geqq y \mid \bar{X} = x\}$ where $(\bar{X}, \bar{Y})$ is an average of n independent bivariate random vectors. A more general version, corresponding to the conditioning on a $p-1$-dimensional linear function of a p-dimensional variable is also included. A separate formula is given for the lattice case. The expansion is a generalization of the Lugannani and Rice (1980) formula, which reappears if $\bar{X}$ and $\bar{Y}$ are independent. As an example an approximation to the hypergeometric distribution is derived.

CONDITIONAL PROBABILITY; HYPERGEOMETRIC DISTRIBUTION; LATTICE VARIABLE; TAIL PROBABILITY; UNIFORM SADDLEPOINT EXPANSION

1. Introduction

The saddlepoint method used to drive asymptotic approximations to integrals of a certain type is known to give remarkably good approximations. In Daniels (1954) it was shown that this technique was applicable to the problem of approximating densities of sums of independent random variables. In fact, to apply the resulting approximation, it is necessary to know the cumulant generating function for the statistic, the density of which is to be approximated, whereas it is immaterial in this sense whether it is a sum of independent random variables. However, in the case of an average, $\bar{X}$ say, of n independent replications, it is known that the relative error to the density of $\bar{x}$ is $O(n^{-1})$ as n tends to ∞, uniformly for $\bar{x}$ in a bounded set. For comparison the Edgeworth expansions typically keep a relative error of order $O(n^{-k/2})$ uniformly only within sets for which $\bar{x}/\sqrt{n}$ grow slowly. In this sense the saddlepoint expansions are large-deviation expansions. Similar expansions were derived by a different method in Esscher (1932) for distribution functions, except that the expansion was in powers of $n^{-1/2}$. By employing a technique outlined in Bleistein (1966), Lugannani and Rice (1980) derived

Received 31 July 1986; revision received 9 September 1986.
* Postal address: Royal Veterinary and Agricultural University, Department of Mathematics, Thorvaldsensvej 40, DK-1871 Frederiksberg C, Denmark.

saddlepoint expansions for distribution functions of a one-dimensional statistic. A review of this and related methods is given in Daniels (1987).

For a conditional density it is straightforward to obtain a large-deviation expansion, simply by approximating the numerator and denominator separately by a saddlepoint expansion. This is the so-called double saddlepoint approximation, see Barndorff-Nielsen and Cox (1979). It is easy to see that this approximation keeps the same properties in terms of the relative error uniformly within sets of large deviations as do the single saddlepoint expansions. It is the purpose of the present paper to derive a saddlepoint expansion for the conditional distribution function of one coordinate of a random vector given the others. A solution to this problem is simple if the conditional cumulant generating function is tractable for further calculations, but this may not always be the case. However, this possibility should be kept in mind as a preferable method whenever feasible, because the expansion derived below is based on a saddlepoint expansion of a multivariate integral, which may be less accurate than for one-dimensional integrals. Also the possibility of combining the two methods is worth considering, i.e. to calculate the conditional cumulant generating function given some of the coordinates directly, and then apply the formula given below to the conditional probability given the remaining conditioning coordinates. The calculation of the expansion for the conditional distribution function requires knowledge of the cumulant generating function for the distribution for the entire vector random variable under study, and the solution of two saddlepoint equations, one for this vector random variable and one for the vector of conditioning coordinates. The expression is given in Section 2 for the expansion of $P\{\bar{Y} \geqq \bar{y} \mid \bar{X} = \bar{x}\}$ where $(\bar{X}, \bar{Y})$ is the average of n independent replications of bivariate random vectors, and for the p-dimensional case where we condition on a linear function of dimension $p - 1$. We stick to the case of n independent replications to clarify the orders of terms, although the approximation may be used for other cases as well. Formally this is done merely by taking $n = 1$. We only state the first-order expansion; further terms may be obtained as described in Bleistein (1966), and they will be in orders of integer powers of n^{-1} relative to the main term uniformly in large deviation sets. Section 3 contains an outline of the proof based on inversion of characteristic functions. A brief sketch of another proof, which is somewhat simpler, is also given, because the method may be of some general interest. This consists of a saddlepoint expansion for the integral of Daniels' saddlepoint approximation to the density, and involves a technique from Bleistein (1966) for a saddlepoint near an endpoint of integration. The two methods lead to the same expansion, just as for unconditional probabilities where they both lead to the Lugannani and Rice expansion. However, the second method holds for the continuous case only, whereas the first one generalizes to the lattice case also, by modifications as in Daniels (1987), as is shown in Section 4. The second method of proof may be compared to the expansion obtained by applying the method of Laplace to the integral of the conditional saddlepoint density approximation. This method yields an expansion in powers of $n^{-1/2}$, and for the unconditional case it was shown by Robinson (1982) that it has an asymptotic behaviour similar to the expansion in Esscher (1932). For further discussion of these methods, see Daniels (1987). Finally, in Section 5 we apply the method to an example, namely the hypergeometric distribution, for which the approximation turns out to give an excellent agreement with the exact values. Since it can hardly be considered a problem to obtain exact values for the hypergeometric distributions, except possibly for magnitudes for which the normal approximation gives satisfactory values, this distribution is included merely as an example. More important applications may be to conditional tests in exponential family models, or to other statistical problems of conditional inference.

2. The expansion for the conditional distribution

Consider first the bivariate continuous case. Let $(X_1, Y_1), \cdots, (X_n, Y_n)$ be independent identically distributed bivariate random variables with density $f(x, y)$ at (x, y) and cumulant generating function κ given by

$$\kappa(s, u) = \ln \int\int f(x, y) \exp\{sx + uy\} du\, ds, \qquad s, u \in \mathbb{C}, \tag{2.1}$$

which we shall assume exists in a neighbourhood of (0, 0). Let $\bar{X} = n^{-1}\Sigma X_i$, $\bar{Y} = n^{-1}\Sigma Y_i$ and fix a point $(\tilde{x}, \tilde{y})$. We want to expand the conditional probability $P\{\bar{Y} \geqq \tilde{y} \mid \bar{X} = \tilde{x}\}$. Denote the derivatives of κ as follows:

$$\dot{\kappa}_s(s, u) = \frac{\partial}{\partial s}\kappa(s, u), \quad \dot{\kappa}_u(s, u) = \frac{\partial}{\partial u}\kappa(s, u),$$
$$\ddot{\kappa}_{ss}(s, u) = \frac{\partial^2}{\partial s^2}\kappa(s, u), \tag{2.2}$$

etc., and let $\ddot{\kappa}$ denote the 2×2 matrix of second derivatives. We need the saddlepoint, $\tilde{s}_0$ say, corresponding to $\bar{X} = \tilde{x}$, defined by the saddlepoint equation

$$\dot{\kappa}_s(\tilde{s}_0, 0) = \tilde{x}, \qquad \tilde{s}_0 \in \mathbb{R} \tag{2.3}$$

and the bivariate saddlepoint, $(\tilde{s}, \tilde{u})$ say, corresponding to $(\bar{X}, \bar{Y}) = (\tilde{x}, \tilde{y})$, defined by

$$\dot{\kappa}_s(\tilde{s}, \tilde{u}) = \tilde{x}, \quad \dot{\kappa}_u(\tilde{s}, \tilde{u}) = \tilde{y}, \qquad \tilde{s}, \tilde{u} \in \mathbb{R}. \tag{2.4}$$

The approximation now becomes

$$P\{\bar{Y} \geqq \tilde{y} \mid \bar{X} = \tilde{x}\}$$

$$\sim 1 - \Phi(\sqrt{n}\tilde{w}) + \phi(\sqrt{n}\tilde{w})\{[\ddot{\kappa}_{ss}(\tilde{s}_0, 0)]^{1/2}/(\sqrt{n}\tilde{u}\,|\ddot{\kappa}(\tilde{s}, \tilde{u})|^{1/2}) - (\sqrt{n}\tilde{w})^{-1}\}, \tag{2.5}$$

where Φ and ϕ are the standard normal distribution and density functions, respectively, $|\ddot{\kappa}(\tilde{s}, \tilde{u})|$ means the determinant of $\ddot{\kappa}(\tilde{s}, \tilde{u})$ and

$$\tilde{w} = \text{sign}(\tilde{u})\{2[\tilde{s}\tilde{x} + \tilde{u}\tilde{y} - \kappa(\tilde{s}, \tilde{u})] - 2[\tilde{s}_0\tilde{x} - \kappa(\tilde{s}_0, 0)]\}^{1/2}. \tag{2.6}$$

The quantity $\sqrt{n}\tilde{w}$ is the signed square root of minus twice the log likelihood ratio statistic, based on the observation $(\bar{X}, \bar{Y}) = (\tilde{x}, \tilde{y})$, for the hypothesis $u = 0$ in the generated exponential family with densities $f(x, y)\exp\{sx + uy\}$ parametrized by (s, u). The error of approximation (2.5) is $O(n^{-1})$ relative to the main term as $n \to \infty$, uniformly for $(\tilde{x}, \tilde{y}) \in K$, where K is some fixed compact neighbourhood of $(E\bar{X}, E\bar{Y})$. As this set K we may take any compact neighbourhood of $(E\bar{X}, E\bar{Y})$ which is interior to the set of points $(\dot{\kappa}_s(s, u), \dot{\kappa}_u(s, u))$ for which $\kappa(s, u)$ exists.

If X_i and Y_i are independent the approximation reduces to the one obtained by Lugannani and Rice (1980).

The generalization of (2.5) to conditional probabilities of the form $P\{\bar{X}_p \geqq \tilde{x}_p \mid (\bar{X}_1, \cdots, \bar{X}_{p-1}) = (\tilde{x}_1, \cdots, \tilde{x}_{p-1})\}$ is quite trivial as well as its proof. Stated in somewhat more generality, although trivially reduced to the case mentioned above, the multivariate version takes the following form, which may be useful for applications. Let $X_1, \cdots, X_n$ be independent identically distributed random vectors in $\mathbb{R}^p$, let A be a $p \times (p-1)$ matrix of rank $p-1$ and $b \in \mathbb{R}^p$ a vector which is linearly independent of the columns of A. With $\bar{X} = n^{-1}\Sigma X_i$ and $\bar{Y} = b'\bar{X}$ we get the approximation analogues to (2.5),

$$P\{\bar{Y} \geqq y \mid A'\bar{X} = \tilde{a}\}$$

$$\sim 1 - \Phi(\sqrt{n}\tilde{w}) + \phi(\sqrt{n}\tilde{w})\{|A'\ddot{\kappa}(\tilde{t}_0)A|^{1/2}/(\sqrt{n}\tilde{u}\,|\ddot{\kappa}(\tilde{t})|^{1/2}|(A, b)|) - (\sqrt{n}\tilde{w})^{-1}\}, \tag{2.7}$$

where $\ddot{\kappa}$ is the $p \times p$ matrix of second derivatives of the cumulant generating function, (A, b) is the $p \times p$ matrix that equals A with b appended as the last column, and the saddlepoints $\tilde{t}_0 = A\tilde{s}_0$ and $\tilde{t} = A\tilde{s} + b\tilde{u}$, both in $\mathbb{R}^p$, are given by the equations

$$A'\dot{\kappa}(A\tilde{s}_0) = \tilde{a}, \quad \tilde{s}_0 \in \mathbb{R}^{p-1}, \tag{2.8}$$

where

$$\dot{\kappa}(t) = \frac{\partial}{\partial t}\kappa(t),$$

and

$$A'\dot{\kappa}(\tilde{t}) = \tilde{a}, \quad b'\dot{\kappa}(\tilde{t}) = \tilde{y}, \tag{2.9}$$

which is equivalent to $\dot{\kappa}(\tilde{t}) = \tilde{x}$, where $\tilde{x}$ is the unique point with $A'\tilde{x} = \tilde{a}$, $b'\tilde{x} = \tilde{y}$. Finally $\tilde{w}$ is

$$\tilde{w} = \operatorname{sign}(\tilde{u})\{2[\tilde{t}\tilde{x} - \kappa(\tilde{t})] - 2[\tilde{s}_0\tilde{a} - \kappa(\tilde{t}_0)]\}^{1/2}. \tag{2.10}$$

If, in particular the columns of (A, b) are orthonormal, the determinant $|(A, b)|$ is 1, and the approximation is recognized as a generalization of (2.5).

The above expressions are all for the case of continuous variables. We return to the lattice case in Section 4.

As in the Lugannani and Rice formula the singularity at $\tilde{u} = 0$ corresponding to $\tilde{w} = 0$ is removable, such that the expressions (2.5) and (2.7) are analytic throughout their domain and should be replaced by their limiting value at $\tilde{u} = 0$, which in the general form corresponding to (2.7) is

$$P\{\bar{Y} \geqq b'\dot{\kappa}(A\tilde{s}_0) \mid A'\bar{X} = \tilde{a}\} \sim \frac{1}{2} - \left(\frac{1}{2\pi n}\right)^{1/2} \{\tfrac{1}{6}\kappa^{(3)}(v^3)/(v'\ddot{\kappa}v)^{3/2} + \tfrac{1}{2}\operatorname{tr}\{A(A'\ddot{\kappa}A)^{-1}A'[\kappa^{(3)}(v)]\}/(v'\ddot{\kappa}v)^{1/2}\}, \tag{2.11}$$

where the derivatives $\ddot{\kappa}$ and $\kappa^{(3)}$ are evaluated at $\tilde{t}_0 = A\tilde{s}_0$, $v \in \mathbb{R}^p$ is the vector $b - A(A'\ddot{\kappa}A)^{-1}A'\ddot{\kappa}b$, while $\kappa^{(3)}(v^3)$ is the three-fold product $\Sigma\Sigma\Sigma\kappa^{(3)}_{ijk}v_iv_jv_k$, where (v_i) and $\kappa^{(3)}_{ijk}$ are the coordinates of v and $\kappa^{(3)}$, respectively, and finally $\kappa^{(3)}(v)$ is the $p \times p$ matrix with (i, j)th coordinate $\Sigma\kappa^{(3)}_{ijk}v_k$. If $A'\bar{X}$ and $b'\bar{X}$ are independent the expression reduces to $\frac{1}{2} - \frac{1}{6}\lambda_3(2\pi n)^{-1/2}$, where λ_3 is the standardized third cumulant of $b'X_i$, in agreement with the Lugannani and Rice formula. The expression (2.11) is identical to the one obtained by the mixed Edgeworth saddlepoint approximation at this particular value, see Barndorff-Nielsen and Cox (1979); just as the Lugannani and Rice formula reduces to the Edgeworth expansion when evaluated at the mean.

It is often convenient for applications to restate the expression (2.7) in terms of the line of support of the conditional distribution. Thus let $c \in \mathbb{R}^p$ be a non-zero vector satisfying $A'c = 0$. Then, if we let $\tilde{x}$ denote the point in $\mathbb{R}^p$ such that $A'\tilde{x} = \tilde{a}$ and $b'\tilde{x} = \tilde{y}$, the conditional distribution of $\bar{X}$ is located on $\tilde{x} + rc$, $r \in \mathbb{R}$, and we may write $\bar{X} = \tilde{x} + \bar{R}c$, where $\bar{R} = (\bar{Y} - \tilde{y})/(b'c)$. Now approximation (2.7) becomes

$$P\{\bar{R} \geqq 0 \mid A\bar{X} = A\tilde{x}\} \sim 1 - \Phi(\sqrt{n}\tilde{w}) + \phi(\sqrt{n}\tilde{w})\{|\ddot{\kappa}(\tilde{t}_0)|^{1/2}(c'\ddot{\kappa}^{-1}(\tilde{t}_0)c)^{1/2}/(\sqrt{n}(c'\tilde{t})|\ddot{\kappa}(\tilde{t})|^{1/2}) - (\sqrt{n}\tilde{w})^{-1}\}, \tag{2.12}$$

where the sign of w should be redefined to equal that of $(c'\tilde{t})$, to take into account the possible change of 'direction' if $b'c < 0$. The saddlepoint equation (2.8) for $\tilde{t}_0$ is in this setting more conveniently stated as the solution to $A'\dot{\kappa}(\tilde{t}_0) = A\tilde{x}$ with the restriction $c'\tilde{t}_0 = 0$, while the saddlepoint $\tilde{t}$ from (2.9) is simply given by $\dot{\kappa}(\tilde{t}) = \tilde{x}$. The limit of (2.12) at $c'\tilde{t} = 0$ is identical to (2.11), except that v should be replaced by $(\ddot{\kappa}(\tilde{t}_0))^{-1}c$.

3. Conditions and proofs

We shall prove the expansion for the bivariate case for which the method is more transparent and, except for notation, practically identical to the one for the multivariate case. The following three conditions are required for the validity of the proof.

Conditions.

(I) Some positive power of the characteristic function for (X_i, Y_i) is integrable.

(II) The cumulant generating function κ for (X_i, Y_i) exists in a neighbourhood of $(0, 0)$.

(III) The variance matrix $\ddot{\kappa}(0, 0)$ is non-singular.

The last condition obviously causes no loss of generality, because we may restrict attention to a subspace if it does not hold.

In the proof below we shall outline the method but not pay much attention to the error term or to the possibility of deriving higher order expansions. That the error is of the form described in Section 2 follows from the methods being used, for which we refer to Daniels (1954), Bleistein (1966) and, in particular, Olver (1974), Chapters 3–4.

Proof. The inversion formula for the density $\bar{f}$ of $(\bar{X}, \bar{Y})$ is

$$(3.1)\qquad \bar{f}(\bar{x}, \bar{y}) = \left(\frac{n}{2\pi i}\right)^2 \int_{-i\infty}^{i\infty} \int_{-i\infty}^{i\infty} \exp\{n[\kappa(s, u) - s\bar{x} - u\bar{y}]\} ds\, du$$

from which we obtain the integral, $Q(\bar{y} \mid \bar{x})$ say, given by

$$(3.2)\qquad \begin{aligned} Q(\bar{y} \mid \bar{x}) &= \int_{\bar{y}}^{\infty} \bar{f}(\bar{x}, \bar{y}) d\bar{y} \\ &= \left(\frac{n}{2\pi i}\right)^2 \int_{c-i\infty}^{c+i\infty} \int_{-i\infty}^{i\infty} \exp\{n[\kappa(s, u) - s\bar{x} - u\bar{y}]\} ds \frac{du}{nu}, \end{aligned}$$

where $c > 0$ indicates a transformation of the path of integration to avoid the singularity at $u = 0$. In the case $c < 0$ we should subtract a certain quantity from the left-hand side, but for simplicity we stick to the case $c > 0$ in the sequel. For the integral in s in (3.2) for fixed u we now use a standard saddlepoint approximation, exactly as for the usual approximation to densities as in Daniels (1954). The saddlepoint, $\tilde{s}_u$ say, is defined by the equation

$$(3.3)\qquad \dot{\kappa}_s(\tilde{s}_u, u) = \bar{x}, \quad \tilde{s}_u \in \mathbb{R},$$

and the approximation to (3.2) becomes

$$(3.4) \qquad Q(\tilde{y} \mid \tilde{x}) \sim \left(\frac{n}{2\pi}\right)^{1/2} \frac{1}{2\pi i} \int_{c-i\infty}^{c+i\infty} \exp\{nh(u)\}(\ddot{\kappa}_{ss}(\tilde{s}_u, u))^{-1/2} \frac{du}{u},$$

where

$$(3.5) \qquad h(u) = \kappa(\tilde{s}_u, u) - \tilde{s}_u\tilde{x} - u\tilde{y}.$$

By differentiation of (3.3) with respect to u we see that $d\tilde{s}_u/du = -(\ddot{\kappa}_{ss}(\tilde{s}_u, u))^{-1}\ddot{\kappa}_{su}(\tilde{s}_u, u)$, and hence that the first two derivatives of h are

$$(3.6) \qquad \dot{h}(u) = \dot{\kappa}_u(\tilde{s}_u, u) - \tilde{y},$$

$$(3.7) \qquad \ddot{h}(u) = \ddot{\kappa}_{uu}(\tilde{s}_u, u) - (\ddot{\kappa}_{su}(\tilde{s}_u, u))^2/\ddot{\kappa}_{ss}(\tilde{s}_u, u).$$

To approximate (3.4) we use another saddlepoint expansion, modified as described in Bleistein (1966) to deal with the singularity at $u = 0$, and, in fact, quite similar to the one used to obtain the Lugannani and Rice formula; see Daniels (1986). The saddlepoint $\tilde{u} \in \mathbb{R}$ is obtained by equating (3.6) to zero, which leads to the equations (2.4). We introduce the new variable w by the equations

$$(3.8) \qquad \tfrac{1}{2}(w - \tilde{w})^2 = h(u) - h(\tilde{u}),$$

$$(3.9) \qquad \tfrac{1}{2}\tilde{w}^2 = h(0) - h(\tilde{u}), \quad \text{sign}(\tilde{w}) = \text{sign}(\tilde{u}),$$

such that $w = 0$ when $u = 0$, and $w = \tilde{w}$ corresponds to $u = \tilde{u}$. We now choose $c = \tilde{u}$ in (3.4), assume $\tilde{u} > 0$ to avoid technicalities, and rewrite (3.4) as

$$(3.10) \quad Q(\tilde{y} \mid \tilde{x}) \sim \left(\frac{n}{2\pi}\right)^{1/2} \exp\{nh(\tilde{u})\} \frac{1}{2\pi i} \int_{\tilde{w}-i\infty}^{\tilde{w}+i\infty} \exp\left\{\frac{n}{2}(w - \tilde{w})^2\right\} g(w) \frac{dw}{w},$$

where

$$(3.11) \qquad g(w) = \frac{du}{dw}\frac{w}{u}\{\ddot{\kappa}_{ss}(\tilde{s}_u, u)\}^{-1/2}$$

is analytic with a removable singularity at $w = 0$. In fact, we have

$$(3.12) \qquad g(0) = \{\ddot{\kappa}_{ss}(\tilde{s}_0, 0)\}^{-1/2}, \quad g(\tilde{w}) = \tilde{w}\tilde{u}^{-1}|\ddot{\kappa}(\tilde{s}, \tilde{u})|^{-1/2},$$

where we have derived the value of dw/du at $u = \tilde{u}$ by differentiating (3.8) twice with respect to u, which gives

$$(3.13) \qquad (w - \tilde{w})\frac{d^2w}{du^2} + \left(\frac{dw}{du}\right)^2 = \ddot{h}(u),$$

and the calculation is finished by insertion of (3.7) and taking $u = \tilde{u}$, $w = \tilde{w}$. Now, following Bleistein (1966), the integral in (3.10) is approximated by the integral obtained by replacing $g(w)$ with the linear function $g(0) + w\tilde{w}^{-1}(g(\tilde{w}) - g(0))$, which agrees with g at the two 'critical points'

$w = 0$ and $w = \tilde{w}$. Bleistein showed by a partial integration that the remainder then is an integral of the same form, but with a factor n^{-1}. Thus (3.10) is approximated by

$$Q(\tilde{y} \mid \tilde{x}) \sim \left(\frac{n}{2\pi}\right)^{1/2} g(0)\exp\{nh(0)\} \frac{1}{2\pi i} \exp\left\{-\frac{n}{2}\tilde{w}^2\right\}$$

$$\times \int_{\tilde{w}-i\infty}^{\tilde{w}+i\infty} \exp\left\{\frac{n}{2}(w-\tilde{w})^2\right\} \{w^{-1} + \tilde{w}^{-1}(g(\tilde{w})/g(0) - 1)\}dw$$

$$(3.14) \quad = \left(\frac{n}{2\pi}\right)^{1/2} \{\ddot{\kappa}_{ss}(\tilde{s}_0, 0)\}^{-1/2}\exp\{-n(\tilde{s}_0\tilde{x} - \kappa(\tilde{s}_0, 0))\}$$

$$\times [1 - \Phi(\sqrt{n}\tilde{w}) + \phi(\sqrt{n}\tilde{w})(\sqrt{n}\tilde{w})^{-1}(g(\tilde{w})/g(0) - 1)],$$

where the final expression is noticed to be valid also when $\tilde{u} < 0$, because to the order considered the same quantity should then be subtracted from this as from $Q(\tilde{y} \mid \tilde{x})$. The factor in front of the square brackets in the last expression is exactly the first-order saddlepoint approximation to the marginal density of $\bar{X}$ at $\tilde{x}$. Hence the result follows by insertion.

A second method of proof starts from the first-order saddlepoint approximation to the bivariate density

$$(3.15) \quad f(\tilde{x}, \tilde{y}) \sim \frac{n}{2\pi} |\ddot{\kappa}(\tilde{s}_y, \tilde{u}_y)|^{-1/2}\exp\{-n[\tilde{s}_y\tilde{x} + \tilde{u}_y\tilde{y} - \kappa(\tilde{s}_y, \tilde{u}_y)]\},$$

where $(\tilde{s}_y, \tilde{u}_y)$ is the bivariate saddlepoint corresponding to $(\tilde{x}, \tilde{y})$. To obtain an approximation to the integral $Q(\tilde{y} \mid \tilde{x})$ of $f(\tilde{x}, \tilde{y})$ from $\tilde{y}$ to ∞, we use a saddlepoint approximation to the integral of the right side of (3.15), modified as described in Bleistein (1966), Section 5 or Temme (1982), to take into account the endpoint of integration which may be close to the saddlepoint. This method appears a bit easier than the one given in the proof above, but it does not easily generalize to the lattice case as does the other method by slight modifications to be discussed in Section 4. To obtain higher-order terms, in the continuous case, by the second method based on integration of (3.15), we have to include higher-order terms in the expansion of $f(\tilde{x}, \tilde{y})$ as well as in the expansion of the integral. For such expansions it is probably easier to use the result in Temme (1982), which is adapted exactly to integrals of this type.

4. The lattice case

With modifications quite analogous to those described in Daniels (1987) for the Lugannani and Rice formula, the proof given in Section 3 carries over to the case when Y_i is a lattice random variable. The conditioning variable X_i may be continuous or lattice, but for the moment consider the bivariate case corresponding to Formula (2.5), except that we assume that $\mathbb{Z}^2$ is a minimal lattice for (X_i, Y_i). Let $n\tilde{x}$ and $n\tilde{y}$ be integers, then Formula (2.5) may be replaced by

$$P\{\bar{Y} \geqq \tilde{y} \mid \bar{X} = \tilde{x}\} \sim 1 - \Phi(\sqrt{n}\tilde{w}) + \phi(\sqrt{n}\tilde{w})\{[\kappa_{ss}''(\tilde{s}_0, 0)]^{1/2}/(\sqrt{n}(1 - \exp(-\tilde{u}))\,|\kappa''(\tilde{s}, \tilde{u})|^{1/2}) - (\sqrt{n}\tilde{w})^{-1}\}, \tag{4.1}$$

where the quantities $\tilde{w}$, $\tilde{s}_0$ and $(\tilde{s}, \tilde{u})$ are still as given in (2.3), (2.4) and (2.6). It may be preferable, for reasons discussed in Daniels (1987) to introduce a continuity correction and hence define

$$\hat{y} = \tilde{y} - \tfrac{1}{2}n^{-1}, \tag{4.2}$$

and the corresponding bivariate saddlepoint $(\hat{s}, \hat{u})$ by Equations (2.4) with $\tilde{y}$ replaced by $\hat{y}$. Furthermore $\hat{w}$ is a redefinition of $\tilde{w}$ in (2.6) with $\tilde{s}$, $\tilde{u}$ and $\tilde{y}$ replaced by $\hat{s}$, $\hat{u}$ and $\hat{y}$, respectively, whereas $\tilde{s}_0$ is unchanged. With this continuity correction, Formula (4.1) is replaced by

$$P\{\bar{Y} \geqq \tilde{y} \mid \bar{X} = \tilde{x}\} \sim 1 - \Phi(\sqrt{n}\hat{w}) + \phi(\sqrt{n}\hat{w})\{[\kappa_{ss}''(\tilde{s}_0, 0)]^{1/2}/(\sqrt{n}\,2\sinh(\tfrac{1}{2}\hat{u})\,|\kappa''(\hat{s}, \hat{u})|^{1/2}) - (\sqrt{n}\hat{w})^{-1}\}. \tag{4.3}$$

The expressions (4.1) and (4.3) are unchanged for the case when X_i is continuous, but Y_i still lattice. The asymptotic properties of the approximations (4.1) and (4.3) are the same as for the continuous case, i.e. the relative error is still $O(n^{-1})$ uniformly for $(\tilde{x}, \tilde{y})$ in a compact set. Also the conditions for validity, given in Section 3, are still the same except that Condition I is replaced by the condition that the lattice considered is a minimal lattice for the distribution in question. If there is a continuous component, such as X_i when X_i is continuous and Y_i lattice, this must still fulfil Condition I.

Concerning the proof the only change, except for minor trivial changes in the inversion formula, stems from the replacement in (3.2) of the integral from $\tilde{y}$ to ∞ by a sum, where ny runs from $n\tilde{y}$ to ∞. This summation results in the factor $(1 - \exp(-u))^{-1}$ instead of $(nu)^{-1}$ on the right side of (3.2), and this change corresponds exactly to the change from the approximation in (2.5) to the one in (4.1). Details concerning the continuity correction are found in Daniels (1987). A further advantage of the continuity correction is that it takes us 'away from the boundary' of the support, where the saddlepoint is undefined, except when the desired probability is trivially equal to 1.

Let us now turn to the general formulation where $\bar{X}$ is a p-dimensional random vector, and we consider the conditional distribution of $\bar{Y} = b'\bar{X}$ given $A'\bar{X}$ as in Formula (2.7). We assume that the conditional distribution of $\bar{Y}$ given $A'\bar{X} = \tilde{a}$ is a lattice distribution and that b has been scaled such that $n\bar{Y}$

has minimal lattice $y_0 + \mathbb{Z}$, where y_0 is some constant that may depend on $\tilde{a}$. If $\tilde{y}$ is a lattice point in this conditional distribution of $\bar{Y}$, the saddlepoint approximation corresponding to (4.3) now becomes

$$P\{\bar{Y} \geqq \tilde{y} \mid A'\bar{X} = \tilde{a}\} \sim 1 - \Phi(\sqrt{n}\hat{w})$$
$$+ \phi(\sqrt{n}\hat{w})\{|A'\ddot{\kappa}(\tilde{t}_0)A|^{1/2}/(\sqrt{n}2\sinh(\tfrac{1}{2}\hat{u})|\ddot{\kappa}(\hat{t})|^{1/2}|(A,\ b)|) - (\sqrt{n}\hat{w})^{-1}\} \tag{4.4}$$

with notation as in Formula (2.7), except for the continuity correction $\hat{y} = \tilde{y} - \frac{1}{2}n^{-1}$ and corresponding changes in the definitions (2.9) and (2.10) leading to $\hat{t}$ and $\hat{w}$ in place of $\tilde{t}$ and $\tilde{w}$, while $\hat{u}$ is still given by the relation $\hat{t} = A\hat{s} + b\hat{u}$. Whether $A'\bar{X}$ is lattice or continuous or a mixture of the two is immaterial for the approximation (4.4). The setup considered here trivially covers the case when $X_i \in \mathbb{Z}^p$ and we condition on a $(p-1)$-dimensional linear function of $\bar{X}$, in which case the remaining component will be either degenerate or lattice. The limit of the expression in (4.4) as $\hat{u} \to 0$ is identical to the expression (2.11). This is not so for the uncorrected version corresponding to (4.1).

As for the continuous case the analogue of version (2.12) of the formula may be useful. Thus, let $c \in \mathbb{R}^p$ satisfy $A'c = 0$, and let us further require that $b'c = 1$, such that $n\bar{X}$ may be written as $n\bar{X} = n\tilde{x} + n\bar{Z}c$, with $\mathbb{Z}$ as minimal lattice for $n\bar{Z} = n(\bar{Y} - \tilde{y})$. The continuity correction included in (4.4) becomes $\hat{x} = \tilde{x} - \frac{1}{2}n^{-1}c$, and the two saddlepoints are given by the equations $\dot{\kappa}(\hat{t}) = \hat{x}$ and $A\dot{\kappa}(\tilde{t}_0) = A\tilde{x}$ with the restriction $c'\tilde{t}_0 = 0$. Then Formula (4.4) takes the form

$$P\{\bar{Z} \geqq 0 \mid A\bar{X} = A\tilde{x}\} \sim 1 - \Phi(\sqrt{n}\hat{w})$$
$$+ \phi(\sqrt{n}\hat{w})\{|\ddot{\kappa}(\tilde{t}_0)|^{1/2}(c'\ddot{\kappa}^{-1}(\tilde{t}_0)c)^{1/2}/(\sqrt{n}2\sinh(\tfrac{1}{2}c'\hat{t})|\ddot{\kappa}(\hat{t})|^{1/2}) - (\sqrt{n}\hat{w})^{-1}\}, \tag{4.5}$$

where $\hat{w}$ is the continuity corrected form of (2.10), i.e.

$$\hat{w} = \operatorname{sign}(c'\hat{t})\{2[\hat{t}\hat{x} - \kappa(\hat{t})] - 2[\tilde{t}_0\hat{x} - \kappa(\tilde{t}_0)]\}^{1/2}. \tag{4.6}$$

The corresponding restatement of the uncorrected version corresponding to (4.1) is apparent from the above expressions.

5. Example: The hypergeometric distribution

The hypergeometric distribution is a one-dimensional distribution with distribution function

$$H_{n,r,N}(k) = \sum_{j=m}^{k} \binom{r}{j}\binom{N-r}{n-j} \Big/ \binom{N}{n}, \qquad m = \max\{0, n+r-N\} \tag{5.1}$$

where $k \leqq r, n$ and $r, n \leqq N$ and all numbers are non-negative integers. An attempt to apply the Lugannani and Rice formula leads to a complicated equation involving the hypergeometric function. The distribution may, however, be represented as a conditional distribution constructed from four independent Poisson random variables, and hence we may apply the result in Section 4, more specifically we shall use the continuity corrected version (4.5).

Let (X_{ij}), $i, j = 1, 2$, be four independent Poisson random variables with $EX_{ij} = 1$. Considered as a two by two table the marginal totals are $X_{i\cdot} = X_{i1} + X_{i2}$ and $X_{\cdot j} = X_{1j} + X_{2j}$, while $X_{\cdot\cdot} = \Sigma\Sigma X_{ij}$ is the total. The conditional distribution of X_{11} given the marginal totals is of the form (5.1) with $N = X_{\cdot\cdot}$, $n = X_{1\cdot}$ and $r = X_{\cdot 1}$. Thus we consider this conditional distribution, or equivalently the conditional distribution of the vector $X = (X_{11}, X_{12}, X_{21}, X_{22})'$ given the marginal totals.

Fix a point $(\breve{x}_{ij})$ and let $\breve{x}_{i\cdot}$, $\breve{x}_{\cdot j}$, $\breve{x}_{\cdot\cdot}$ denote the corresponding margins, and consider the conditional probability of $\{X_{11} \geqq \breve{x}_{11}\}$ given these. The conditional distribution of X is concentrated on the set $\{\breve{x} + zc;\, z \in \mathbb{Z}\}$, where c is the vector $(1, -1, -1, 1)'$, and the set is a minimal lattice for the distribution. Notice that in the notation we have chosen, the number of replications from the previous sections is $n = 1$. The continuity correction replaces $\breve{x} = (\breve{x}_{11}, \breve{x}_{12}, \breve{x}_{21}, \breve{x}_{22})'$ by the vector $\hat{x} = \breve{x} - \frac{1}{2}c$. The cumulant generating function and its first two derivatives are

$$
\begin{aligned}
\kappa(t) &= \sum_{i,j} (\lambda_{ij} - 1), \quad \lambda_{ij} = \exp(t_{ij}), \\
\dot{\kappa}(t) &= (\lambda_{11}, \lambda_{12}, \lambda_{21}, \lambda_{22})', \qquad (5.2)\\
\ddot{\kappa}(t) &= \operatorname{diag}(\lambda_{11}, \lambda_{12}, \lambda_{21}, \lambda_{22}),
\end{aligned}
$$

where $t = (t_{11}, t_{12}, t_{21}, t_{22})'$. Thus the saddlepoint $\hat{t}$, cf. (4.5), has coordinates $\hat{t}_{ij} = \ln \hat{\lambda}_{ij}$, say, where $\hat{\lambda}_{ij} = \hat{x}_{ij}$. The other saddlepoint $\tilde{t}_0$ with coordinates $\ln \tilde{\lambda}_{ij}$, say, must satisfy $c'\tilde{t}_0 = 0$ which is equivalent to $\tilde{\lambda}_{11}\tilde{\lambda}_{22}/(\tilde{\lambda}_{12}\tilde{\lambda}_{21}) = 1$, and it is easily seen that the solution is

$$
\tilde{\lambda}_{ij} = \breve{x}_{i\cdot}\breve{x}_{\cdot j}/\breve{x}_{\cdot\cdot}, \qquad (5.3)
$$

which in statistical jargon is 'the expected value' of X_{ij} in the model of independence between rows and columns. It is now a matter of insertion to obtain the expression corresponding to (4.5) for which we note that

$$
\begin{aligned}
|\ddot{\kappa}(\tilde{t}_0)|^{1/2} &= (\Pi \breve{x}_{i\cdot})(\Pi \breve{x}_{\cdot j})/\breve{x}_{\cdot\cdot}^2 \\
|\ddot{\kappa}(\hat{t})|^{1/2} &= \Big(\prod_{i,j} \hat{x}_{ij}\Big)^{1/2}, \\
c'\hat{t} &= \ln(\hat{x}_{11}\hat{x}_{22}/(\hat{x}_{12}\hat{x}_{21})) \qquad (5.4)\\
(c'\ddot{\kappa}^{-1}(\tilde{t}_0)c)^{1/2} &= \Big(\sum_{i,j} \tilde{\lambda}_{ij}^{-1}\Big)^{1/2}, \\
\hat{w} &= \operatorname{sign}(c'\hat{t})\Big\{2\Big[\sum_{i,j} \hat{x}_{ij}\ln \hat{x}_{ij} - \sum_{i} \breve{x}_{i\cdot}\ln \breve{x}_{i\cdot} - \sum_{j} \breve{x}_{\cdot j}\ln \breve{x}_{\cdot j} + \breve{x}_{\cdot\cdot}\ln \breve{x}_{\cdot\cdot}\Big]\Big\}^{1/2},
\end{aligned}
$$

which, inserted in (4.5), give the approximation

$$P\{X_{11} \geqq \tilde{x}_{11} \mid X_{1\cdot} = \tilde{x}_{1\cdot}, X_{2\cdot} = \tilde{x}_{2\cdot}, X_{\cdot 1} = \tilde{x}_{\cdot 1}\}$$

$$\sim 1 - \Phi(\hat{w}) + \phi(\hat{w}) \left\{ \left(\frac{\tilde{x}_{1\cdot}\tilde{x}_{2\cdot}\tilde{x}_{\cdot 1}\tilde{x}_{\cdot 2}}{\tilde{x}_{\cdot\cdot}\hat{x}_{11}\hat{x}_{12}\hat{x}_{21}\hat{x}_{22}} \right)^{1/2} (2 \sinh(\tfrac{1}{2}c'\hat{t}))^{-1} + \hat{w}^{-1} \right\}.$$

(5.5)

The approximation is only undefined when one of the marginal totals is 0 in which case the probability is 1. The quantity $\hat{w}$ is the signed square root of minus twice the log likelihood ratio statistic for the test of independence in the 2×2 table, except that it has been calculated with a continuity correction.

TABLE 1
Hypergeometric distribution. Approximation (5.5) to $P\{X_{11} \geqq x_{11} \mid x_{1\cdot}, x_{2\cdot}, x_{\cdot 1}\}$ compared to the exact value, for four independent Poisson variables (X_{ij}) with the same mean

x_{11} x_{21}	x_{12} x_{22}	Exact probability	Approximation (5.5)	Relative error (%)
85	5			
75	35	1.505×10^{-6}	1.501×10^{-6}	−0.3
14	6			
8	12	0.05548	0.05541	−0.1
5	3			
1	9	0.03167	0.03127	−1.3
5	1			
1	5	0.04004	0.03929	−1.9
6	0			
0	6	1.082×10^{-3}	0.976×10^{-3}	−9.9

Some numerical examples of the approximation (4.5) to $P\{X_{11} \geqq x_{11} \mid x_{1\cdot}, x_{2\cdot}, x_{\cdot 1}\}$ are given in Table 1. It is seen that for these examples the relative error never exceeds 10%. Because the example is merely included to give an impression of the quality of the approximation, it is not compared to alternative approximations. An extensive study of various kinds of normal, binomial and Poisson approximations to the hypergeometric distribution, is given in Molenaar (1970). If approximations are needed, however, it is clearly an advantage to have a single approximation that works well over the major range of parameters and distributions.

References

BARNDORFF-NIELSEN, O. AND COX, D. R. (1979) Edgeworth and saddle-point approximations with statistical applications. *J.R. Statist. Soc. B* **41**, 279–312.

BLEISTEIN, N. (1966) Uniform asymptotic expansions of integrals with stationary point near algebraic singularity. *Comm. Pure Appl. Math.* **19**, 353–370.

DANIELS, H. E. (1954) Saddlepoint approximations in statistics. *Ann. Math. Statist.* **25**, 631–650.

DANIELS, H. E. (1987) Tail probability approximations. *Internat Statist. Rev.* **55**, 37–48.

ESSCHER, F. (1932) On the probability function in the collective theory of risk. *Skand. Akt. Tidsskr.* **15**, 175–195.

LUGANNANI, R. AND RICE, S. O. (1980) Saddlepoint approximation for the distribution of the sum of independent random variables. *Adv. Appl. Prob.* **12**, 475–490.

MOLENAAR, W. (1970) *Approximations to the Poisson, Binomial and Hypergeometric Distribution Functions*. Mathematical Centre Tracts, No. 31. Mathematisch Centrum, Amsterdam.

OLVER, F. W. J. (1974) *Asymptotics and Special Functions*. Academic Press, New York.

ROBINSON, J. (1982) Saddlepoint approximation for permutation tests and confidence intervals. *J.R. Statist. Soc. B* **44**, 91–101.

TEMME, N. M. (1982) The uniform asymptotic expansion of integrals related to cumulative distribution functions. *Siam J. Math. Anal.* **13**, 239–253.

Chapter 5

Saddlepoint expansions for directional test probabilities

Introduction by Nicola Sartori
Università di Padova

The paper develops an expansion for the conditional tail probability of a multivariate continuous random variable, given its direction of departure from the expected value. The result is used to give an approximation for a directional test probability for a simple hypothesis, specified by a fixed value θ_0 of a vector parameter θ of a statistical model. This is achieved by applying the general result to the score function $D_1(\theta_0)$ of the model. In particular, the directional tail probability is defined as the conditional probability of the event $\{\|D_1(\theta_0)\| \geq \|D_1(\theta_0)_{\text{obs}}\|\}$ given the direction of $D_1(\theta_0)$, which is $D_1(\theta_0)/\|D_1(\theta_0)\|$, where $\|\cdot\|$ denotes the Euclidean norm and $D_1(\theta_0)_{\text{obs}}$ is the observed value of $D_1(\theta_0)$.

The directional tail probability proposed by Ib is essentially the same as the observed level of significance obtained with conical tests in Fraser and Massam (1985), although the original motivation in the latter is somewhat different. Fraser (1993) acknowledges that the term "directional" introduced by Ib is more appropriate and adopts it in subsequent work, which is mentioned at the end of this introduction.

The motivation for the use of a directional significance level is briefly discussed in §1. While conditioning on ancillary statistics or conditioning for the elimination of nuisance parameters are well-established and quite generally accepted routes to inference, it is not so obvious why one should condition on the observed direction of departure. In this respect, Ib avoids any argument for an inferential justification of this choice, as well as any consideration about power, and takes a more pragmatic point of view: the directional test is first order equivalent to the usual tests, such as the likelihood ratio test. Therefore, if standard asymptotic results have to be used for inference, the advantage of the directional test is that the result in the paper provides a saddlepoint approximation for the directional tail probability which guarantees a uniform relative error for large deviations. Without this conditioning, the saddlepoint method is more difficult to apply, because integration of the density over the level surface of the test statistic is required. The large deviations property of the proposed approximation can be particularly important for highly significant results and is the reason why Ib suggests that this method may be preferred also to the Bartlett corrected likelihood ratio test.

The use of this "conditioning for convenience" argument, as defined in Reid (1995, §6.1), is also present in the original proposal of Fraser and Massam (1985), although here the convenience is more apparent because of the availability of the expansion for the directional tail probability.

§2 contains the main result of the paper. Starting from $X_1, \ldots, X_n$ independent and identically distributed random vectors in $\mathbb{R}^p$ with distribution P, the exponential family generated by P is considered. Conditions are given, but are not too restrictive. Basically it is assumed that all moments of X_i exist. Theorem 1 then gives an expansion for the conditional distribution of $\bar{X} = n^{-1}(X_1 + \cdots + X_n)$ given its direction $U = \mathrm{dir}(\bar{X}) \in \mathbb{R}^p$, defined for $\bar{X} \neq x_0$ by $\bar{X} = x_0 + RU$, with $R = \|\bar{X} - x_0\|$ and $x_0 = E(X_i)$. A by-product of the calculation of the conditional distribution is an expansion for the density of U with respect to the uniform distribution on the unit sphere in $\mathbb{R}^p$. However, the main result is the expansion for $P\{R > \tilde{r} \mid U = \tilde{u}\}$, where $\tilde{x} = x_0 + \tilde{r}\tilde{u} \neq x_0$, which involves cumulants of X_i and chi-squared distribution functions.

The proof starts with a transformation from the density f_n of $\bar{X}$ to that of (R, U), giving the density

$$2\pi^{p/2}\Gamma(p/2)r^{p-1}f_n(x_0 + ru), \quad r > 0, \quad u \in V \subset \mathbb{R}^p, \quad \|u\| = 1\,. \tag{5.1}$$

Then it expresses the target tail probability $P\{R > \tilde{r} \mid U = \tilde{u}\}$ as a ratio of two integrals with integrand function (5.1). Such integrals are approximated using first a saddlepoint expansion for $f_n(x)$ and then, after a change of variable, using a method developed by Bleistein (1966). This method is designed to deal with a singularity or an end point near the saddlepoint, and essentially replaces the part of the integrand with singularities with a linear function that agrees with the original function at the singularity points.

There are three important aspects of the approximation for $P\{R > \tilde{r} \mid U = \tilde{u}\}$. First, the relative error is of order $O(n^{-1})$ in large deviations, as inherited from the saddlepoint approximation and the method of Bleistein. Second, when $p = 1$, the expansion agrees with that of Lugannani and Rice (1980), whose accuracy is well known. Finally, although the results are expressed in terms corresponding to the chosen basis, the approximation does not depend on this choice.

§3 particularizes the general result to full exponential families with distribution $\{P_\theta, \theta \in \Theta\}$. In particular, if $X_1, \ldots, X_n$ are independent random vectors from the distribution P_{θ_0}, where $\theta_0 \in \text{int}(\Theta)$, the conditional tail probability gives a significance level for testing the simple hypothesis that θ is equal to θ_0. Moreover, in this case the conditional test on $\|\bar{X} - x_0\|$ given the direction $\bar{X} - x_0$ is equivalent to the conditional likelihood ratio test given the direction of the score function, and is also first order equivalent to the likelihood ratio test.

Although the paper does not consider composite hypotheses, it covers the case of affine hypotheses in exponential families. This is done by using the result of §2 in the conditional model given an estimator under the hypothesis. This conditional model is still an exponential family in which the hypothesis is simple. Ib adapts the general formula without the explicit use of the conditional density, since this is proportional to the marginal density, restricted in a suitable subspace, and the normalizing constant cancels when conditioning on the direction. This version of the approximation is then used in the examples of §4.

The first example considers a normal linear regression model. With known variance, the directional tail probability approximation of any linear hypothesis on the mean is exact and equal to the significance level computed using the exact chi-squared distribution of the likelihood ratio test. This is because the length and the direction are independent. With unknown variance this equivalence is no longer true and the approximation for the conditional tail probability can be compared with the exact test based on the F distribution. A numerical comparison is carried out between these two and also with the usual chi-squared approximation for the likelihood ratio test and its Bartlett corrected version. For low-dimensional hypotheses the saddlepoint approximation has the usual accuracy, even with extremely small sample sizes, and outperforms the other approximations. On the other hand, with higher-dimensional hypotheses the accuracy of all approximations is unsatisfactory for practical purposes, unless the sample size is sufficiently large. In this respect, the second example about comparison of gamma random variables is less interesting, since it does not consider high-dimensional hypotheses.

Analytical approximations for directional tail probabilities have been considered in subsequent literature in Cheah *et al.* (1994) and Fraser and Reid (2006). The first paper is restricted to exponential families and proposes an approximation similar to Ib's, though harder to calculate. From the examples in the paper it is not clear if the additional computational effort is worth the limited gain in accuracy. The second paper introduces another analytical approximation for the directional tail probability which is valid for a general parametric model. It is based on a tangent exponential model and, like Ib's approximation, involves chi-squared distribution functions. The paper does not give numerical evidence and comparison with the other approximations. However, it is likely that the accuracy of all these analytical approximations might be strongly affected by high-dimensional parameter under the null hypothesis. This seems to be the case also for the modified likelihood ratio test proposed by Skovgaard (2001), as showed by the numerical evidence in Davison *et al.* (2014) and Fraser *et al.* (2016). In these papers, the directional tail probability is still computed using saddlepoint approximations for densities, but analytical approximations for integrals are replaced by two one-dimensional numerical integrations. The very accurate results of this method, also in extremely high-dimensional settings, seem to suggest that the lack of accuracy of the previous proposals is mostly due to the analytical approximations of the integrals, rather than to a lack of accuracy of the saddlepoint approximation itself.

J. R. Statist. Soc. B (1988)
50, *No.* 2, *pp.* 269–280

Saddlepoint Expansions for Directional Test Probabilities

By Ib M. SKOVGAARD†

Royal Veterinary and Agricultural University, Copenhagen, Denmark

[Received September 1986, Revised June 1987]

SUMMARY

An expansion is derived for the conditional tail probability of a multivariate random variable given its direction from the expected value. In particular, if the score function of a statistical model is chosen as this variable, such a conditional probability gives a test for a simple hypothesis, which is asymptotically equivalent to the likelihood ratio test. The expansion is of the large deviation type and is derived through the use of saddlepoint methods. Thus the relative error of the test probability is $O(n^{-1})$ uniformly in a bounded set for an average of n independent replications. The approximation is based on the cumulant transform for the random variable, but is otherwise a simple expression in terms of chi-squared distributions. In the one-dimensional case it reduces to the expansion obtained by Lugannani and Rice. Numerical examples show an excellent fit, comparable with other saddlepoint expansions, when the dimension is low, even for very small sample sizes, but for higher dimensions more replications are required to give a similar approximation.

Keywords: CONDITIONAL TAIL PROBABILITY; LARGE DEVIATION EXPANSION; SADDLEPOINT METHOD

1. INTRODUCTION

When testing a statistical hypothesis it is often convenient to use an asymptotic approximation to the test probability. The standard method is the chi-squared approximation, for example applied to the likelihood ratio test, but this method may be inaccurate for small or moderate sample sizes. An equally important problem is that it will usually not be known whether the result is sufficiently accurate. In the search for more accurate methods it is natural to consider saddlepoint approximations which are known to give remarkably accurate results even for small sample sizes: see, for example, Esscher (1932), Daniels (1954) and Lugannani and Rice (1980).

The classical application of the saddlepoint approximation is to sums of independent random variables, but the technique may, in modified form, be applied to densities of estimators when these are in a local one-to-one correspondence with the score function, which is usually the case: see Field (1982), Daniels (1983) and Skovgaard (1985). With test statistics, which are often asymptotically quadratic in the score function, the saddlepoint method is more difficult to apply, because the density has to be integrated over the level surface of the test statistic.

An attempt to overcome this difficulty is to modify the test by conditioning on the direction of the score function from the origin. More precisely, for a simple hypothesis specified by a fixed value θ_0 of a vector parameter θ, let $D_1(\theta_0)$ denote the derivative at θ_0 of the log-likelihood function. Then the 'directional test probability' is computed as the conditional probability of the event $\{\, \| D_1(\theta_0) \| \geqslant \| D_1(\theta_0)_{\text{obs}} \| \,\}$ given the

† *Address for correspondence*: Department of Mathematics, The Royal Veterinary and Agricultural University, Thorvaldsensvej 40, DK-1871 Frederiksberg C, Denmark.

 0035–9246/88/50269 $2.00

direction of $D_1(\theta_0)$, i.e. $D_1(\theta_0)/\| D_1(\theta_0) \|$, where $\| \ \|$ denotes the Euclidean norm and the suffix 'obs' refers to the observed value. This kind of test can only be applied to continuous distributions of $D_1(\theta_0)$, and so can the approximations to this test probability derived in this paper.

It is not the purpose here to claim that such a directional test should be preferred on any inferential grounds, but it is first order equivalent to the usual tests, such as the likelihood ratio test, and since it can be approximated by a saddlepoint approximation, as is shown in Section 2, it may be preferred to the other tests, if the standard asymptotic results were to be used otherwise. The saddlepoint method gives an approximation with a fairly constant relative error to the tail probability and consequently the method may be of value for highly significant results. Because it is not the point to choose a directional test because of any kind of optimality, power considerations are not included in the paper. A similar kind of directional test has been suggested in regression models for somewhat different reasons in Fraser and Massam (1985).

An alternative way of improving the chi-squared approximation to the likelihood ratio test is to use a Bartlett correction (Bartlett, 1954), which scales the chi-square distribution by a factor chosen to make the expectation correct to order $O(n^{-1})$: see Lawley (1956), Chandra and Ghosh (1979) and Barndorff-Nielsen and Cox (1984). Although this method is known to improve the approximation substantially, it does not possess the good asymptotic properties of the large deviation approximations like the saddlepoint expansions, namely of keeping a uniform relative error to the tail probability for large deviations. Even a factor of 2 is generally of no great practical importance for test probabilities, whereas an absolute error of 0.01, say, may be important if the correct result is 0.001.

For composite hypotheses the obvious generalisation of the directional test is as follows. Let θ parameterise the model and β the hypothesis, such that we have a mapping $\beta \to \theta$. Define $D_1(\theta)$ as before and let $\hat{\beta}$ denote the maximum likelihood estimator. Then we may calculate the conditional tail probability of $D_1(\hat{\theta})$ given $\hat{\beta}$ and the direction of $D_1(\hat{\theta})$, where $\hat{\theta}$ is the point corresponding to $\hat{\beta}$. The saddlepoint technique can be applied for this case also, but we shall not be concerned with composite hypotheses in this paper, except for cases that trivially reduce to simple hypotheses.

In Section 2 we derive the basic saddlepoint expansion for the tail probability of an average $\bar{X}$ given its direction from the expected value. As a by-product we obtain an expansion for the distribution of the direction. The relative error of the approximation is bounded by $O(n^{-1})$ uniformly for $\bar{x}$ in a bounded set. Further terms may be computed by the same technique, but they are hardly of practical interest.

In Section 3 we state the result in the notation of exponential families. There is no loss of generality in considering exponential families as long as we only consider the null distribution of $D_1(\theta_0)$; we only need to consider the exponential family generated by this statistic. However, for an exponential family the test is equivalent to the conditional test on the likelihood ratio given the direction of the score function. Finally, in Section 4, we consider some numerical examples.

The methodology used to derive the expansion is a combination of standard saddlepoint techniques (see, for example, Olver (1974)), saddlepoint expansions of densities (see Daniels (1954)) and uniform saddlepoint expansions with two nearby critical points (see Bleistein (1966) and references therein).

2. EXPANSION OF DIRECTIONAL TAIL PROBABILITIES

Let $X_1, \ldots, X_n$ be independent, identically distributed random vectors in $\mathbb{R}^p$. The distribution of X_i is assumed to be absolutely continuous with respect to the Lebesgue measure on $\mathbb{R}^p$. The cumulant transform of this distribution, denoted P, is

$$\kappa(t) = \log E \exp(\langle t, X_i \rangle), \quad t \in T \subset \mathbb{R}^p, \tag{2.1}$$
$$T = \{t \in \mathbb{R}^p \mid \int \exp(\langle t, x \rangle) \, \mathrm{d}P(x) < \infty\},$$

where $\langle t, x \rangle = \Sigma t^j x_j = t^j x_j$ is the inner product of $t = (t^1, \ldots, t^p)$ and $x = (x_1, \ldots, x_p)$, and we use the summation convention implying that summation over any index appearing twice in a term is understood. Superscripts are used instead of subscripts for co-ordinates in the parameter space. Note that

$$\| t \| = \langle t, t \rangle^{1/2}.$$

Although the results are expressed in terms corresponding to the chosen basis, the approximations that are obtained may be seen not to depend on this choice. A convenient choice may simplify the appearance of the results, as will be seen later.

Concerning the distribution P we shall assume that the following conditions are fulfilled.

Condition 1. Some power of the characteristic function of X_i is integrable.

Condition 2. The set $T \subset \mathbb{R}^p$ defined in equation (2.1) contains zero as an inner point.

Condition 3. The variance matrix of X_i is non-singular.

The last assumption is introduced for notational convenience only, since we might otherwise consider the observations as elements of a vector space of lower dimension. The second assumption implies that all moments of X_i exist.

We shall need the associated exponential family generated by P. It contains the distributions with densities with respect to P

$$g_t(x) = \exp(\langle t, x \rangle - \kappa(t)), \quad t \in T \subset \mathbb{R}^p. \tag{2.2}$$

The cumulants of the distribution with density g_t are the derivatives $\kappa'(t), \kappa''(t), \ldots, \kappa^{(k)}(t), \ldots$, for $t \in \operatorname{int}(T)$. In particular we denote the variance matrices and their inverses by

$$\Sigma(t) = \kappa''(t), \tag{2.3}$$
$$\Delta(t) = \Sigma(t)^{-1},$$

and for the distribution P corresponding to $t = 0$ we write Σ_0 for $\Sigma(0)$, Δ_0 for $\Delta(0)$, etc. Of particular importance is the *saddlepoint* $t_x \in \mathbb{R}^p$ corresponding to $x \in V$, defined by the saddlepoint equation

$$\kappa'(t_x) = x, \tag{2.4}$$

whenever this has a solution. From the theory of exponential families it follows that one such solution exists whenever

$$x \in M(P) = \kappa'(\text{int}(T)), \tag{2.5}$$

which contains $x_0 = EX$ as an inner point because of condition 2. The solution is always unique.

We want to expand the conditional distribution of $\bar{X} = n^{-1}(X_1 + \ldots + X_n)$ given its direction $U = \text{dir}(\bar{X}) \in \mathbb{R}^p$ defined for $\bar{X} \neq x_0$ by the equation

$$\bar{X} = x_0 + RU, \qquad R = \| \bar{X} - x_0 \|, \tag{2.6}$$

where $x_0 = EX$. As a by-product of the calculation we obtain an expansion of the density of U with respect to the uniform distribution on the unit sphere in $\mathbb{R}^p$. The expansions are given in the following theorem in a co-ordinate-free notation which is clarified by co-ordinate versions of some of the expressions. Note that, for example, $\Delta_0(u)$ for $u \in \mathbb{R}^p$ is a vector in $\mathbb{R}^p$, in co-ordinates $[\Delta_0]^{ij}u_i$, while $\Delta_0(u^2)$ denotes the scalar $[\Delta_0]^{ij}u_i u_j$, and correspondingly $\kappa_0^{(3)}(t^3)$, $t \in \mathbb{R}^p$, denotes the scalar $[\kappa_0^{(3)}]_{ijk} t^i t^j t^k$.

Theorem 1. The density of U with respect to the uniform distribution on the unit sphere in $\mathbb{R}^p$ is

$$|\Delta_0|^{1/2}[\Delta_0(\tilde{u}^2)]^{-p/2} h_n(\tilde{u})[1 + O(n^{-1})] \tag{2.7}$$

as $n \to \infty$ uniformly in $\tilde{u} \in \mathbb{R}^p$, $\|\tilde{u}\| = 1$, where

$$\begin{aligned} h_n(\tilde{u}) &= \{1 + c_p n^{-1/2}[\tfrac{1}{6}(p+1)[\Delta_0(\tilde{u}^2)]^{-3/2}\kappa_0^{(3)}\{[\Delta_0(\tilde{u})]^3\} \\ &\quad - \tfrac{1}{2}[\Delta_0(\tilde{u}^2)]^{-1/2}\,\text{tr}(\Delta_0\{\kappa_0^{(3)}[\Delta_0(\tilde{u})]\})]\}, \\ c_p &= \sqrt{2}\,\Gamma(\tfrac{1}{2}p + \tfrac{1}{2})/\Gamma(\tfrac{1}{2}p), \end{aligned} \tag{2.8}$$

and the co-ordinate version of the matrix within tr(...) in this expression for $h_n(\tilde{u})$ is

$$(\Delta_0\{\kappa_0^{(3)}[\Delta_0(\tilde{u})]\})^i_j = [\Delta_0]^{ia}[\kappa_0^{(3)}]_{abj}[\Delta_0(\tilde{u})]^b.$$

With F_k denoting the chi-squared distribution function on k degrees of freedom, we have the following expansion of the conditional tail probability for $\tilde{x} = x_0 + \tilde{r}\tilde{u} \neq x_0$:

$$\begin{aligned} &P\{R > \tilde{r} \mid U = \tilde{u}\} \\ &\quad = h_n(\tilde{u})^{-1}\{1 - F_p(n\tilde{s}^2) + c_p[1 - F_{p+1}(n\tilde{s}^2)]G(\tilde{x})/(n^{1/2}\tilde{s})\}[1 + O(n^{-1})], \end{aligned} \tag{2.9}$$

as $n \to \infty$ uniformly in $\tilde{x}$ in some compact neighbourhood of x_0, where

$$G(\tilde{x}) = |\Sigma_0|^{1/2}|\tilde{\Sigma}|^{-1/2}\{\Delta_0[(\tilde{x} - x_0)^2]\}^{p/2}\tilde{s}^{-p+2}(\langle \tilde{t}, \tilde{x} - x_0\rangle)^{-1} - 1, \tag{2.10}$$

$$\tilde{s} = [2(\langle \tilde{t}, \tilde{x}\rangle - \tilde{\kappa})]^{1/2}, \tag{2.11}$$

$\tilde{t} = t_{\tilde{x}}$ is the saddlepoint corresponding to $\tilde{x}$, and we have used the abbreviations $\tilde{\kappa} = \kappa(\tilde{t})$, $\tilde{\Sigma} = \Sigma(\tilde{t})$.

Proof. Let f_n denote the density of $\bar{X}$ with respect to the Lebesgue measure on

$\mathbb{R}^p$. Then the transformation $X \to (R, U)$ transforms this density into the density

$$A_{p-1} r^{p-1} f_n(x_0 + ru), \quad r > 0, \quad u \in V, \quad \| u \| = 1, \tag{2.12}$$

with respect to the Lebesgue measure on $\mathbb{R}_+$ by the uniform distribution on the unit sphere in $\mathbb{R}^p$, the measure of which is

$$A_{p-1} = 2\pi^{p/2}\Gamma(\tfrac{1}{2}p).$$

Consider now the 'tail integral'

$$J(\tilde{x}) = \int_{\tilde{r}}^{\infty} A_{p-1} r^{p-1} f_n(x_0 + r\tilde{u})\, dr. \tag{2.13}$$

The conditional tail probability to be approximated is

$$P\{R \geqslant \tilde{r} \mid U = \tilde{u}\} = J(\tilde{x})/\tilde{J}(0), \tag{2.14}$$

where

$$\tilde{J}(0) = \lim_{r \to \infty} J(x_0 + r\tilde{u}). \tag{2.15}$$

Thus the problem is to expand integrals of the form (2.13). The first step is to replace $f_n(x_0 + r\tilde{u})$ in equation (2.13) by its saddlepoint approximation, given in Daniels (1954) for $p = 1$,

$$f_n(x) \sim (n/2\pi)^{p/2} |\Sigma_x|^{-1/2} \exp[-n(\langle t_x, x\rangle - \kappa_x)], \tag{2.16}$$

where t_x is the saddlepoint corresponding to x and $\kappa_x = \kappa(t_x)$, etc. This approximation involves a relative error which is $O(n^{-1})$ uniformly in $x \in K$, say, where K is any compact subset of $M(P)$, defined in equation (2.5). By insertion of equation (2.16) into equation (2.13) we obtain the expansion

$$J(\tilde{x}) \sim \int_{\tilde{r}}^{r_{\max}} [2(\tfrac{1}{2}n)^{p/2}/\Gamma(\tfrac{1}{2}p)] r^{p-1} |\Sigma_x|^{-1/2} \exp[-n(\langle t_x, x\rangle - \kappa_x)]\, dr, \tag{2.17}$$

where $r_{\max} = \sup\{r;\, x_0 + r\tilde{u} \in K\}$. The relative error in equation (2.17) is also $O(n^{-1})$ uniformly in any compact $K_1 \subset \mathrm{int}(K)$ because the truncation of the integral results in an exponentially decreasing error only, relative to the main term. The saddlepoint corresponding to $x = x_0$ is zero, and the exponent in equation (2.17) is seen to attain its maximum value at this point and to decrease strictly as r increases. Thus we may change variables from r to $s > 0$ given by

$$\tfrac{1}{2}s^2 = \langle t_x, x_0 + r\tilde{u}\rangle - \kappa_x, \quad s > 0, \tag{2.18}$$

and rewrite the integral as

$$J(\tilde{x}) \sim \int_{\tilde{s}}^{s_{\max}} k_p(s)\, g(s)\, ds, \tag{2.19}$$

where

$$g(s) = (r/s)^{p-1}(dr/ds) |\Sigma_x|^{-1/2}, \tag{2.20}$$

and

$$k_p(s) = [2(\tfrac{1}{2}n)^{p/2}/\Gamma(\tfrac{1}{2}p)] s^{p-1} \exp(-\tfrac{1}{2}ns^2) \tag{2.21}$$

is the density of the square root of a chi-squared distribution with p degrees of freedom, scaled by the factor n^{-1}. The function g is analytic in the interval $[0, s_{max})$ with a removable singularity at zero. To obtain an expansion of the integral (2.19), that holds uniformly as x approaches zero, we use a method developed by Bleistein (1966), designed to cope with a singularity or an endpoint near the saddlepoint, and based on a method of partial integration. This method consists of the approximation of integral (2.19) by the corresponding integral with g replaced by a linear function that agrees with g at the two 'critical points' $s=0$ and $s=\tilde{s}$. Higher order expansions may be obtained by repeated use of such approximations, but that is not necessary for our purpose. Thus in equation (2.19) we replace $g(s)$ by

$$g(s) \sim g(0)\{1 + s\tilde{s}^{-1}[g(\tilde{s})/g(0) - 1]\}, \tag{2.22}$$

integrate from $\tilde{s}$ to infinity and obtain the expansion

$$J(\tilde{x}) \sim g(0)\{1 - F_p(n\tilde{s}^2) + c_p n^{-1/2}\tilde{s}^{-1}[g(\tilde{s})/g(0) - 1][1 - F_{p+1}(n\tilde{s}^2)]\} \tag{2.23}$$

where c_p is defined in equation (2.8). It only remains to calculate $g(0)$, $g(\tilde{s})$ and the limiting value of $J(\tilde{x})$ as $\tilde{r} \to 0$,

$$\tilde{J}(0) = g(0)\{1 + c_p n^{-1/2}[g'(0)/g(0)]\}. \tag{2.24}$$

To calculate the desired values of g and g' we differentiate equation (2.18) three times with respect to s and obtain the equations

$$s = \langle t_x, \tilde{u}\rangle(\mathrm{d}r/\mathrm{d}s),$$

$$1 = \langle t_x, \tilde{u}\rangle(\mathrm{d}^2r/\mathrm{d}s^2) + \Delta_x(\tilde{u}^2)(\mathrm{d}r/\mathrm{d}s)^2,$$

$$0 = \langle t_x, \tilde{u}\rangle(\mathrm{d}^3r/\mathrm{d}s^3) + 3\Delta_x(\tilde{u}^2)(\mathrm{d}^2r/\mathrm{d}s^2)(\mathrm{d}r/\mathrm{d}s) - \kappa_x^{(3)}\{[\Delta_x(\tilde{u})]^3\}(\mathrm{d}r/\mathrm{d}s)^3,$$

from which we obtain

$$\mathrm{d}r/\mathrm{d}s = s/\langle t_x, \tilde{u}\rangle \to [\Delta_0(\tilde{u}^2)]^{-1/2} \quad \text{as} \quad s \to 0,$$

$$\mathrm{d}^2r/\mathrm{d}s^2 \to \tfrac{1}{3}(\kappa_0^{(3)}\{[\Delta_0(\tilde{u})]^3\})[\Delta_0(\tilde{u}^2)]^{-2} \quad \text{as} \quad s \to 0.$$

It is now straightforward to calculate $g(0) = |\Sigma_0|^{-1/2}[\Delta_0(\tilde{u}^2)]^{-p/2}$ and $g(\tilde{s})$, while

$$\begin{aligned} g'(0)/g(0) &= -\tfrac{1}{2}\frac{\mathrm{d}}{\mathrm{d}s}\log|\Sigma_x| + \tfrac{1}{2}(p+1)(\mathrm{d}^2r/\mathrm{d}s^2)/(\mathrm{d}r/\mathrm{d}s) \\ &= \tfrac{1}{6}(p+1)[\Delta_0(\tilde{u}^2)]^{-3/2}\kappa_0^{(3)}\{[\Delta_0(\tilde{u})]^3\} - \tfrac{1}{2}\operatorname{tr}(\Delta_0\{\kappa_0^{(3)}[\Delta_0(\tilde{u})]\})[\Delta_0(\tilde{u}^2)]^{-1/2} \end{aligned} \tag{2.25}$$

where all derivatives are calculated at $s=0$. It is now a matter of insertion to obtain the result, except for the statement about the error which follows from the method of Bleistein along the same lines, as the analogous result is shown for the saddlepoint approximation (2.16). □

If we choose the basis such that Σ_0, and hence Δ_0, is the identity matrix, then $\Delta_0(\tilde{u})^2 = 1$ and the approximation in equation (2.7) to the density of U becomes $h_n(\tilde{u})$. Thus the factor in front of $h_n(\tilde{u})$ is a correction factor originating from a change of basis and, in consequence, of the inner product.

In the one-dimensional case, $p=1$, the expansion reduces to that obtained by

Lugannani and Rice (1980) (see also Daniels (1987)). Whether this agreement holds if the expansion is carried to higher order is unclear.

When computing the approximation in applications, it may be convenient to consider the entire set of observations as one (vector) observation and consequently to put $n = 1$. For n independent replications this will lead to the expansion given.

3. APPLICATIONS TO EXPONENTIAL FAMILIES

Let $\{P_\theta, \theta \in \Theta)$ be a full exponential family on $\mathbb{R}^p$, with densities

$$\exp[\langle \theta, x\rangle - \psi(\theta)], \quad \theta \in \Theta \subset \mathbb{R}^p, \tag{3.1}$$

with respect to some absolutely continuous measure on $\mathbb{R}^p$. If $X_1, \ldots, X_n$ are independent random vectors from the distribution $P_0 = P_{\theta_0}$, where $\theta_0 \in \operatorname{int}(\Theta)$, the cumulant transform of X_i is

$$\kappa(t) = \log \int \exp\langle t, x\rangle \, \mathrm{d}P_0(x) = \psi(\theta_0 + t) - \psi(\theta_0) \tag{3.2}$$

for $\theta_0 + t \in \Theta$. Thus the saddlepoint equation is

$$\kappa'(t_x) = x \Leftrightarrow t_x = \theta_x - \theta_0 \tag{3.3}$$

where θ_x denotes the maximum likelihood estimator corresponding to $\bar{X} = x$, which is given by the equation $\psi'(\theta_x) = x$, whenever this has a solution.

The associated exponential family, see equation (2.2), is the family with densities (3.1) with $\theta = \theta_0 + t \in \Theta$, and with the notation from Section 2 we have the identities

$$\Sigma_0 = \psi''(\theta_0) = \kappa''(0), \; \Delta_0 = \Sigma_0^{-1}, \; \kappa_0^{(3)} = \psi^{(3)}(\theta_0), \; \tilde{\Sigma} = \psi''(\tilde{\theta}) = \kappa''(\tilde{t}), \; \tilde{t} = t_{\tilde{x}} = \tilde{\theta} - \theta_0, \tag{3.4}$$

such that corresponding to the value $\tilde{x}$ of $\bar{X}$ we have the maximum likelihood estimate $\tilde{\theta}$ and the saddlepoint $\tilde{t} = \tilde{\theta} - \theta_0$. The expansion for the conditional tail probability of $\bar{X}$ given the direction from the mean $x_0 = E\bar{X} = \psi'(\theta_0)$ is now obtained directly from equations (2.9) and (2.8) where we note that

$$h_n(\tilde{u}) = h_n(\tilde{x} - x_0),$$

and

$$n\tilde{s}^2 = -2 \log Q = 2n[\langle \tilde{\theta} - \theta_0, x\rangle - \psi(\tilde{\theta}) + \psi(\theta_0)], \tag{3.5}$$

is minus twice the log-likelihood ratio test for the hypothesis $\theta = \theta_0$ within the family (3.1). The transformation from $\| \tilde{x} - x_0 \| = \tilde{r}$ to $\tilde{s}$ is monotone in any fixed direction, such that the conditional test on $\| \bar{X} - x_0 \|$ given the direction of $\bar{X} - x_0$ is equivalent to the conditional likelihood ratio test given this direction. The dependence of the approximate test probability for a fixed value of $-2 \log Q$ on the direction is clearly seen from equations (2.8) and (2.9). In particular the two tails in the one-dimensional case differ by a term of order $O(n^{-1/2})$ if the third cumulant is non-zero, as noted by McCullagh (1984). Also, it may be seen that, because the limiting distribution of the direction is uniform when Δ_0 is the identity matrix, the directional test is first order equivalent with the likelihood ratio test, and hence with several other commonly used tests.

If we want to test an affine hypothesis in the exponential family (3.1), i.e. a hypothesis of the form $\theta = \theta_0 + A\beta$, where A is a $p \times q$ matrix of rank q and $\beta \in \mathbb{R}^q$, then the result of Section 2 may also be used in slightly modified form. The reason is that if we condition on the estimator under the hypothesis, or equivalently on $A^t\bar{X}$, where A^t is the transpose of A, we obtain an exponential family for which the hypothesis is simple. We do not need to compute the densities of this conditional family, or rather the cumulant transform for the conditional null density, although this may be numerically a better solution when it is feasible. We want to approximate the conditional distribution of $\| \bar{X} - \hat{\psi}_0' \|$ given its direction and given $\hat{\theta}_0$, where $\hat{\psi}_0' = \psi'(\hat{\theta}_0)$ and $\hat{\theta}_0$ is the maximum likelihood estimate of θ under the hypothesis. Conditioned on $\hat{\theta}_0$, or equivalently on $A^t\bar{X}$, the distribution of $\bar{X}$ is restricted to a $(p-q)$-dimensional subspace, and its density on this space is proportional to the unconditional density. The normalisation constant cancels when we condition on the direction, and therefore the approximation to the directional test becomes the same as for the simple hypothesis except that p should be replaced by $p-q$ throughout. Thus for this case the expansion becomes

$$P\{ \| \bar{X} - \hat{\psi}_0' \| \geqslant \| \tilde{x} - x_0 \| \mid (\bar{X} - \hat{\psi}_0')/\| \bar{X} - \hat{\psi}_0' \| = (\tilde{x} - x_0)/\| \tilde{x} - x_0 \|, \hat{\psi}_0' = x_0\}$$
$$\sim [h_n(\tilde{x} - x_0)]^{-1}\{1 - F_{p-q}(n\tilde{s}^2) + c_{p-q}[1 - F_{p-q+1}(n\tilde{s}^2)]G(\tilde{x})/(n^{1/2}\tilde{s})\}, \tag{3.6}$$

where

$$h_n(y) = 1 + c_{p-q}n^{-1/2}[\tfrac{1}{6}(p+1-q)[\hat{\Delta}_0(y^2)]^{-3/2}\hat{\kappa}_0^{(3)}\{[\Delta_0(y)]^3\}$$
$$-\tfrac{1}{2}[\hat{\Delta}_0(y^2)]^{-1/2}\,\mathrm{tr}(\hat{\Delta}_0\{\hat{\kappa}_0^{(3)}[\Delta_0(y)]\})],$$
$$G(\tilde{x}) = |\hat{\Sigma}_0|^{1/2}|\tilde{\Sigma}|^{-1/2}\{\hat{\Delta}_0[(\tilde{x} - x_0)^2]\}^{(p-q)/2}\tilde{s}^{-p+q+2}\langle\tilde{\theta} - \hat{\theta}_0, \tilde{x} - x_0\rangle^{-1} - 1,$$

while $\tilde{s}$ is given in equation (3.5), where θ_0 means $\hat{\theta}_0$, c_p is given in equation (2.8) and the values of $\hat{\Delta}_0$, $\hat{\Sigma}_0$ etc. are evaluated at the estimate $\hat{\theta}_0$ under the hypothesis.

This form of the expansion, which has the same asymptotic properties as that for the simple hypothesis, will be used for the examples in Section 4. The possibility of computing the cumulant transform for the conditional distribution given A^tX does not lead to tractable expressions for these cases. The 'intermediate' possibilities of conditioning not on A^tX itself, but on some linear function of it, for which the conditional cumulant transform is sufficiently simple, before applying the approximation (3.6), is worth noting.

4. NUMERICAL EXAMPLES

In the one-dimensional case our problem reduces to the problem of approximating distribution functions, in which case the astonishing accuracy of the saddlepoint approximations is well established. Hence we shall deal only with multivariate cases and, to consider examples of some practical relevance, choose two examples of canonical hypotheses in exponential families as discussed in Section 3.

4.1. *Example 1: Normal Linear Models*

In the case of known variance it is easy to see that the approximation to the directional test of any linear hypothesis is exact, and that this test is equivalent to the chi-squared test based on the mean squared deviations.

If the variance is unknown the results are no longer exact as will be seen here. Recalling that the results are not exact for the chi-squared approximation to the likelihood ratio test either, this is not so surprising. Since the exact solution, in the form of the F test, is well known, the only purpose of this example is to gain some insight into the behaviour of the expansion.

Let $X_1, \ldots, X_n$ be independent, X_i normally distributed with mean μ_i and variance σ^2. Assume that the vector $\mu \in \mathbb{R}^n$ with co-ordinates (μ_i) belongs to some linear subspace $M \subset \mathbb{R}^n$ of dimension $m < n$. We test the hypothesis that $\mu \in H \subset M$, where H is a linear subspace of dimension $h < m$. Let $p_M: \mathbb{R}^n \to M$ denote the orthogonal projection of $\mathbb{R}^n$ on to M and p_H the projection on H. The model is an exponential family with the density of $X = (X_1, \ldots, X_n)$ at $x = (x_1, \ldots, x_n)$ given as

$$(2\pi\sigma^2)^{-n/2} \exp[\theta_1 \| x \|^2 + \langle \theta_2, p_M(x) \rangle - \psi(\theta)], \tag{4.1}$$

$$\psi(\theta) = \| \theta_2 \|^2/(-4\theta_1) - \tfrac{1}{2}n \log(-\theta_1), \quad \theta_1 < 0, \quad \theta_2 \in M, \tag{4.2}$$

where $\theta_1 = (-2\sigma^2)^{-1}$ and $\theta_2 = \mu/\sigma^2$. The minimal sufficient statistic is $(\| x \|^2, p_M(X))$, and the estimates under the hypothesis are $\hat{\mu}_0 = p_H(X)$ and $\hat{\sigma}_0^2 = \| X - p_H(X) \|^2/n$. In this estimated distribution corresponding to an observed value x, the mean of $\| X \|^2$ is $\| x \|^2$, while the mean of $p_M(X)$ is $p_H(x)$. Thus the test statistic becomes the length of $p_M(x) - p_H(x)$ and we shall condition on its direction and on $\| x - p_H(x) \|^2$. It is easy to see that this test is equivalent to the usual F test, but we shall continue to derive its approximation.

Let

$$G = \| p_M(x) - p_H(x) \|^2/\| x - p_M(x) \|^2 = (m-h)F/(n-m) \tag{4.3}$$

where F is the usual F statistic. It is a trivial matter to compute the expansion (3.6) in which $p = m - h$. The expansion becomes

$$1 - F_{m-h}(n \log(1+G)) + \sqrt{2}[\Gamma(\tfrac{1}{2}(m-h+1))/\Gamma(\tfrac{1}{2}(m-h))][1 - F_{m-h+1}(n \log(1+G))] \times \{[G/\log(1+G)]^{(m-h-2)/2}(1+G)^{h/2+1} - 1\}[n \log(1+G)]^{-1/2}, \tag{4.4}$$

which is an approximation to the distribution of $n \log(1+G)$, for which we may work out the exact distribution from the F distribution. Thus the expansion may be viewed as an approximation to the F distribution. Its numerical behaviour, however, should not be judged as such because the specific example has not been used to derive the form of the expansion, but it cannot be excluded that the behaviour is special for this choice of example. As an approximation to the F distribution with $(m-h, n-m)$ degrees of freedom it may seem odd that it depends on h and not only on $m-h$ and $n-m$, but this is so because the likelihood ratio test statistic depends on h (through n). This problem is avoided by conditioning on $p_H(X)$ before calculating the expansion (3.6), in which case the problem essentially reduces to the case $h = 0$. Thus in the numerical examples we only consider $h = 0$, while n and m may vary.

In Table 1 some examples of values of the approximation are given for varying m and $n - m$. They are listed as approximations to the F distribution which has then been transformed to the distribution of $n \log(1+G)$ according to equation (4.3) before equation (4.4) has been calculated. For comparison we also list the approximations based on the usual chi-squared approximation to the likelihood ratio test and the Bartlett corrected chi-squared approximation. The Bartlett approximation is obtained by multiplying minus twice the log-likelihood ratio statistic, $-2 \log Q$, by a factor

TABLE 1
Numerical values for example 1†

m	$n-m$	x	$P\{F \geqslant x\}$	*Saddlepoint*	χ^2	*Bartlett*
1	1	405300	0.001	0.000886	3.74×10^{-7}	0.00012
1	1	161.45	0.05	0.0443	0.0014	0.0159
1	1	5.83	0.25	0.224	0.050	0.138
1	1	161.45^{-1}	0.95	0.945	0.912	0.933
1	3	167.0	0.001	0.00094	0.000059	0.00061
1	3	10.13	0.05	0.0476	0.0151	0.0382
1	3	216^{-1}	0.95	0.949	0.937	0.947
1	9	22.86	0.001	0.00098	0.00038	0.00092
1	9	5.12	0.05	0.0495	0.0338	0.0478
10	1	241.88	0.05	0.85	3.8×10^{-14}	2.6×10^{-8}
10	10	2.98	0.05	0.19	0.0021	0.019
10	20	2.35	0.05	0.115	0.0097	0.035
10	80	1.95	0.05	0.066	0.033	0.0485

†Three approximations to the likelihood ratio test for linear normal models are given: the saddlepoint approximation (4.4), the χ^2 approximation to $-2 \log Q$ and the Bartlett corrected χ^2 approximation. All are compared with the exact distribution based on an F distribution with $(m, n-m)$ degrees of freedom.

to make its expectation equal to $p-q=m$ to second order. In this example the expected value of $-2 \log Q$ may be calculated exactly, but that would not be a reasonable comparison, because we want to compare more general methods. Therefore we use the general approximation, obtained by the delta method, which for this case becomes

$$E\{-2 \log Q\} = E\{n \log(1+G)\} \sim m[1+(m+2)/2n], \tag{4.5}$$

which means that the chi-squared approximation is applied to $(-2 \log Q)/[1+(m+2)/2n]$.

For $m=1$ the saddlepoint approximation (4.4) matches the well-known good behaviour of the usual saddlepoint expansions, even down to $n=2$. Even though the hypothesis is only one dimension smaller than the model, such that the test statistic is one dimensional in the conditional distribution given the estimator, the expansion had to be derived through a two-dimensional approximation, and hence it is not identical with the Lugannani and Rice (1980) expansion in this case. For $m=2$ it can be worked out analytically that the relative error of the saddlepoint approximation is maximal at $n=3$ and never exceeds 7.6% throughout the tail of the distribution. When m is large and $n-m$ is relatively small the expansion turns out to be useless. However, the case $m=10$, $n-m=10$, corresponds to two independent observations of a 10-dimensional random vector from a distribution with 11 parameters, while the case $m=10$, $n-m=80$, corresponds to nine replications.

It is concluded that for high dimensional models relatively more replications are needed to give a reasonable accuracy. This is probably not surprising and is also true for the usual saddlepoint approximation to densities. For these cases the chi-squared approximations, with or without Bartlett corrections, also require a higher number of replications, although the Bartlett corrected approximation does slightly better than the saddlepoint approximation in some cases in terms of relative accuracy. In the far tail the saddlepoint approximation does not increase its relative error to any great extent, as is the case for the uncorrected and corrected chi-squared approximation. This is the usual pattern for large deviation approximations compared with the 'central' approximations such as the two chi-squared approximations.

4.2. *Example 2: Testing Homogeneity of Variances*

Let $Y_1, \ldots, Y_k$ be independent scaled gamma variables with known shape parameters $\alpha_1, \ldots, \alpha_k$ and expectations $EY_i = \alpha_i \xi_i$. We want to test the hypothesis $\xi_1 = \ldots = \xi_k$. Let $\theta_i = -1/\xi_i$: then the model is an exponential family with $\theta = (\theta_1, \ldots, \theta_k)$ as the canonical parameter and $Y = (Y_1, \ldots, Y_k)$ as the corresponding sufficient statistic. Furthermore the logarithm of the normalising constant in equation (3.1) is $\psi(\theta) = \Sigma[-\alpha_i \log(-\theta_i)]$. The estimates of ξ_i with and without the hypothesis are $\bar{y} = (\Sigma y_i / \Sigma \alpha_i)$ and $\bar{y}_i = y_i/\alpha_i$ respectively. Simple computations lead to the following expressions for the quantities used in equation (3.6),

$$|\hat{\Sigma}_0|^{1/2} = \Pi(\alpha_i^{1/2}\bar{y}),$$
$$|\hat{\Sigma}|^{1/2} = \Pi(\alpha_i^{1/2}\bar{y}_i),$$
$$\hat{\Delta}_0[(\bar{y} - \hat{\psi}_0')^2] = \Sigma\alpha_i(y_i - \bar{y})^2/\bar{y}^2,$$
$$\langle \hat{\theta} - \hat{\theta}_0, \bar{y} - \hat{\psi}_0' \rangle = \Sigma\alpha_i(\bar{y} - 1)/\bar{y}_i,$$
$$\tilde{s}^2 = 2\Sigma\alpha_i(\log \bar{y} - \log \bar{y}_i),$$

where $\tilde{s}^2$ is recalled to be minus twice the log-likelihood ratio. The conditional distribution to be approximated is that obtained by conditioning on $\bar{Y}$ and on the direction of the vector with co-ordinates $Y_i - \alpha_i \bar{Y} = U_i R$, say, where the vector U_i is of unit length as measured by the $\hat{\Delta}_0$ inner product, i.e. $\Sigma U_i^2/\alpha_i = \bar{y}^2$. If we fix the value of $\bar{Y}$ at $\bar{y}$, and U_i at $u_i = v_i \bar{y}$, say, we may write down the approximation to the tail probability of R for fixed $\bar{y}$ and (v_i), $\Sigma v_i = 0$, $\Sigma v_i^2/\alpha_i = 1$, as

$$P\{R \geqslant r \mid \bar{Y} = \bar{y}, U_i = v_i\bar{y}, i = 1, \ldots, k\}$$
$$\sim (1 + cA)^{-1}\{[1 - F_{k-1}(\tilde{s}^2)] + cB[1 - F_k(\tilde{s}^2)]\}, \qquad (4.6)$$

where

$$c = \sqrt{2}[\Gamma(\tfrac{1}{2}k)/\Gamma(\tfrac{1}{2}(k-1))],$$
$$A = \tfrac{1}{3}k(\Sigma r^3 v_i^3/\alpha_i^2)r^{-3} - (\Sigma r v_i/\alpha_i)r^{-1/2},$$
$$B = \tilde{s}^{-1}\{[\Pi((\alpha_i + rv_i)/\alpha_i)]^{-1} r^{k-1}\tilde{s}^{-k+3}[-\Sigma r v_i \alpha_i/(\alpha_i + rv_i)]^{-1} - 1\},$$
$$\tilde{s}^2 = -2\Sigma\alpha_i \log(1 + rv_i/\alpha_i),$$

and the distribution of R is concentrated on the interval $0 < r < \min\{\alpha_i/|v_i|;\ v_i < 0\}$.

It turns out to be easy to work out the exact distribution. The exact density is proportional to

$$r^{k-2}\Pi(1 + rv_i/\alpha_i)^{\alpha_i - 1}, \qquad (4.7)$$

which, as a polynomial in r, is easily integrated when all the α_is are integers.

TABLE 2
Numerical values for example 2 for the case $k = 2$, $\alpha_1 = \alpha_2 = \alpha$†

α	r	*Exact*	*Saddlepoint*
0.5	$1-1.23\times10^{-6}$	0.00100	0.00123
0.5	$1-1.23\times10^{-4}$	0.0100	0.0121
0.5	0.9877	0.100	0.116
0.5	0.7010	0.500	0.525
0.5	0.0785	0.950	0.953
1	1.4128	0.00100	0.00110
1	1.4001	0.0100	0.0108
1	1.2728	0.100	0.104
1	0.7010	0.500	0.506
1	0.0707	0.950	0.951

† Exact tail probabilities and the saddlepoint approximation (4.8) are given.

We restrict the numerical examples to the case $k = 2$, $\alpha_1 = \alpha_2 = \alpha$, for which the exact distribution of R^2 is a beta distribution with $(1, 2\alpha)$ degrees of freedom, scaled to the interval $(0, 2\alpha)$. The approximation for this case is

$$P\{R \geqslant r \mid \bar{Y}\} \sim 1 - F_1(\tilde{s}^2) + (2/\pi)^{1/2}(r^{-1} - \tilde{s}^{-1})(1 - r^2/2\alpha)^{\alpha}, \tag{4.8}$$
$$\tilde{s}^2 = -2\alpha \log(1 - r^2/2\alpha).$$

The example (Table 2) shows a very similar behaviour of the approximation to that of the previous example, namely that of a limited relative error throughout the tail, even for samples as small as those in the table. Results for higher dimensional models are not given for this case, because the limitations of the approximation in that sense were shown in example 1.

REFERENCES

Barndorff-Nielsen, O. E. and Cox, D. R. (1984) Bartlett adjustments to the likelihood ratio statistic and the distribution of the maximum likelihood estimator. *J. R. Statist. Soc.* B, **46**, 483–495.

Bartlett (1954) A note on multiplying factors for the various χ^2 approximations. *J. R. Statist. Soc.* B, **16**, 296–298.

Bleistein, N. (1966) Uniform asymptotic expansions of integrals with stationary point near algebraic singularity. *Commun. Pure Appl. Math.*, **19**, 353–370.

Chandra, T. K. and Ghosh, J. K. (1979) Valid asymptotic expansions for the likelihood ratio statistic and other perturbed chi-squared variables. *Sankhya* A, **41**, 22–47.

Daniels, H. E. (1954) Saddlepoint approximations in statistics. *Ann. Math. Statist.*, **25**, 631–650.

——(1983) Saddlepoint approximations for estimating equations. *Biometrika*, **70**, 89–96.

——(1987) Tail probability approximations. *Int. Statist. Rev.*, **55**, 37–48.

Esscher, F. (1932) On the probability function in the collective theory of risk. *Skand. Akt. Tidsskr.*, **15**, 175–195.

Field, C. (1982) Small sample asymptotic expansions for multivariate *M*-estimates. *Ann. Statist.*, **10**, 672–689.

Fraser, D. A. S. and Massam, H. (1985) Conical tests: observed levels of significance and confidence regions. *Statist. Hefte*, **26**, 1–17.

Lawley, D. N. (1956) A general method for approximating to the distribution of the likelihood ratio criteria. *Biometrika*, **43**, 295–303.

Lugannani, R. and Rice, S. O. (1980) Saddlepoint approximation for the distribution of the sum of independent random variables. *Adv. Appl. Probabil.*, **12**, 475–490.

McCullagh, P. (1984) Local sufficiency. *Biometrika*, **71**, 233–244.

Olver, F. W. J. (1974) *Asymptotics and Special Functions.* New York: Academic Press.

Skovgaard, I. M. (1985) Large deviation approximations for maximum likelihood estimators. *Probabil. Math. Statist.*, **6**, 89–107.

Chapter 6

A review of higher order likelihood inference

Introduction by Nancy Reid

University of Toronto

The meetings of the International Statistical Institute, now the World Statistics Congress, are remarkable for their breadth in many dimensions, including location, scientific content, international participation, and special events. The 47th Session in Paris in 1989 was particularly memorable for a spectacular banquet held in the Conciergerie of the Palais du Justice. The secretary-general of the National Organizing Committee for this Session was Dr. Jean-Louis Bodin, later President of the International Statistical Institute (Saporta, 2015).

In those days, on checking in at the registration desk, one received a large and heavy bundle of four or five volumes containing all the papers to be presented at the meeting, and usually a very sturdy briefcase in which to transport them home. And since they were books, hence valued, one did find room in the suitcase and then room on one's office shelves for these tomes. About ten years ago I made a fairly clean sweep of my over-crowded office, and these volumes, I confess, made it to the recycle bin. But not before I had extracted two papers: Richard Smith's excellent review of non-regular likelihood inference (Smith, 1989) and Ib's review of higher order likelihood inference. This is a small measure of my view of the quality of this paper. I remember the delivery of the paper as well as being exceptionally clear and interesting, whilst being delivered in Ib's unassuming and low-key manner.

The study of higher order likelihood inference was very active throughout the 1980's, accompanied by a great sense of excitement that new ground was being broken in the theory of inference. This built on the development in the 1970s of the theory of exponential families (Efron, 1975; Efron and Hinkley, 1978), but was launched I think by the publication in the August 1980 issue of *Biometrika* of a remarkable series of papers that all seemed to be converging on the same approximation to the distribution of the maximum likelihood estimator: what we now call the p^* formula (Barndorff-Nielsen, 1980; Cox, 1980; Durbin, 1980; Hinkley, 1980). There were intriguing links to the literature on the differential geometry of statistical models, developed in some generality in Amari (1982), building on Efron (1975) and Dawid (1975).

While a great deal of work in asymptotic theory throughout the 1980's and 1990's addressed accurate approximation, usually of p-values, Ib here takes a slightly different, and more classical point of view. As emphasized in the introduction, the goal is to consider classes of procedures for inference with good, or even optimal, properties, rather than taking a given statistic, for example the maximum likelihood estimator, and finding good approximations to its distribution. This is perhaps clearer from the point of view of estimation: in regular models the minimum variance of a consistent estimator is given by the inverse of the expected Fisher information. An asymptotically efficient estimator is then any estimator whose mean-squared error converges in the limit to the minimum value. Among a collection of such asymptotically efficient estimators, it is natural to then turn to an investigation of which ones get there faster, by looking at an asymptotic expansion of the mean-squared error and comparing higher order terms in the expansion. A similar approach can be taken for testing, for which the relevant optimality criterion is power. As uniformly most powerful tests do not generally exist outside of special settings, it is natural to consider asymptotic expansions of the power function of asymptotically optimal tests with a view to identifying higher order properties.

The main emphasis on Ib's review paper is on the power function, although questions of efficient estimation are addressed because of their prominence in the literature at the time. In connection with this he emphasizes that consideration of power is quite compatible with the use of p-values to assess statistical significance, and does not constrain one to the accept/reject paradigm of Neyman and Pearson. This is a subtle point that I have not seen discussed in the literature on higher order asymptotics. Suppose for simplicity that we have a well-defined test statistic, $T = t(Y)$, with Y from a parametric model with density $f(y; \beta)$. The p-value function,

$p(\beta)$ in modern usage, is

$$p(\beta) = p(\beta; t^{obs}) = \Pr(T \geq t^{obs}; \beta) = \int_{t^{obs}}^{\infty} f(t; \beta) dt, \tag{6.1}$$

where t^{obs} is the observed value of the test statistic, and it is assumed that large values of T indicate departure from the hypothesized value of β. The power function in this notation is

$$p(\beta; t_\alpha) = \Pr(T \geq t_\alpha; \beta), \tag{6.2}$$

where t_α is defined by the requirement that $\Pr(T \geq t_\alpha; \beta_0) = \alpha$ for a particular choice β_0. In Ib's review $p(\beta; t_\alpha)$ is written $\lambda(\beta; \alpha) = \Pr_\beta(C_\alpha)$, so that the critical region of the test can be left unspecified, and not necessarily determined by large values of a particular test statistic. In much of the recent work on confidence distributions and significance functions, the emphasis is on $p(\beta; t^{obs})$. But if we view the p-value function as depending on two arguments $p(\beta; t)$, then both concepts are subsumed. A cross-section on a particular value of t can identify either the significance level, or the power, depending on the choice of t. Similarly a cross-section on a particular value of β determines the critical value of t for a given size.

Ib concludes his discussion of these and related points in §2 by emphasizing the central role of the power functions of tests as the basic ingredient to be studied further. The heart of the review is in the long Section 3, which addresses first, second, and third order theory for testing, and includes a survey of related results for estimation, in the context of a scalar or vector parameter of interest with no nuisance parameters. He begins the section with a very clear discussion of what are conventionally called composite alternatives, the null being $H_0 : \beta = \beta_0$. If β is a scalar parameter, then composite alternatives are two-sided, $\beta \neq \beta_0$, and uniformly most powerful tests do not exist, even in the simple case of regular exponential families. Skovgaard describes unbiasedness of the test, power function never less than the size, as a "theoretical favourite", and equi-tailed tests as having "conquered the majority of the market of applications", which I would say remains true today. He then describes the equi-tailed conditional test, which conditions on the direction of the test statistic from its mean. This approach was suggested for some problems in regression by Fraser and Massam (1985), and developed further for both scalar and vector parameters in Skovgaard (1988): see Chapter 5 in this volume. In recent work (Davison *et al.*, 2014; Fraser *et al.*, 2016) we simplified the approach to calculating these conditional p-values, and the resulting inference seems to be remarkably accurate.

The more conventional approach to the problem of lack of uniformly most powerful tests is to investigate the power function of different tests, and Ib reviews the work of Pfanzagl in this direction. He illustrates the well-known result that tests based on the score function $\partial \log(\beta; y)/\partial\beta$ are locally most powerful, and that any test asymptotically equivalent to a test based on the score function will have the same first order power. His approach to this result is via his development of approximate exponential families, illustrated here in (1). The nature of this approximation does need some clarification, which is provided in his book Skovgaard (1989b). His summary of these well-known asymptotic results is "any of the commonly used test statistics behave to first order as in the case of an exponential family". This is an interesting approach closely related to his work on analytical statistical models and the proof of the p^* formula in Skovgaard (1990), as most elementary treatments of first order asymptotics would replace exponential family by normal model. The asymptotic efficiency of the maximum likelihood estimator, and any estimators not very different from the maximum likelihood estimator, is briefly summarized.

In view of the inability of limiting theory to provide a choice among test statistics, it was natural for researchers to consider an expansion of the power function, to assess whether tests equivalent in the limit could be distinguished by their higher order properties. There were several results developed in the 1970's and 1980's to address this question, but they all in some sense failed. First, somewhat surprisingly, it turned out that any test statistic with a power function that can be expanded in decreasing powers of $n^{1/2}$ has the property that equality of the leading term implies equality of the second term, and the only distinction arises in the third term, that of $O(n^{-1})$, where there is no uniform domination. Ib describes the two lines of work that led to the same result, and how they built on the asymptotic theory presented in Pfanzagl (1980). The source papers, and Pfanzagl's book, are very technical and detailed, and the summary of the key issues here is masterfully concise. There were similar, technically somewhat easier, results available about the efficiency of point estimators, which are briefly summarized. As a side remark he points out that defining a point estimator as the 0% point of a confidence region cannot be justified by asymptotic arguments like these, for rather subtle reasons.

In §3.4 Ib provides a very clear and concise discussion of the conflict between conditioning and power. Although this point is raised in several places, possibly first in Cox (1958), it continues to be overlooked in discussions of theoretical statistics. Most people who think about it agree that in the famous weighing machine example, it is correct to conduct inference conditionally on the machine used, if that machine is chosen at random. What is not appreciated, but emphasized here, is that a version of this problem is disguised in many applications of regression: Ib emphasizes non-linear normal theory regression (Skovgaard, 1985b), but the argument also applies to linear, non-normal regression. A brief discussion of Birnbaum's result leads to a proposal to use a weaker form of conditioning on ancillary statistics, given at (4). He relates this to McCullagh (1987) on approximate ancillarity, leading to the conclusion that the likelihood ratio statistic, or its signed square root in the case of a scalar parameter, is the best choice among asymptotically equivalent test statistics, on the grounds that its power properties are not unreasonable, and it obeys a form of conditioning on approximately ancillary statistics. This conclusion has been born out by later work on the accuracy of inference based on an adjustment to the likelihood ratio statistic and to its signed square root; see also Skovgaard (2001).

Rather less is provided in §4 in the setting of inference for a real-valued parameter $\psi(\beta)$, with the remaining components of β treated as nuisance parameters. In the case that the dimension of the nuisance parameter is fixed, but the sample size $n \to \infty$, the approach using asymptotic expansions of the power function are similar to those reviewed in §3. More interesting is the class of so-called Neyman–Scott problems, where each observation, or possibly set of observations, introduces a new nuisance parameter. In Ib's notation the model for y_i is governed by β_i, with the only restriction being the hypothesis that $\psi(\beta_i)$ is constant. He links this to the work on semi-parametric models obtained by postulating an unknown mixing distribution for the β_i, here called $Q(\cdot)$. The comments here are more speculative, suggesting links to Owen's empirical likelihood, which was presented in the same invited paper session as this paper. Murphy and van der Vaart (2000) carried this approach much further, giving general conditions for the profile likelihood approach to be valid in this setting, but there is as yet little higher order asymptotic theory for semi-parametric models.

§5 briefly highlights a different line of work, which I think was more active at the time and certainly in the years that followed, of improving the approximation of the usual test quantities, especially the standardized maximum likelihood estimator and the likelihood ratio test. A review paper on higher order likelihood inference could be based entirely on this class of results, indeed I have written some myself, and I have always admired Ib's choice in this review to step back and consider the asymptotics involved in the choice of test statistics. The concluding remarks are, as always, interesting, precise and concise. While Ib says at the outset that in some places he is “deliberately provocative”, his personal style and clarity of exposition really does not permit this!

A REVIEW OF HIGHER ORDER LIKELIHOOD INFERENCE

Ib M. Skovgaard

Royal Veterinary and Agricultural University
Department of Mathematics
Thorvaldsensvej 40, DK 1871 Frederiksberg C
Denmark

Published in *Bulletin of the International Statistical Institute, 47th Session*, Paris 1989
Book 3: 331 – 350.

Key words: ancillarity principle, asymptotic expansions, efficiency, nuisance parameters, powers of tests, second order, third order.

1 INTRODUCTION

The purpose of the present paper is to review the applications of second and third order asymptotics to likelihood based statistical inference. Among the numerous applications of asymptotic approximations and expansions the review is confined to the methodological issues of deriving inferential procedures and choosing between these. Maybe the most important role of asymptotics has been to derive distributional approximations for given inferential procedures chosen by other criteria. While such applications are closely related to the current theme they are not of primary concern here. Only independent replications from 'regular' parametric statistical models are considered and almost exclusively *general inferential procedures*, i.e., procedures that do not depend on special features such as the possibility of elimination of nuisance parameters by marginalization or conditioning. Thus, important results concerning, for example, parametrizations, or inferential techniques for special cases or special classes of models, are not discussed. It is not implied by these limitations that such results are less important than those discussed here. There may well be other omissions, within the scope of the paper, simply because of the vast amount of literature on the subject and my lack of knowledge of major parts of it. Furthermore it should be noted that it has been attempted to review the 'state of art', not the historical development. Therefore, references are given primarily to recent accounts, and several authors not mentioned here deserve credit for the development of the theory.

The paper centers on second and third order asymptotic results. Some first order results need to be reviewed as a basis for the more refined asymptotics. These first order results are considered here to be well-known and explicit references are not given. General accounts may be found, for example, in the books by Cox & Hinkley (1974), and Lehmann (1959, 1983).

The review is deliberately stripped of any mathematical detail. Thus, regularity conditions are not discussed and most of the results are formulated in an imprecise manner. The reader should refer to the original papers for more thorough descriptions. An attempt to provide a basis for results of the kind described here, in the form of a class of regular models, has been described in Skovgaard (1989).

Some rather categorical viewpoints are taken as the basis concerning the type of inferential procedures and criteria of optimality, especially in the next section and in connection with the nuisance parameter setup and conditional inference. In practice there may be several reasons for choosing different priorities and different methods of statistical analyses, but theory can hardly be advanced without some more definite setting. A somewhat provocative style of presentation is used in places, mainly to promote discussion.

2 INFERENTIAL PROCEDURES AND OPTIMALITY CRITERIA

The basic aims of parametric statistical inference are:

(1) Point estimation of parameters.
(2) Testing hypotheses.
(3) Providing confidence regions for parameters.

Methods of model checking, which are at least as important, are rarely based (entirely) on parametric methods and are not discussed here. The context in which these aims are discussed is that of scientific investigations in which knowledge is accumulated by experiments or other statistical investigations, as opposed to situations in which 'actions' have to be taken as a result of the analysis as, for example, in control theory or in some financial analyses. Thus, in particular, the obvious relevance of decision theory to analyses where the loss function has a definite interpretation as a 'real' loss (or gain), falls outside the present framework. Whether decision theory is of importance in other cases is a different issue.

Most statisticians agree on the basic probabilistic criterion for evaluation of the quality of a test, namely that of maximizing the power. It is rarely possible to maximize the power uniformly, and then other (asymptotic) criteria come into play, for example, local maximization of power to a certain order of approximation. Such criteria will be discussed below, but the fact remains that power maximization in some form is the goal. This remains true when a test is considered in its 'modern' form by ascribing a p-value to any given set of data, measuring the '(dis)-agreement' of data with the hypothesis, rather than testing at a fixed level of rejection and then deciding to accept or reject the hypothesis, cf. Cox (1977). In scientific investigations it is usually not reasonable to make such a decision. The p-value is, of course, obtained if we have a nested family of tests in the classical sense, one at each possible level.

One way to approach the evaluation of confidence regions is to employ the usual correspondence between tests and confidence regions. From the confidence set at a certain level for a, possibly multi-dimensional, parameter β we deem the test for each particular value of that parameter significant at that level if the given value is not in the confidence set. Similarly, if we have a family of tests at a certain level, one for each simple hypothesis for the parameter, then a confidence region at this level is defined as the non-significant parameter points. In this way the criterion of power for tests at a certain level is converted into the criterion of coverage probabilities for confidence regions, and there is no longer any reasons to consider the optimality of confidence regions separately. If, however, other criteria are used, such as that of minimizing the length of confidence intervals or, more generally, the measure of confidence sets, or if we insist that the confidence set must be an interval, then a separate issue arises.

The criterion of minimal length seems to be somewhat dubious as a general criterion, compared to coverage probabilities, because of the lack of transformation invariance. At any rate, a specific parametrization has to be chosen for which the length is required to be minimal. For this reason, and also because it is not obvious if length is the quantity to minimize, this criterion is not considered further here.

The property that the confidence set for a one-dimensional parameter becomes an interval is definitely desirable because it simplifies the interpretation of the results. This requires, however, that the given interval corresponds to a set of parameter values that comply reasonably with the data. If this is not the case, it does not appear sensible to insist that the confidence region must be an interval. Thus, the fact that the region is an interval is a desirable *result*, not necessarily a desirable property of the confidence region *procedure*.

For example, consider the problem of estimating the intercept of the regression line with the x-axis in a linear regression of Y on x based on the Normal distribution. If the estimate of the slope is comparatively close to zero, with a confidence interval containing zero, while the mean of the Y's is clearly positive, then the only possible conclusion concerning the intercept with the x-axes is that it is not contained in an interval around the mean of the x's. Hence the confidence region for the intercept will be to disjoint half-lines. This result may well lead to the conclusion that it is not reasonable to infer about the intercept, and we would usually prefer to see results leading to a single (bounded) interval. To insist on a procedure giving a single confidence interval for the intercept in any case would, however, be misleading.

For these reasons the problem of deriving confidence regions is not considered to require special methodological considerations and is not considered further in the present paper.

In statistical applications point estimates play an important role. Often the important conclusions from an experiment concern whether some significant differences are found, for example, between treatments, and if so the estimates of treatment effects will be at the focus of attention. However, we also know from practical investigations that such estimates should not be taken too seriously without some accompanying measure of their accuracy. In other words, point estimates are convenient summaries, but confidence sets, possibly in the form of estimate plus/minus twice the standard error of the estimate, are necessary for careful interpretations.

Much work concerning comparison of statistical procedures has been devoted to the comparison of point estimators. The criterion used for such comparisons is usually that of minimizing the expected loss in terms of some loss function, if possible uniformly under certain conditions, otherwise by the minimax criterion, or by the expected loss with respect to some prior distribution. Quadratic loss functions are often considered, but a general adoption of quadratic loss just transfers the burden of choice of loss function to an approximately equivalent burden of choice of parametrization.

Based on the view of point estimates as convenient summaries of confidence regions it seems more reasonable to make sure that the point estimates reflect the confidence regions as well as possible rather than attempting to minimize some expected distance between the point estimate and the 'true' parameter value. Suppose that for each given level we have a procedure giving a confidence region for the parameter at that level, and that we are satisfied that this procedure provides 'optimal' confidence regions to the best of our knowledge and abilities. Then from this point of view it may be reasonable to use as the point estimate, the intersection of all confidence regions at all positive confidence levels. Briefly described this defines the point estimate as a zero level confidence region.

There is a number of open problems related to this procedure. First of all the confidence regions at different levels should be nested. This is so if the confidence regions are constructed from a class of tests (one at each specified level for each specific parameter value) which are nested in the usual sense that significance at one level implies significance at any more extreme level. Secondly, it is not certain that the 'point estimate' constructed in this way exists, because the intersection of the confidence regions may be empty. However, if each of the confidence regions is non-empty, then under mild regularity conditions it follows from Cantor's intersection theorem that the intersection is non-empty, still provided that the confidence regions constitute a class of nested sets. The condition that each confidence region is non-empty imposes a genuine restriction on the class of tests used to construct these, namely the condition that no possible data set make us reject all parameter values within the model. It seems reasonable to require that tests are constructed in accordance with this condition.

In this connection it should be emphasized that model checking is not within the scope of this discussion. Thus, the model is assumed to contain hypothesis and alternatives. In many situations, one may prefer not to specify the alternatives and instead write down a test statistic directed towards the relevant, but unspecified, alternatives. While such a procedure may be perfectly adequate, it is from the present point of view regarded as an ad hoc procedure used in a case of an unspecified model.

In conclusion, we adopt the power functions of tests as the basic criterion for the construction of statistical procedures in parametric models. Directly, as a criterion to choose between possible tests, indirectly to convert these into confidence sets, and, in turn, to point estimates defined as zero level confidence sets. Therefore, the discussion centers on results concerning powers of tests. However, some attention is given to results on optimality of estimators in terms of expected loss. This kind of results has received so much attention that it can hardly be neglected in the review. The discussion is structured according to the absence or presence of nuisance parameters. A special section is devoted to the discussion of the schism between conditional and optimal tests.

3 NO NUISANCE PARAMETERS

3.1 Exact and first order results

We consider a statistical model indexed by a parameter β taking values in a subset B of $\mathbf{R}^p$, $p \in \mathbf{N}$. The data $Y = (Y_1, \ldots, Y_n)$ are assumed to consist of independent replications. The density function corresponding to the parameter $\beta \in B$ for Y at y with respect to some underlying measure on the sample space is denoted $f(y; \beta)$.

We want to draw inference on the parameter β. In the framework of hypothesis testing we want to construct tests for each of the hypotheses of the form $\beta = \beta_0$, $\beta_0 \in B$. These tests may then be used to construct confidence regions and point estimates for β. Only simple hypotheses are considered; composite hypotheses may be viewed as instances of nuisance parameters, discussed separately.

The optimal test of the hypothesis $\beta = \beta_0$ against some particular simple alternative $\beta = \beta_1$ is known from the Neyman-Pearson lemma to be the likelihood ratio test (or the Neyman-Pearson test) with critical regions of the form

$$\{y : f(y; \beta_1)/f(y; \beta_0) > c\}.$$

In general, this test is not the most powerful test against other alternatives. For a one-dimensional exponential family, with densities of the form

$$f(y; \beta) = a(y) \exp\{t(y)\beta - \kappa(\beta)\},$$

where a and t are measurable functions, the Neyman-Pearson test against the alternative $\beta_1 > \beta_0$ is uniformly most powerful against all one-sided alternatives $\beta > \beta_0$. In the two-sided case the problem is how to balance the power against alternatives in the two directions from β_0, once it has been realized that no uniformly most powerful test exists. Three different general principles exist, although adoption of any of these remains a matter of choice and taste. The favorite among authors of theoretical comparisons of tests seems to be the principle of *unbiased tests*, i.e., tests with power functions satisfying the unbiasedness condition

$$\lambda(\beta; \alpha) \geq \lambda(\beta_0; \alpha)$$

for any $\beta \in B$, where

$$\lambda(\beta; \alpha) = \Pr_\beta\{C_\alpha\},$$

is the power function for a the test at level α with the critical region C_α. The second principle, which seems to have conquered the majority of the market of applications, is that of *equi-tailed rejection probabilities*, i.e.,

$$\Pr_{\beta_0}\{t(Y) \geq \bar{t}_\alpha\} = \Pr_{\beta_0}\{t(Y) \leq \underline{t}_\alpha\} = \alpha/2,$$

where $\bar{t}_\alpha$ and $\underline{t}_\alpha$ are the upper and lower critical values at level α for the test based on the canonical sufficient statistic $t(Y)$. Notice that the tests with critical regions of the form $\{t(Y) \geq c\}$ and $\{t(Y) \leq c\}$, respectively, are uniformly most powerful against one-sided alternatives. While this principle is intuitively appealing it is not defined in terms of the distributional properties of the test, but has to be defined in terms of a particular test statistic. A third possibility is that of *equi-tailed conditional rejection probabilities*, i.e.,

$$\Pr_{\beta_0}\{\, t(Y) \geq \bar{t}_\alpha \mid t(Y) > t_0 \,\} = \Pr_{\beta_0}\{\, t(Y) \leq \underline{t}_\alpha \mid t(Y) < t_0 \,\} = \alpha/2,$$

where $t_0 = \mathrm{E}_{\beta_0}\{t(Y)\}$. In specific connections this idea was promoted (in its multivariate analogue) as 'conical tests' in Fraser & Massam(1985). Also this principle is defined in terms of a specific test statistic rather than in terms of the properties of the critical regions.

In the more general case of a multi-dimensional parameter the power of the test has to be balanced against alternatives in all possible directions from the hypothesis $\beta = \beta_0$. The condition of unbiasedness generalizes immediately to this case, imposing the mathematically convenient condition on the power function that its first derivative in all possible directions from β_0 vanishes. However, this condition is not sufficiently restrictive to determine a unique most powerful test, even in the simplest cases. The criteria of equi-tailed rejection probabilities and its conditional analogue generalize to multi-dimensional parameters only in the case of a continuous sample space; namely by equating either the conditional density of the direction of the test statistic from its mean given the critical region, or the conditional probability of the critical region given the direction of the test statistic. Combined with the criterion of maximal power any of these two principles leads to uniquely defined tests for exponential families.

Perhaps a more consequent attitude towards the balancing of power of the test in different directions is that adopted, i.a., by Pfanzagl (see, e.g., Pfanzagl, 1980); namely that we have to recognize the fact that no uniformly most powerful test exists. Then, we may investigate properties of different tests and classes of tests in terms of their power, for example, whether a test is admissible, i.e., has a power function which agrees with the power function of any test which is uniformly as least as powerful. The choice of test is then left to specific applications.

For other models than exponential families the situation is generally not even so fortunate that uniformly most powerful tests exist against all alternatives in a specific direction form the hypothesis. In fact, it is only for exponential families that this happens for any number of independent replications (Pfanzagl, 1968). However, to first order of approximation any regular parametric model behaves like an exponential family. A heuristic argument showing that this is so is that for any (sufficiently regular) statistical model we may approximate the densities for β in some neighborhood of β_0 by

$$f_1(y;\beta) = f(y;\beta_0)\exp\{D_1^T(\beta-\beta_0) - \frac{1}{2}(\beta-\beta_0)^T I(\beta_0)(\beta-\beta_0)\}, \qquad (1)$$

where $D_1 = D_1(\beta_0)$ is the (column) vector of first partial derivatives of $\log f(y;\beta)$ with respect to coordinates of β at β_0, and $I(\beta_0)$ is the Fisher information matrix at β_0. The family of approximate densities $\{f_1(y,\beta)\}$ has the form of an exponential family, except that the integral of $f_1(y,\beta)$ over the sample space may not be exactly one. Any error arising from the approximation of $f(y,\beta)$ by $f_1(y,\beta)$ may, however, be shown to be of order $O(n^{-1/2})$ as $n \to \infty$. A precise version of this result and its higher order analogues is given in Skovgaard (1989). The first order approximations for the test theory are, of course, well known. A careful account, also including higher-order results, may be found in Pfanzagl (1980, Chapter 9).

From (1) it is seen that the test statistic which, to first order of approximation, has the properties of the canonical sufficient statistic in an exponential family, is the score statistic $D_1(\beta_0)$. Thus, tests based on this statistic may be constructed to be first order efficient (i.e., of optimal power) against alternatives in any particular direction. As is well known, this is true for numerous test statistics that are asymptotically equivalent to $D_1(\beta_0)$, for example, the likelihood ratio test against a sequence of contiguous alternatives $\beta_n = \beta_0 + \delta/\sqrt{n}$ (for a fixed vector δ), or the test based on the maximum likelihood estimator or on some other efficient estimator sequence. Briefly speaking, any of the commonly used test statistics behave to first order as in the case of an exponential family.

Concerning estimation and minimization of loss functions, it is hardly necessary to review first order theory here. Numerous estimator sequences $\hat{\beta}_n$ are first order efficient in the sense that $\sqrt{n}(\hat{\beta}_n - \beta_0)$ converges to a (Normal) distribution with mean zero and variance $I(\beta_0)^{-1}$. More precisely, disregarding problems of super-efficiency, first order efficiency requires that for any β the mean squared error converges to the inverse of the Fisher information per observation. It should be noted that the choice of quadratic loss function is not crucial here, because first order efficient estimators transform to first order efficient estimators of a parameter obtained by a (smooth) one-to-one transformation.

If, as the starting point, we take the maximum likelihood estimator $\hat{\beta}_n$ of β, then any estimator of the form $\hat{\beta}_n + \rho_n(Y)$ with $\rho_n(Y) = O_P(n^{-1/2})$ is efficient to first order (still provided sufficient regularity). The same result is obtained through the definition of the estimator as a zero-level confidence region, if the confidence region is given in terms of the test based on the score statistic (or on any asymptotically equivalent statistic), combined with any of the three mentioned criteria for balancing the test against alternatives in different directions. These choices are, admittedly, somewhat arbitrary as presented here, and the lack of existence of optimal tests makes the derivation of the estimator in this way a little dubious. To some extent this is a natural consequence of the lack of choice of test statistic/confidence regions.

3.2 Second order results

We consider again n independent replications and expand distributions of statistics to second order, i.e., we include the term of order $n^{-1/2}$ when the statistic has been standardized so that it has a limiting non-degenerate distribution. Among the many procedures that are optimal in terms of their first order asymptotic behaviour we might expect to be able to discriminate on the basis of the second order terms. This expectation turns out, however, to be false. Neither power functions of first order efficient tests nor expected losses for estimators differ in their second order terms. Some reservations are necessary here, but only mild ones. For example, the loss function considered must be genuinely 'two-sided'. A precise version exists for bowl-shaped loss functions, i.e., loss functions that increase in all directions from β_0, cf. Ghosh et al. (1980), but the result holds more generally. Expansions for expected losses are usually in powers of n^{-1}, because the linear term in $\hat{\beta}_n - \beta_0$ has asymptotic expectation of order n^{-1}. Hence the second-order term (of order $n^{-1/2}$) vanishes and differences between first order efficient estimators have to be found in the third order expansions of the expected losses. In the terminology of efficiency of estimators, second-order terms usually refer to the terms of order n^{-1}, but here we stick to the terminology corresponding to powers of $n^{-1/2}$.

For powers of tests, the result that first order efficiency implies efficiency to second order is more surprising, because the second order terms do not vanish. An explanation and review of the phenomenon is given in Bickel et al. (1981). The result was developed in a number of papers (in varying generality), i.a., Pfanzagl (1973, 1979), Chibisov (1974) and Pfanzagl & Wefelmeyer (1978). Other references may be found in Bickel et al. (1981) and in Pfanzagl (1980). More precisely, consider the power function of a test against a sequence of contiguous (simple) alternatives

$$\lambda^{(n)}(\beta_0 + \delta/\sqrt{n};\alpha) = \lambda_0(\delta;\alpha) + \lambda_1(\delta;\alpha)/\sqrt{n} + \lambda_2(\delta;\alpha)/n + o(n^{-1}), \tag{2}$$

where $\lambda^{(n)}$ denotes the power based on n independent replications, and sufficient regularity is assumed to admit the expansion (cf., e.g., Pfanzagl, 1980). The third order term λ_2 is included for later reference. Let

$$\bar{\lambda}^{(n)}(\beta_0 + \delta/\sqrt{n};\alpha) = \bar{\lambda}_0(\delta;\alpha) + \bar{\lambda}_1(\delta;\alpha)/\sqrt{n} + \bar{\lambda}_2(\delta;\alpha)/n + o(n^{-1}) \tag{3}$$

denote the corresponding expansion for the sequence of optimal tests against $\beta_0 + \delta/\sqrt{n}$, i.e., for the Neyman-Pearson tests. Then the result is that

$$\lambda_0(\delta;\alpha) = \bar{\lambda}_0(\delta;\alpha) \quad\Rightarrow\quad \lambda_1(\delta;\alpha) = \bar{\lambda}_1(\delta;\alpha).$$

We conclude that second order asymptotics fail to discriminate between tests and estimators that are optimal to first order, and consequently we incorporate the third order terms in our investigations.

3.3 Third order results

3.3.1 Hypothesis testing

Turning now to the third order term $\lambda_2(\delta;\alpha)$ in (2), differences between first order optimal procedures do arise. It is no longer possible to find tests that are asymptotically optimal to third order against all alternatives in a specific direction, unless the model agrees with an exponential family to the order considered (Pfanzagl, 1980, Section 9.5). Pfanzagl continues in the one-dimensional case to prove that the class of Neyman-Pearson tests is an asymptotically complete class to third order, in the sense that any third order admissible test sequence has power functions which are asymptotically bounded to third order by the power functions of the Neyman-Pearson tests against some sequence of simple alternatives (Pfanzagl, 1980, Section 9.5).

In two (essentially independent) lines of development, one by Chandra and co-authors (Chandra & Joshi, 1983, Chandra & Mukerjee, 1984, 1985, Chandra & Samanta, 1988), another by Amari and Kumon (Amari, 1983, Kumon & Amari, 1983, 1985, cf. also Amari, 1985, 1987), further investigations of third order powers are carried out. In both of these the behaviour of some commonly used test statistics are compared, including the score test (or Rao's test), the Wald test (based on the maximum likelihood estimator), and the likelihood ratio test. Also, lacking the existence of a uniformly most powerful test against alternatives in a specified direction, attention is drawn to the locally most powerful test, i.e., the test which maximizes $\lambda_2(\delta,\alpha)$ in a neighborhood of $\delta=0$ limited to a specific direction, among the first (and hence second) order efficient tests. Notice that when only alternatives in a specific direction are considered it does not matter whether we consider one-dimensional or multi-dimensional parameters. It is found that the score test is locally most powerful to third order against one-sided alternatives (cf. Chandra & Samanta, 1988, Amari, 1983, 1985). Amari (1983) explicitly points out that for any third order admissible test, the entire function $\lambda_2(\delta,\alpha)$ is determined by a single quantity and continues to compare these functions for the 'classical' tests mentioned above. It follows from Pfanzagl's work that the locally most powerful test must fall short of other tests for some values of δ. Indeed, Amari (1987, Figure 6) demonstrates that the likelihood ratio test does comparatively well for a wide range of δ's whereas the other tests (including the locally most powerful test) have rather large deficiencies for some values of δ.

In the case of two-sided tests for a one-dimensional parameter, Chandra and co-authors as well as Kumon and Amari impose the restriction of unbiasedness (to third order). This restriction does not alter the fact that uniformly most powerful tests exist to first and second order, but not to third order except in the case of an exponential family. However, the tests considered are, of course, different from the ones in the one-sided case where there is no condition of unbiasedness. Chandra & Samanta (1988) show that the score test (based on the score statistic with a two-sided critical region) is again the locally most powerful test, but Amari (1983, 1985, 1987) finds the score test to be inadmissible! These results may not, in themselves, be contradictory, because the score test might locally agree with the locally most powerful test but fall short of this for larger values of δ. Coupled with the fact that the function is determined by its local behaviour (Amari, 1983, 1985) the results do, however, seem to be contradictory. It would be interesting to have clarified if there is a contradiction and if so, which result is correct. None of the authors comment on this apparent disagreement. Amari's results (Amari, 1987, Figure 7) again suggest the likelihood ratio test as one with a good overall behaviour, although it does, of course, have a lower local power than the locally most powerful test.

In the multi-dimensional case where we want a test balanced against alternatives in all possible directions from the (simple) hypothesis β_0, most is left to the choice of method of balancing the power. Even with the condition of unbiasedness, Kumon & Amari (1985), cf. also Amari (1985), have to impose a severe restriction on the shapes of the critical regions to compare various test. Furthermore, the power functions they compare are averaged over all directions by integration over spheres with respect to the information metric. While this is an intuitively appealing measure of power it may not be the only criterion of relevance, but first of all because of the restriction on the critical regions considered, their results are less convincing than in the one-dimensional case.

3.3.2 Estimation

To third order of approximation differences occur between expected losses of first order efficient estimators. An estimator sequence is called third order efficient (second order efficient in traditional terminology, because the relevant expansions are in powers of n^{-1}), if it minimizes the third order (n^{-1}) term of the expected loss at any parameter value, among all first order efficient estimators. It has been shown (see, i.a., Rao, 1961, Efron, 1975) that the bias corrected version of the maximum likelihood estimator is third order efficient when quadratic loss is considered. The bias correction is performed by subtraction of the bias at the estimated parameter value. A stronger result by Ghosh et al. (1980) provides a 'complete class' of third order efficient estimators in the form of the maximum likelihood estimator with different 'bias' corrections according to the competing estimators considered. More precisely, for any first order efficient estimator sequence T_n, a function of the maximum likelihood estimator exists which is 'centered' in the same way as T_n, and which has at least as small expected loss for any bowl-shaped loss function, i.e., for any loss function which increases in any direction from the true parameter value. A strict version of the result is found in Ghosh et al. (1980). It should be noted, however, that because of the problem of centering it is not possible to choose a specific estimator which is uniformly better than any other with respect to all loss functions. For the specific choice of quadratic loss in some parametrization, it remains true that the maximum likelihood estimator has to be corrected for bias in that parametrization to become third order efficient. Also other estimators are efficient to third order, see, for example, Eguchi (1983).

The definition of an estimator as a zero level confidence region does not lead to equally precise answers. Letting the level tend to zero in any of the third order efficient tests derived by Kumon & Amari (1983, Section 4.2) we obtain again the bias corrected maximum likelihood estimator, seemingly in agreement with the results mentioned above. This is, however, an artifact produced by an unjustified asymptotic operation. Their asymptotic considerations are concerned with increasing sample size for a fixed level of the test, and the alternative limiting operation should not lead to a definite point. The size of the confidence region at any level is asymptotically of order $n^{-1/2}$ so a 'bias correction' (which is of order n^{-1}) to the the point estimate cannot be determined by this method. In any case the result concerning the point estimate must be independent of the parametrization when defined in terms of power considerations of the tests, and consequently a bias correction in a specific (arbitrary) parametrization cannot be preferred on this basis. Thus, again this definition of an estimate leads to (perhaps) less precise conclusion. Whether this is reasonable is another matter.

3.4 Conditional vs. powerful tests

So far we have considered only the optimality of tests in terms of power. A restriction on the class of available tests may be imposed by insisting that any test should be evaluated in the conditional distribution given an ancillary statistic if such one exists. An ancillary statistic is here defined in the narrow sense of a statistic with a distribution that is independent on the parameter (i.e., constant under the model considered). In higher order likelihood theory the idea is extended to asymptotically ancillary statistics (Efron & Hinkley, 1978), but the general ideas and problems related to the exact ancillaries have to be confronted first.

A choice of priorities is necessary here, because conditional tests are not generally most powerful. Consider, for example, the famous case of two measuring instruments (Cox, 1958, Cox & Hinkley, 1974, Example 2.33). By flipping a coin it is decided whether to use a precise instrument resulting in a measurement with distribution $N(\mu, 1)$, say, where μ is the (unknown) quantity to be estimated, or whether to use an imprecise instrument resulting in the distribution $N(\mu, 1000)$. The random variable indicating which instrument is chosen is ancillary, and the principle of conditionality informs us to condition on the outcome of this randomization, i.e., to treat the experiment as if the instrument used was chosen in advance. While this seems intuitively reasonable, the most powerful test against a simple hypothesis, $\mu = \mu_0$, is such that the hypothesis, if true, is rejected with higher probability when the precise instrument is used. Heuristicly, this is seen to be so because most is gained by allocation of the critical region to the part of the experiment with the higher power.

Despite the fact that this example is entirely unrealistic the same dilemma occurs in disguise in, for example, any non-linear Normal regression problem. Here the role of the measurement instrument is played by the observed Fisher information (suitably standardized), or, equivalently, by the distance from the model to the observation in the mean value space, see, e.g., Skovgaard (1985). For these models the statistic is only approximately ancillary except in special cases. It so happens that the score test approximates the (locally) most powerful test whereas the likelihood ratio test approximates the conditional test (Kumon & Amari, 1983). Thus, the choice between these two tests corresponds closely to the choice of test in the example with the two measuring instruments.

In a more accurate formulation the principle of ancillarity implies that the same conclusions should be drawn from an experiment with a given observation, as from the experiment obtained from this by conditioning on any ancillary statistic, i.e., by the experiment (with the same outcome) obtained be fixing in advance the value of the ancillary statistic at its observed value (Birnbaum, 1962, see also Cox & Hinkley, 1974, Chapter 2).

However intuitive the principle of ancillarity may seem it turns out to be virtually impossible to obey in non-Bayesian inference. The main reason is that together with either the principle of sufficiency or the even weaker invariance principle (Basu, 1973) it may be shown that the strong likelihood principle is implied (Birnbaum, 1962), requiring inference to be drawn solely on the basis of the likelihood function (not on its distribution). Review's of these principles may be found in Dawid (1976) and Cox & Hinkley (1974, Chapter 2). Another way in which this problem is encountered is by the observation that no unique maximal ancillary statistic exists, i.e., there may be several ancillary statistics and we cannot condition on all of them (Basu, 1959). The following example from Basu's paper illustrates the problem. *Example:* Let $(X_1, Y_1), \ldots, (X_n, Y_n)$ be independent identically distributed binary Normal variables with zero means and unit variances. The unknown parameter is the correlation coefficient ρ. The vector $(X_1, \ldots, X_n)$ is ancillary and hence the ancillarity principle requires us to condition on this statistic. But also $(Y_1, \ldots, Y_n)$ is ancillary, and the two statistics together constitute the entire sample of observations. How should we then choose between the two ancillary statistics and what remains of the principle if it only applies to one of the ancillaries? •

A number of authors discuss criteria for choosing which ancillaries to condition on, i.a., Cox (1971), Fraser (1972) and Kalbfleisch (1975), but since any such choice is incompatible with the principle and the basic argument of ancillarity, convincing solutions to the problem are not obtained.

Perhaps the principle to *condition* on any ancillary is too strong. After all, the fact that a statistic is ancillary merely suggests that it should be impossible to draw any conclusions based on the observation of such a statistic (we can't get something for nothing). In terms of test theory this condition may be formulated in the following way. Let $p(Y)$ denote the p-value obtained from a test of some hypothesis based on the observation Y and let $A(Y)$ denote an ancillary statistic. Then we impose the condition that

$$\mathrm{Pr}_0\{\, p(Y) \leq \alpha \mid A(Y) \,\} \leq \alpha \tag{4}$$

for any $\alpha \in (0, 1)$ and any ancillary $A(Y)$, where the probability is based on any distribution within the null-hypothesis. Without the conditioning this is the basic inequality for a test, and it seems reasonable to demand that we cannot increase the chance of obtaining significant results on the basis of observations that tell us nothing about the parameter. Thus, tests that do not satisfy the condition (4) may be considered logically deficient.

In the example concerning the correlation coefficient above, it may be noted that it causes no problem to construct a test for $\rho = 0$ compatible with the condition (4), because the distribution of the sample correlation coefficient is independent of either of the two mentioned ancillaries. The drastic difference between conditioning on *either* or on *both* of the ancillaries is apparent here. When the null-hypothesis ρ_0 for the correlation coefficient is non-zero the problem becomes more difficult (and more interesting), but in any case the possibility of obtaining tests obeying (4) exists, although we may be forced to use somewhat conservative tests.

In asymptotic theory we want to extend the requirement (4) to imply that conditioning on approximate ancillaries must lead to an approximate compliance with the inequality. This means that based on a statistic with little information we should not be able to draw strong conclusions. This natural requirement touches some of the basic problems with the Neyman-Pearson test theory which does not rule out the possibility of a strong rejection of a hypothesis in favour of alternatives that don't fit much better. How these problems should be overcome is not clear; for now we take the point of view that if a statistic $A(Y)$ is ancillary to a certain order of asymptotic approximation, then the inequality (4) must hold to the same order of approximation. An investigation to second or third order of powers of tests satisfying this condition has not been carried out, but several closely related results exist.

In McCullagh (1984) an approach was taken, compatible with the arguments above. Corresponding to each fixed (simple) hypothesis for a multi-dimensional parameter, a statistic exists which to second order is independent of any ancillary statistic. This is the 'orthogonal statistic' referred to in McCullagh (1987, Section 8.5). It was then suggested to base tests and confidence procedures on these statistics. The choice of such statistics is not unique and it remains an open problem whether tests constructed in this way are optimal among tests satisfying the approximate version of condition (4). However, the approximation (1) may be extended to show that to second order the (arrays of) first two derivatives, $D_1(\beta_0)$ and $D_2(\beta_0)$, are sufficient to this order of approximation (Skovgaard, 1989), and then it is seen from the results in Cox (1980) that conditioning on any ancillary statistic to this order leads to the same result. It therefore seems reasonable to conjecture that McCullagh's method, which is equivalent to tests based on the 'signed' likelihood ratio statistic, leads to optimal tests obeying (4) to second order.

McCullagh (1987, Section 8.4) furthermore shows that the likelihood ratio test statistic is independent to third order of any ancillary statistic; the signed version of this statistic only to second order. Thus, until further results are obtained it seems that the likelihood ratio test might be preferred, because it obeys the condition (4) to high order of accuracy, especially when the general existence of an accurate approximation, in the form of the Bartlett corrected chi-squared approximation, is taken into account (see, e.g., McCullagh, 1987, Section 7.4.4). Barndorff-Nielsen (1984, 1985, 1986) shows that a signed version of the likelihood ratio statistic is independent of ancillary statistics to third order, but his results concern specific ancillary statistics.

4. NUISANCE PARAMETERS

Problems of testing and estimation in the presence of nuisance parameters arise when some function $\psi(\beta)$ of the parameter $\beta \in \mathbf{R}^p$ is of interest. Here we limit ourselves to the case of a one-dimensional interest parameter $\psi(\beta) \in \mathbf{R}$. The nuisance parameter is often given explicitly as a supplementary parameter $\xi \in \mathbf{R}^{p-1}$ and the parametrization chosen such that $\beta = (\psi, \xi)$, i.e., such that the interest parameter is a coordinate of β. For theoretical considerations we prefer the formulation in terms of a parameter function $\psi(\beta)$ of interest, rather than the (in some contexts, at least) somewhat arbitrary representation $\beta = (\psi, \xi)$. This has the advantage that it prevents us from investigation of methods based on an estimate of the nuisance parameter for unspecified value of ψ. A different issue is that some methods may be based on choices of parametrizations with specific properties, see, for example, Cox & Reid (1987).

There are two, principle distinct, types of asymptotics for inference about parameter functions (i.e., problems with nuisance parameters). Both are relevant, but usually to clearly different types of applications. In the first, here referred to as 'fixed nuisance parameter', independent replications are considered from the model with a fixed parameter vector β. In terms of hypothesis testing this corresponds to composite hypotheses, formulated as $H_0 : \psi(\beta) = \psi_0$, based on i.i.d. observations. The second type of asymptotics, referred to as 'a sequence of nuisance parameters', consists of independent observations $Y_1, \ldots Y_n$, where the distribution of Y_i is given by the (unknown) parameter value β_i, but with the restriction that $\psi(\beta_i)$ is constant. In terms of the nuisance parameters, the full parameters would be $(\psi, \xi_1), \ldots, (\psi, \xi_n)$. The latter type of asymptotics is the 'genuine' nuisance parameter situation, known to cause problems for likelihood based inference methods(cf. Neyman & Scott, 1948). We consider the two types of asymptotics separately.

4.1 Fixed nuisance parameter

In this case we have independent replications from a model with parameter $\beta \in \mathbf{R}^p$ and want to infer about the parameter $\psi(\beta) \in \mathbf{R}$. Provided sufficient regularity, the maximum likelihood estimator is consistent (and first order efficient) and the fact that a nuisance parameter is present dose not cause serious problems, asymptotically speaking.

Consider the problem of testing $\psi(\beta) = \psi_0$, with ψ_0 fixed. Tests based on the maximum likelihood estimator $\hat{\psi}_n = \psi(\hat{\beta}_n)$ are first order uniformly most powerful against one-sided alternatives ($\psi > \psi_0$ or $\psi < \psi_0$), and so are other asymptotically equivalent tests (Pfanzagl, 1980, Section 13), for example the test based on the score statistic calculated at the estimate $\hat{\beta}_0$ of β restricted to the level surface $\psi(\beta) = \psi_0$. As in the case of no nuisance parameter, first order efficiency implies second order efficiency (Pfanzagl, 1979, 1980, Section 14, Pfanzagl & Wefelmeyer, 1978). In general, no third order uniformly most powerful test against one-sided alternatives exists; an exception again being the exponential families. Pfanzagl (1980, Section 15) constructs a third order complete class of tests, and notes that generally the likelihood ratio test obtained by a replacement of the nuisance parameter by its maximum likelihood estimate, is not even admissible. I feel uncertain whether this is because an estimate of the explicitly parametrized nuisance parameter ξ is used, or whether it is true also when the estimates $\hat{\beta}$ (without the hypothesis) and $\hat{\beta}_0$ are used in this substitution.

Also for estimators of $\psi(\beta)$, the results are not much different from those without nuisance parameters. Because any bowl-shaped loss function which depends only on $\psi(\beta)$ is also bowl-shaped as a function of β, the result from Ghosh et al. (1980) is directly applicable. Thus, any third order efficient estimator of ψ has as expected loss which is bounded below by that of a suitably centered function of the maximum likelihood estimator (of β).

4.2 A sequence of nuisance parameters

More serious difficulties arise when the parameters $\beta_1, \beta_2, \ldots,$ need not be identical, but are restricted only to a single level surface of the form $\psi(\beta_i) = \psi$, say, in usual sloppy notation. Even the problem of finding a general method leading to a consistent estimator of ψ (in regular cases) is far from easy, and equally difficult is the construction of tests with reasonable first order properties for the hypothesis $\psi = \psi_0$. Methods based on marginal or conditional likelihoods exist when it is possible to reduce the problem to one involving only the parameter ψ, but in general no such reductions exist. Also approximate methods of marginal or conditional likelihoods have been suggested, but none of these methods have been shown to lead to consistent estimators in general. When consistency fails, first order properties fail and there is no point in attempting an investigation of higher order asymptotics.

One path remains open, and it seems quite logical to enter that in view of the type of asymptotics considered. We restrict the problem by considering the sequence of nuisance parameters to be drawn from some (unknown) distribution. Given the exchangability of the observations in the model, this step should not be alarming to statisticians, used to population considerations. In the present framework it means that $\beta_1, \beta_2, \ldots,$ are assumed to constitute a sequence of (unobservable) i.i.d. random variables drawn from some unknown distribution Q_ψ, say, with support on the level surface $\psi(\beta) = \psi$. Thus, $Y_1, Y_2, \ldots,$ become a sequence of independent replications from the mixture distribution with density

$$\int f(y;\beta)\, dQ_\psi(\beta)$$

with respect to some underlying measure. In this way the problem is converted to a problem with i.i.d. observations from a semi-parametric model.

This approach was adopted by Kiefer & Wolfowitz (1956) who proved the consistency of the maximum likelihood estimator under regularity conditions. Improvements of their conditions have been obtained later, and contributions have been made to the theory of mixture families, but I am not aware of any major developments of asymptotic likelihood theory along this line of approach. It would be interesting to know whether first order asymptotic results of the kind discussed in the previous sections hold, and in that case whether higher order refinements may be obtained. The recent results on 'empirical likelihoods' (Owen, 1988) give some hope of further developments. Pfanzagl (1982) develops first order theory, expanded to second order in Pfanzagl (1985), for 'general statistical models', and the results are similar to the ones discussed above for the parametric models, but I am not sure whether his framework covers the present situation. A particular problem is whether a semi-parametric likelihood ratio test is tractable for this situation, with a limiting chi-squared distribution, maybe even with the possibility of a Bartlett correction. Notice that it does not hurt in the single nuisance parameter case to consider the nuisance parameter as a random variable, because it makes no difference whether we estimate a parameter value or a degenerate distribution. Thus, it is not necessary in applications to decide which of the two types of asymptotics to lean to.

4.3 Profile likelihoods

Important contributions to the theory of nuisance parameters are given in relation to (modified) profile likelihood functions. The profile likelihood is the function of the interest parameter obtained by maximization over the induced level surface in the parameter space. Modifications to this profile likelihood are obtained as approximations to marginal or conditional likelihoods that approximately eliminate the dependence on the nuisance parameters (Barndorff-Nielsen, 1983, Cox & Reid, 1987, Davison, 1988, Tibshirani & McCullagh, 1988). Although asymptotic approximations are heavily involved in these methods, they are primarily justified on other grounds than those considered here, and it is somewhat unclear where to place the methods in terms of the two types of asymptotics considered above. The profile likelihood methods do not in general lead to consistent estimators in the type of asymptotics with an infinite sequence of nuisance parameters, and for a fixed nuisance parameter no optimality properties have been derived. This does not imply that the profile likelihood methods are not relevant, and they have been shown to possess good statistical properties in concrete examples.

5 DISTRIBUTIONAL APPROXIMATIONS

Much work on asymptotics in relation to likelihood inference deals with the derivation of asymptotic approximations to distributions of estimators and test statistics. Although such approximations are not really the theme of this review they are behind most of the results discussed here.

A slightly different angle of approach starts from the statistical procedure and is concerned with accurate approximations to the related distributions. In particular, it is of interest to see to what extent approximations of the large deviation type (such as the saddlepoint approximations) can be used for this purpose. This type of approximation is represented by Barndorff-Nielsen's (1980) formula for the conditional distribution of the maximum likelihood estimator. Barndorff-Nielsen (1989) approaches convenient and accurate methods for assigning confidence limits, based on this formula and a method related to the distributional saddlepoint expansion (Lugannani & Rice, 1980).

Another important achievement concerning accurate approximations is the proof of the general existence of a Bartlett correction factor for the chi-squared approximation to the distribution of the likelihood ratio test (Lawley, 1956, cf. also McCullagh, 1987, Section 7.4.4).

The idea of choosing among the many first order equivalent test statistics, one for which a saddlepoint approximation to the null-distribution exists, rather than seeking a test with certain higher order properties, was promoted in Skovgaard (1988). A review of saddlepoint methods is found in Reid (1988).

6 CONCLUDING REMARKS

With the results available so far it seems sensible to base statistical procedures on the likelihood ratio test, because of its good conditional properties given (approximately) ancillary statistics, coupled with its first order properties (shared with many other procedures), and its high overall third order power. Reservations are necessary here concerning sufficient regularity of the problem considered. When it is not computationally feasible to base, for example, confidence sets on the likelihood ratio tests, it may be reasonable to look for good (e.g., higher order) approximations to this procedure, rather than to start from an efficient estimator and its distribution as is commonly done in practice.

Whether other statistical procedures exist, more powerful than those based on the likelihood ratio test, but with equally good conditional properties, is an open problem. To second order of approximation this does not seem to be the case.

In the case of a sequence of (unknown) nuisance parameters it is still an open problem whether likelihood methods (based on the approach by Kiefer & Wolfowitz, 1956) with reasonable general asymptotic properties are available.

BIBLIOGRAPHY

Amari, S.-I. (1983). Comparisons of asymptotically efficient tests in terms of geometry of statistical structures. *Bulletin of the International Statistical Institute. Proceedings of the 44th Session* **2**. ISI, Madrid.

Amari, S.-I. (1985). *Differential-Geometrical Methods in Statistics. Lecture Notes in Statistics* **28**. Springer, Berlin.

Amari, S.-I. (1987). Differential geometrical theory of statistics — towards new developments. *Differential Geometry in Statistical Inference. Lecture Notes, Monograph Series* **10,** pp. 19-94. Inst. Math. Statist., Hayward, Calif..

Barndorff-Nielsen, O. E. (1980). Conditionality resolutions. *Biometrika* **67**, 293-310.

Barndorff-Nielsen, O. E. (1983). On a formula for the distribution of the maximum likelihood estimator. *Biometrika* **70**, 343-356.

Barndorff-Nielsen, O. E. (1984). On conditionality resolution and the likelihood ratio for curved exponential models. *Scand. J. Statist.* **11**, 157-170. Corr. (1985), **12**, 191.

Barndorff-Nielsen, O. E. (1985). Confidence limits from $c|\hat{j}|^{\frac{1}{2}}\bar{L}$ in the one-parameter case. *Scand. J. Statist.* **12**, 83-87.

Barndorff-Nielsen, O. E. (1986). Inference on full or partial parameters, based on the standardized signed log likelihood ratio. *Biometrika* **73**, 307-322.

Barndorff-Nielsen, O. E. (1989). Approximate interval probabilities. Manuscript. Dept. Theor. Statist., Aarhus Univ.

Basu, D. (1959). The family of ancillary statistics. *Sankhya* **21**, 247-256.

Basu, D. (1973). Statistical information and likelihood. *Proceedings on Conference on Foundational Questions on Statistical Inference.* Memoirs, **1**. Ed: Barndorff-Nielsen, O.E., Blæsild, P. & Schou, G., pp. 139-247. Dept. Theor. Statist., Aarhus Univ..

Bickel, P. J., Chibisov, D. M. & van Zwet, W. R. (1981). On efficiency of first and second order. *Int. Statist. Rev.* **49**, 169-175.

Birnbaum, A. (1962). On the foundations of statistical inference. (With discussion). *J. Amer. Statist. Assoc.* **57**, 269-326.

Chandra, T. K. & Joshi, S. N. (1983). Comparison of the likelihood ratio, Rao's and Wald's tests and a conjecture of C. R. Rao. *Sankhya A* **45**, 226-246.

Chandra, T. K. & Mukerjee, R. (1984). On the optimality of Rao's statistic. *Comm. in Statist. A. Theory Methods* **13**, 1507-1515.

Chandra, T. K. & Mukerjee, R. (1985). Comparison of the likelihood ratio, Wald's and Rao's tests. *Sankhya A* **47**, 271-284.

Chandra, T. K. & Samanta, T. (1988). On the second order local comparison between perturbed maximum likelihood estimators and Rao's statistic as test statistics. *J. Mult. Anal.* **25**, 201-222.

Chibisov, D. M. (1974). Asymptotic expansions for some asymptotically optimal tests. *Proc. Praque Symp. Asymptotic Statist.* **2**, pp. 37-68 . Academia, Praque.

Cox, D. R. (1958). Some problems connected with statistical inference. *Ann. Math. Statist.* **29**, 357-372.

Cox, D. R. (1971). The choice between alternative ancillary statistics. *J. R. Statist. Soc. B* **33**, 251-255.

Cox, D. R. (1977). The role of significance tests. (With discussion). *Scand. J. Statist.* **4**, 49-70.

Cox, D. R. (1980). Local ancillarity. *Biometrika* **67**, 279-286.

Cox, D. R. & Hinkley, D. V. (1974). *Theoretical Statistics.* Chapman and Hall, London.

Cox, D. R. & Reid, N. (1987). Parameter orthogonality and approximate conditional inference (with discussion). *J. R. Statist. Soc. B* **49**, 1-39.

Davison, A. (1988). Approximate conditional inference in generalized linear models. *J. R. Statist. Soc. B* **50**, 445-461.

Dawid, A. P. (1976). Conformity of inference patterns. Paper presented at the European Meeting of Statisticians, Grenoble.

Efron, B. (1975). Defining the curvature of a statistical problem (with applications to second order efficiency). (With discussion). *Ann. Statist.* **3**, 1189-1242.

Efron, B. & Hinkley, D. V. (1978). Assessing the accuracy of the maximum likelihood estimator: Observed versus expected Fisher information. *Biometrika* **65**, 457-487.

Eguchi, S. (1983). Second order efficiency of minimum contrast estimators in a curved exponential family. *Ann. Statist.* **11**, 793-803.

Fraser, D. A. S. (1972). The elusive ancillary. *Multivariate Statistical Inference..* Ed: Kabe & Gupta, pp. 41-48.

Fraser, D. A. S. & Massam, H. (1985). Conical tests: observed levles of significance and confidence regions. *Statist. Hefte* **26**, 1-17.

Ghosh, J. K., Sinha, B. K. & Wieand, H. S. (1980). Second order efficiency of the MLE with respect to a bounded bowl-shaped loss function. *Ann.Statist.* **8**, 506-521.

Kalbfleisch, J. D. (1975). Sufficiency and conditionality. *Biometrika* **62**, 251-259.

Kiefer, J. & Wolfowitz, J. (1956). Consistency of the maximum likelihood estimator in the presence of infinitely many incidental parameters. *Ann. Math. Statist.* **27**, 887-906.

Kumon, M. & Amari, S.-I. (1983). Geometrical theory of higher-order asymptotics of test, interval estimator and conditional inference. *Proc. R. Soc. Lond. A* **387**, 429-458.

Kumon, M. & Amari, S.-I. (1985). Differential geometry of testing hypothesis — A higher order asymptotic theory in multi-parameter curved exponential family. Tech. report 85-1, Univ. of Tokyo.

Lawley, D. N. (1956). A general method for approximating to the distribution of the likelihood ratio criteria. *Biometrika* **43**, 295-303.

Lehmann, E. L. (1959). *Testing Statistical Hypotheses*. Wiley, New York.

Lehmann, E. L. (1983). *Theory of Point Estimation*. Wiley, New York.

Lugannani, R. & Rice, S. O. (1980). Saddlepoint approximation for the distribution of the sum of independent random variables. *Adv. Appl. Prob.* **12**, 475-490.

McCullagh, P. (1984). Local sufficiency. *Biometrika* **71**, 233-244.

McCullagh, P. (1987). *Tensor Methods in Statistics*. Chapman and Hall, London.

Neyman, J. & Scott, E. L. (1948). Consistent estimates based on partially consistent observations. *Econometrica* **16**, 1-32.

Owen, A. (1988). Empirical likelihood ratio confidence intervals for a single functional. *Biometrika* **75**, 237-249.

Pfanzagl, J. (1968). A characterization of the one parameter exponential family by existence of uniformly most powerful tests. *Sankhya A* **30**, 147-156.

Pfanzagl, J. (1973). Asymptotic expansions related to minimum contrast estimators. *Ann. Statist.* **1**, 993-1026.

Pfanzagl, J. (1979). First order efficiency implies second order efficiency. *Contributions to statistics.* (ed. J. Jureckova), pp. 167-196. Academia, Praque.

Pfanzagl, J. (1980). Asymptotic expansions in parametric statistical theory. *Developments in Statistics,* vol 3 (ed.: P. R. Krishnaiah), pp. 1-97. Academic Press, New York.

Pfanzagl, J. (1982). *Contributions to a General Asymptotic Statistical Theory.* Springer, Berlin.

Pfanzagl, J. (1985). *Asymptotic Expansions for General Statistical Models.* Springer, Berlin.

Pfanzagl, J. & Wefelmeyer, W. (1978). An asymptotically complete class of tests. *Z. Wahrsch. Verw. Geb.* **45**, 49-72.

Rao, C. R. (1961). Asymptotic efficiency and limiting information. *Proc. 4th Berkeley Symp.* **1**, 531-545.

Reid, N. (1988). Saddlepoint methods and statistical inference. (With discussion). *Statist. Sci.* **3**, 213-238.

Skovgaard, I. M. (1985). A second-order investigation of asymptotic ancillarity. *Ann. Statist.* **13**, 534-551.

Skovgaard, I. M. (1988). Saddlepoint expansions for directional test probabilities. *J. R. Statist. Soc. B* **50**, 269-280.

Skovgaard, I. M. (1989). Analytic statistical models. Preprint, Royal Vet. Agric. Univ., Copenhagen.

Tibshirani, R. & McCullagh, P. (1988). A simple method for the adjustment of profile likelihoods. Tech. Report no. 9, University of Toronto, Dept. of Statistics.

SUMMARY

A review is given of some results of second and third order asymptotic approximations to likelihood based statistical inference for regular statistical models. The paper is concerned with the use of asymptotics as a tool for choosing between general statistical inference procedures. For reasons given in the paper, mainly powers of tests are considered. The choice between conditional and optimal tests is discussed, and it is suggested that tests should obey a weakened form of ancillarity principle. As a general method, the likelihood ratio test seems to be preferable on the basis of the criteria chosen, but this conclusion is not definite. Further work, along the line initiated by Kiefer & Wolfowitz (1956), is required for the case of an infinite sequence of nuisance parameters.

RESUME:

Ce papier passe en revue quelques résultats d'approximations asymptotiques de deuxième et troisième ordre appliqués à la vraisemblance basée sur l'inférence statistique pour des modèles statistiques réguliers. Les résultats asymptotiques sont vus comme un instrument de choix parmi les procédures statistiques générales. On considère essentiellement les puissances des tests, on discute du choix entre les tests conditionnels et les tests optimaux et on suggère de faire suivre aux tests le principe d'ancillarité.

En tant que méthode générale, le test du ratio de vraisemblance semble préférable vis-à-vis des critères choisis mais cette conclusion n'est pas définitive.

Il est nécessaire de poursuivre ce travail, à la suite de Kiefer et Wolfowitz (1956), dans le cas d'une suite infinie de paramètres de nuisance.

Chapter 7

On the density of minimum contrast estimators

Introduction by Alessandra Salvan and Luigi Pace

University of Padova, University of Udine

Most citations in the literature mention this contribution as a key reference for a general derivation and a deeper understanding of Barndorff-Nielsen's p^*-formula for the asymptotic conditional density of the maximum likelihood estimator. However, the main result of the work is, more broadly, about existence of the density of minimum constrast estimators together with its exact formula, simply expressed as a function of the joint density of derivatives of the objective function up to order two. Barndorff-Nielsen's p^*-formula can be recovered as an easy consequence, both when it is exact and when it provides an asymptotic approximation.

The results of the paper refer to vector parameters and continuous distributions. Multiple root problems are eschewed by studying the intensity function of the point process of local maxima of the negative contrast function. The intensity function is the density of the minimum contrast estimator if multiple maxima can be excluded, as is often the case in a neighbourhood of the true parameter value, except for an exponentially decreasing probability.

§1 accurately describes the elements of novelty in the paper and the relationship with previous contributions. The aim of providing conditions for existence of the density of the estimator, together with its formula, differs from the objective of related results, from the 1980s, quoted in the paper. In these works, asymptotic approximations are derived for a density assumed to exist. The paper instead extends to a general setting the exact result for a smooth subfamily of an exponential family obtained in Skovgaard (1985a).

§2 states regularity conditions sufficient for the existence of the intensity function and the main Theorem establishes its structure. Ib points out that his conditions are by no means necessary, and indicates some ways to relax them.

To get a closer look at the result, consider data y in a sample space E with statistical model $\{P_\beta\,; \beta \in B \subseteq \mathbb{R}^p\}$. Let $\gamma(y;\beta)$ be the negative contrast function, maximized by $\hat{\beta}$. The intensity function is

$$g(b;\beta_0) = \lim_{\varepsilon\to 0} \frac{P_{\beta_0}\{\gamma(y;\beta) \text{ has a local maximum in } U_\varepsilon(b)\}}{\text{vol}\{U_\varepsilon(b)\}},$$

where $b \in \text{int}(B)$, $U_\varepsilon(b)$ is an ε-neighbourhood of b and $\text{vol}(\cdot)$ its volume.

Let D_1 and D_2 be the gradient and the Hessian, respectively, of $\gamma(y;\beta)$ as a function of β evaluated at $\beta = b$. The assumed conditions are

(C1) For almost all y, $\gamma(y;\beta)$ is three times differentiable with respect to β at $\beta = b$.

(C2) Conditional densities $f(d_1 \mid d_2;\beta_0)$ of D_1 given $D_2 = d_2$ with respect to Lebesgue measure exist and are uniformly bounded.

(C3) Certain moment requirements on derivatives of $\gamma(y;\beta)$ up to third order are satisfied.

The theorem states that, under conditions (C1) to (C3), the limit $g(b;\beta_0)$ exists and is given by formula (2.3) in the paper, in two equivalent forms. The first is the expectation of a function of D_2 under β_0. The function is the product of $f(d_1 \mid D_2;\beta_0)$ at $d_1 = 0$ times $|D_2| I(D_2 < 0)$, the absolute value of the determinant of D_2 when D_2 is negative definite (otherwise zero). Let $f_j(\cdot;\beta_0)$ be the marginal density of D_j, $j = 1, 2$. In the second form, $g(b;\beta_0)$ is written as the product of the marginal density of D_1 at $d_1 = 0$, under β_0, and of the conditional expectation (under β_0), given $D_1 = 0$, of $|D_2| I(D_2 < 0)$, i.e. as

$$g(b;\beta_0) = f_1(0;\beta_0)\, E_{\beta_0}\{|D_2| I(D_2 < 0) \mid D_1 = 0\}. \qquad (7.1)$$

The equivalence of the two forms is seen by writing $f(0 \mid d_2;\beta_0) = f_1(0;\beta_0) f(d_2 \mid d_1 = 0;\beta_0)/f_2(d_2;\beta_0)$. One of the virtues of Ib's exact formula is that it behaves coherently under smooth reparameterizations. We conjecture that this behaviour ensures that derivatives of order higher than two do not appear in the formula.

Condition (C1) allows quadratic approximation of $\gamma(y;\beta)$ near $\beta = b$, as $\tilde{\gamma}(y;\beta) = \gamma(y;b) + D_1^{\mathsf{T}}(\beta - b) + (\beta - b)^{\mathsf{T}} D_2 (\beta - b)/2$. The straightforward part of the proof is to show that (2.3) holds for $\hat{\beta}_1$, the maximizer of $\tilde{\gamma}(y;\beta)$. Indeed, supposing D_2 negative definite, $\hat{\beta}_1 = b - D_2^{-1} D_1$. From the existence part of (C2), the conditional density of $\hat{\beta}_1$ given D_2 is $f(d_1 \mid D_2; \beta_0)|D_2|$ evaluated at d_1 such that $\hat{\beta}_1 = b - D_2^{-1}d_1$. But at $\hat{\beta}_1 = b$, d_1 is 0. The marginal density of $\hat{\beta}_1$ is then obtained by taking the expectation of the conditional density with respect to D_2 under β_0. This leads to the first form of formula (2.3) in the paper and then to formula (7.1) for the intensity function of maxima of the quadratic approximation,

$$\tilde{g}(b; \beta_0) = \lim_{\varepsilon \to 0} \frac{P_{\beta_0}\{\tilde{\gamma}(y; \beta) \text{ has a local maximum in } U_\varepsilon(b)\}}{\text{vol}\{U_\varepsilon(b)\}}.$$

The arduous part of the proof is to show that $g(b; \beta_0) = \tilde{g}(b; \beta_0)$. Boundedness in (C2) and (C3) are needed for the highly technical steps involved.

§3 is avowedly written without attention to mathematical rigour. It illustrates through a heuristic discussion what light the exact formula sheds on the p^*-formula. When $\gamma(y; \beta)$ is the log likelihood function, $\ell(\beta; y)$, the intensity function of local maxima depends on the data through the (minimal) sufficient statistic. If $\hat{\beta}$ is not sufficient, it is supposed that an auxiliary statistic $A(y)$ exists such that $(\hat{\beta}, A)$ recovers sufficiency. In the conditional model for y given $A = a$, the exact density or intensity function from (7.1), $g(b \mid a; \beta_0)$, may be recast as the product of three factors

$$g(b \mid a; \beta_0) = f(0 \mid a; b) \frac{f(0 \mid a; \beta_0)}{f(0 \mid a; b)} E_{\beta_0}\{|D_2| I(D_2 < 0) \mid D_1 = 0, A = a\}. \tag{7.2}$$

For a fixed b, (D_1, A) is a one-to-one function of $(\hat{\beta}, A)$, at least around $\hat{\beta} = b$, so it is sufficient as well. Then D_2 is a function of (D_1, A), so that the third factor in (7.2) is given by $|\jmath(b; b, a)|$, where $\jmath(\beta; \hat{\beta}, a)$ is the observed information function. When A is exactly ancillary, the second factor in (7.2) is the likelihood ratio $\exp\{\ell(\beta_0; y) - \ell(b; y)\}$ at y where $D_1 = 0$ and $A = a$. Hence, (7.2) becomes

$$g(b \mid a; \beta_0) = f(0 \mid a; b) \exp\{\ell(\beta_0; b, a) - \ell(b; b, a)\} \, |\jmath(b; b, a)|. \tag{7.3}$$

In transformation models with A the maximal invariant, the product $f(0 \mid a; b)|\jmath(b; b, a)|^{1/2}$ does not depend on b, so that

$$g(b \mid a; \beta_0) = c(a) \exp\{\ell(\beta_0; b, a) - \ell(b; b, a)\} \, |\jmath(b; b, a)|^{1/2}. \tag{7.4}$$

The right-hand side of (7.4) is Barndorff-Nielsen's p^*-formula, whose exactness in transformation models is a direct consequence of exactness of (7.1). In the other cases where (7.3) holds, a normal approximation for the first factor is $(2\pi)^{-p/2} \, |\text{var}_b(D_1 \mid a)|^{-1/2}$. There is no conditioning in full exponential families, where $\ell(\beta; y) = \ell(\beta; b)$ and $\text{var}_b(D_1) = \jmath(b; b)$. Outside full exponential families, $\text{var}_b(D_1 \mid a) = E_b(-D_2 \mid a)$ and expansion (3.6) in the paper gives $\text{var}_b(D_1 \mid a) = j(b; b, a) + O(1)$. In both cases, the p^*-formula is a large deviation type approximation for $g(b|a; \beta_0)$.

It is a part of the 'magic' of the p^*-formula that it is valid even when the auxiliary A is only approximately ancillary. This generality is recovered by Ib considering further approximations in (7.2) with the last factor replaced by $|j(b; b, a)|$. In a normal deviation region for A, the second factor differs from the likelihood ratio by a multiplicative error of order $O(n^{-1/2})$, or $O(n^{-1})$, if A is first, or second, order ancillary. The approximation $(2\pi)^{-p/2} |j(b; b, a)|^{-1/2}$ for the first factor holds with a similar order of error. Even with an approximate ancillary, the p^*-formula is a large deviation type approximation with respect to b, because it keeps a uniform relative error of order $O(n^{-1/2})$ at worst.

Ib's exact formula is considered in subsequent literature trying to validate it under weaker assumptions (Jensen and Wood, 1998; Almudevar *et al.*, 2000). Skovgaard (2001), §§5.2, 5.3 and 6, considers a generalization to estimators from estimating equations and outlines some perspectives of application outside maximum likelihood estimation, having in mind in particular restricted maximum likelihood estimation in linear mixed models. Almudevar (2016) brings out a relation of formula (2.3) with a formula by Rice for the intensity function of crossings of a level by a stationary process. The problem is related to the characterization of roots of polynomials with random coefficients, or more generally systems of random equations, Brillinger (1972).

The elegance of Ib's exact formula parallels that of Newton's method for finding a zero of a differentiable function. In addition to its role as a source of asymptotic approximations of the large deviation type, a direct numerical use of the formula through simulation when p is small may appear nowadays conceivable.

The Annals of Statistics
1990, Vol. 18, No. 2, 779–789

ON THE DENSITY OF MINIMUM CONTRAST ESTIMATORS

BY IB M. SKOVGAARD

Royal Veterinary and Agricultural University

Conditions for the existence of the density of a minimum contrast estimator in a parametric statistical family are given together with a formula for this density. The formula is exact if multiple local minima cannot occur; otherwise the formula is an exact expression for the point process of local minima of the contrast function. Although it is not in general feasible to compute the expression for the density, the formula can be used as a basis for further expansion of the large deviation type. When the estimate is sufficient, either in the original model or after conditioning on an approximate or exact ancillary, the formula simplifies drastically. In particular, it is shown how Barndorff-Nielsen's formula for the density of the maximum likelihood estimator given an ancillary statistic is derived from the formula given here. In this way the nature of Barndorff-Nielsen's formula as an asymptotic approximation and its appearance as an exact formula for certain cases are demonstrated.

1. Introduction. The theorem given in this paper provides conditions for the existence of the density of the maximum likelihood estimator of a (vector) parameter, together with a formula for this density. The formula is an exact expression of the intensity of the point process of local maxima of the likelihood function, which is the same as the density of the maximum likelihood estimator if multiple local maxima cannot occur; otherwise it exceeds this density by an amount stemming from cases when a local maximum occurs that is not global. Typically, in the case of n independent replications, this difference decreases at an exponential rate in n. The formula was given as Lemma 4.2 in Skovgaard (1985b) under much stronger conditions, within the framework of exponential families. There it was used as an intermediate step toward a large deviation expansion of the density of the maximum likelihood estimator, which turned out to be identical to the one derived by Field (1982) by quite different methods. Field, working with the more general class of M-estimators, assumed the existence of the density of the estimator and proceeded to derive the large deviation expansion by a combination of the methods of Edgeworth expansions from Bhattacharya and Ghosh (1978) and a saddlepoint expansion for the score statistic evaluated at the point at which the density of the estimator was to be calculated. Pazman (1984) derived the same formula for the case of nonlinear regression along a third line of reasoning [cf. Hougaard (1985)]. In the present paper we shall consider the kind of continuity conditions required for the existence of the density of the estimator, a problem that was not addressed for more general families of distributions in any of the

Received April 1988; revised February 1989.
AMS 1980 *subject classifications*. Primary 62F12; secondary 62E15.
Key words and phrases. Barndorff-Nielsen's formula, conditional inference, large deviation expansion, minimum contrast estimator, maximum likelihood estimator, saddlepoint approximation.

papers mentioned above. Another purpose of the present paper is to consider the exact formula in itself, because it may be used to derive more refined approximations for special classes of models, in particular in connection with conditional inference. Thus, as a main example, Barndorff-Nielsen's formula (3.1) [cf. Barndorff-Nielsen (1980, 1983)] is shown to be an easy consequence and its exactness properties may be more transparent in view of the formula given here.

In Section 2 the theorem on the existence and the formula for the density for the estimator are given. The result is stated in terms of minimum contrast estimators, because the method of proof does not rely on any particular properties of the likelihood function. The proof of the theorem is given at the end of that section. Some readers may want to skip this and go directly to the next section.

In Section 3 we discuss the implications of the formula, mainly for conditional inference, in which case it simplifies drastically. In particular, Barndorff-Nielsen's formula is discussed in this section.

2. A formula for the density. Let $\{P_\beta; \beta \in B \subseteq \mathbb{R}^p\}$ be a family of probability measures on some measurable space E, where the parameter space B is some subset of $\mathbb{R}^p$. Consider an estimator $\hat{\beta}$ of the parameter β that minimizes the contrast function

$$-\gamma(y;\beta), \qquad y \in E, \beta \in B,$$

as a function of β when $y \in E$ is the observed data point. For analogy with the special case when γ is the log-likelihood function we prefer to consider the maximization of the negative of the contrast function. The function γ: $E \times \mathbb{R}^p \to \mathbb{R}$ is required to be measurable with respect to the product of P_β and the Borel measure on $\mathbb{R}^p$ for any β. For any $\varepsilon > 0$ and $v \in \mathbb{R}^p$, define

$$(2.1) \qquad U_\varepsilon(v) = \{x \in \mathbb{R}^p; \|x - v\| < \varepsilon\}$$

as the ε-neighbourhood of v in terms of the usual Euclidean norm and let $A_p = \text{vol}(U_1(0))$ denote the volume of the unit ball. Given a fixed parameter value $\beta_0 \in B$, we shall consider the existence of the β_0-density of $\hat{\beta}$ at another fixed point $b \in \text{int}(B)$. Thus all probabilities, expectations, densities, etc., in the sequel are, unless otherwise stated, computed with respect to P_{β_0}. Define

$$(2.2) \quad g(b;\beta_0) = \lim_{\varepsilon \to 0} \left(\varepsilon^p A_p\right)^{-1} P\{\gamma(y;\beta) \text{ has a local maximum in } U_\varepsilon(b)\},$$

which is the intensity of the process of local maxima of the function $\gamma(y;\beta)$ at the point $\beta = b$. It may also be interpreted as the density of a local minimum contrast estimator or, if multiple maxima can be ruled out, the density of the global minimum contrast estimator. Let $D_1(\beta)$ denote the column vector of first partial derivatives of $\gamma(y;\beta)$ with respect to the coordinates of β and $D_2(\beta)$ denote the matrix of second derivatives. The abbreviations D_1 and D_2

are used for the values at the point $\beta = b$. We shall need the following conditions.

(C1) With probability 1, $\gamma(y;\beta)$ is three times differentiable at $\beta = b$.

(C2) Conditional densities $f(d_1|d_2;\beta_0)$ of D_1 given D_2 with respect to the Lebesgue measure exist and satisfy

$$f(d_1|d_2;\beta_0) \le F < \infty,$$

for almost all d_1 and d_2, where F is some constant.

(C3) Let

$$M(\varepsilon) = \sup\left\{\left|(d/dh)^3\gamma(y;\beta + hv)\right|; \beta \in U_\varepsilon(b), v \in \mathbb{R}^p, \|v\| = 1\right\},$$

where the derivative is evaluated at $h = 0$, be the maximal norm of the third differential of $\gamma(y;\beta)$ in the ε-neighbourhood of b. Then two constants, $\eta > 0$ and $\zeta > 0$ exist, such that $0 < \zeta + (p-1)\eta < 1$ and that for any $v \in \mathbb{R}^p$,

$$E\left\{\left|v^{\mathsf{T}} D_2 v\right|^{p/\eta}\right\} < \infty,$$

where v^{T} denotes the transpose of v and

$$E\{[M(\varepsilon)]^{p/\zeta}\} < \infty.$$

THEOREM. *Assume that the conditions* C1, C2 *and* C3 *hold. Then* $g(b;\beta_0)$ *exists and*

$$\begin{aligned} g(b;\beta_0) &= E\{f(0|D_2;\beta_0)|D_2|I_{\text{neg}}(D_2)\} \\ &= f_1(0;\beta_0)E\{|D_2|I_{\text{neg}}(D_2)\,|\,D_1 = 0\}, \end{aligned} \tag{2.3}$$

where $|D_2|$ *denotes the absolute value of the determinant of* D_2, $I_{\text{neg}}(D_2)$ *is the indicator function that equals one if* D_2 *is negative definite and zero otherwise, and* f_1 *is the marginal density of* D_1 *with respect to the Lebesgue measure.*

The conditions as stated in the theorem are not optimal. Some minor technical improvements have been sacrificed to avoid obscuring the theorem too much. For example, $M(\varepsilon)$ needs to be well defined only with probability $1 - o(\varepsilon^p)$ as $\varepsilon \to 0$ and the conditional density in C2 needs only to be uniformly bounded for d_1 in some neighbourhood of zero. In fact, the existence of this density with respect to the Lebesgue measure is required only in such a neighbourhood. It is easy to see from the proof that these relaxed conditions are sufficient.

PROOF OF THE THEOREM. Define the function

$$\tilde{\gamma}(y;\beta) = \gamma(y;b) + D_1^{\mathsf{T}}(\beta - b) + \tfrac{1}{2}(\beta - b)^{\mathsf{T}} D_2(\beta - b) \tag{2.4}$$

and the events

$$(2.5)\qquad L(\varepsilon) = \{y \in E; \gamma(y;\beta) \text{ has a local maximum in } U_\varepsilon(b)\},$$

$$(2.6)\qquad \tilde{L}(\varepsilon) = \{y \in E; \tilde{\gamma}(y;\beta) \text{ has a local maximum in } U_\varepsilon(b)\}.$$

The quantity we seek is given in (2.2), but it will appear from the proof that it can also be calculated as the limit

$$(2.7)\qquad \tilde{g}(b;\beta_0) = \lim_{\varepsilon\to 0} (\varepsilon^p A_p)^{-1} P\{\tilde{L}(\varepsilon)\},$$

which is identical to (2.2) except that $L(\varepsilon)$ is replaced by $\tilde{L}(\varepsilon)$. It follows from the condition C3 by use of Chebyshev's inequality that the event

$$(2.8)\qquad S(\varepsilon) = \{y \in E; \|D_2\| \le \varepsilon^{-\tilde{\eta}} \text{ and } M(\varepsilon) \le \varepsilon^{-\tilde{\zeta}}\}$$

has probability $1 - o(\varepsilon^p)$ as $\varepsilon \to 0$, where $\tilde{\eta} > \eta$ and $\tilde{\zeta} > \zeta$ are two fixed constants chosen to satisfy $\tilde{\zeta} + (p-1)\tilde{\eta} < 1$ and $\|D_2\|$ is the largest absolute eigenvalue of D_2. Within the event $S(\alpha)$ we have the estimate

$$(2.9)\qquad |\gamma(y;\beta) - \tilde{\gamma}(y;\beta)| \le \tfrac{1}{6}\alpha^3 M(\alpha) \le \tfrac{1}{6}\alpha^3\alpha^{-\tilde{\zeta}},$$

for any $\beta \in U_\alpha(b)$. Therefore, if $y \in \tilde{L}(\varepsilon) \cap S((1+\delta)\varepsilon)$ such that the function $\tilde{\gamma}(y;\beta)$ has a local maximum at β_1, say, in $U_\varepsilon(b)$, then for some sufficiently small $\delta > 0$ and $v \in \mathbb{R}^p$ with $\|v\| = 1$ we have

$$(2.10)\qquad \begin{aligned} \gamma(y;\beta_1) - \gamma(y;\beta_1 + \delta\varepsilon v) &= -\tfrac{1}{2}\delta^2\varepsilon^2 v^{\mathsf{T}} D_2 v + (\gamma(y;\beta_1) - \tilde{\gamma}(y;\beta_1)) \\ -(\gamma(y;\beta_1+\delta\varepsilon v) - \tilde{\gamma}(y;\beta_1+\delta\varepsilon v)) &= -\tfrac{1}{2}\delta^2\varepsilon^2 v^{\mathsf{T}} D_2 v \pm \tfrac{1}{3}(1+\delta)^3\varepsilon^{3-\tilde{\zeta}}, \end{aligned}$$

where we have used (2.9) with $\alpha = \varepsilon(1+\delta)$. Notice that on the set $\tilde{L}(\varepsilon)$ the matrix D_2 must be negative semidefinite and hence the first term on the right in (2.10) is nonnegative. We want to infer that (2.10) is, in fact, positive for any v when ε is sufficiently small and hence that $\gamma(y;\beta)$ is larger at β_1 than anywhere on the boundary of $U_{\delta\varepsilon}(\beta_1)$, implying that $L(\varepsilon(1+\delta))$ has occurred. To do this we must be able to exclude the event

$$(2.11)\qquad R_c(\varepsilon) = \bigcup_{\|v\|=1} \{y \in E; |v^{\mathsf{T}} D_2 v| \le c\varepsilon^{1-\tilde{\zeta}}\}$$

for $c = c(\delta) = \frac{2}{3}\delta^{-2}(1+\delta)^3$. On this set the smallest absolute eigenvalue of D_2 is bounded by $c\varepsilon^{1-\tilde{\zeta}}$, while the remaining eigenvalues are bounded on the set $S((1+\delta)\varepsilon)$ by $\varepsilon^{-\tilde{\zeta}}$. Hence, if $y \in \tilde{L}(\varepsilon) \cap S((1+\delta)\varepsilon) \cap R_c(\varepsilon)$, then

$$(2.12)\qquad |D_2| \le c\varepsilon^{1-\tilde{\zeta}}(\varepsilon^{-\tilde{\zeta}})^{p-1} = c\varepsilon^{1-\tilde{\zeta}-(p-1)\tilde{\eta}} = o(1)$$

as $\varepsilon \to 0$ for any $c > 0$. If $\tilde{\gamma}(y;\beta)$ has a local maximum in $U_\varepsilon(b)$ we must have

$$(2.13)\qquad D_1 \in -D_2(U_\varepsilon(0)),$$

which is the set of vectors $-D_2 v$ with $v \in U_\varepsilon(0)$. This set has the volume

$$(2.14)\qquad \operatorname{vol}\{-D_2(U_\varepsilon(0))\} = |D_2|\varepsilon^p A_p.$$

By use of the assumption that the conditional density of D_1 given D_2 is bounded, it follows immediately from (2.12), (2.13) and (2.14) that

$$P\big(\tilde{L}(\varepsilon) \cap S((1+\delta)\varepsilon) \cap R_c(\varepsilon)\big) = o(\varepsilon^p) \tag{2.15}$$

as $\varepsilon \to 0$ for any $c > 0$. Together with the fact that $S((1+\delta)\varepsilon)$ has probability $1 - o(\varepsilon^p)$ as $\varepsilon \to 0$, this implies that

$$\begin{aligned} \tilde{g}(b;\beta_0) &= \lim_{\varepsilon \to 0} (\varepsilon^p A_p)^{-1} P\{\tilde{L}(\varepsilon)\} \\ &= \lim_{\varepsilon \to 0} (\varepsilon^p A_p)^{-1} P\{(\tilde{L}(\varepsilon) \cap S((1+\delta)\varepsilon)) \setminus R_c(\varepsilon)\}. \end{aligned} \tag{2.16}$$

It now follows from (2.10) that

$$\big(\tilde{L}(\varepsilon) \cap S((1+\delta)\varepsilon)\big) \setminus R_{c(\delta)}(\varepsilon) \subseteq L(\varepsilon(1+\delta))$$

and hence that

$$\begin{aligned} \tilde{g}(b;\beta_0) &\le \lim_{\varepsilon \to 0} (\varepsilon^p A_p)^{-1} L(\varepsilon(1+\delta)) \\ &= (1+\delta)^p \lim_{\varepsilon \to 0} \big(\varepsilon^p (1+\delta)^p A_p\big)^{-1} L(\varepsilon(1+\delta)) \\ &= (1+\delta)^p g(b;\beta_0). \end{aligned}$$

Since this holds for any (sufficiently small) $\delta > 0$ we have proved that

$$\tilde{g}(b;\beta_0) \le g(b;\beta_0). \tag{2.17}$$

The other inequality is proved along the same lines although some of the arguments are slightly different. Thus, assume that $y \in L(\varepsilon)$ and let $\beta_1 \in U_\varepsilon(b)$ be a local maximum point of $\gamma(y;\beta)$. Then the derivative of γ with respect to β at β_1 is zero, i.e.,

$$\begin{aligned} 0 &= \frac{d}{d\beta}\gamma(y;\beta_1) \\ &= D_1 + D_2(\beta_1 - b) \pm \tfrac{1}{2}\varepsilon^2 M(\varepsilon) \\ &= D_1 + D_2(\beta_1 - b) \pm \tfrac{1}{2}\varepsilon^{2-\bar{\zeta}}, \end{aligned}$$

if $y \in S(\varepsilon)$, where the notation $\pm$ is used for any vector with a length limited by the quantity indicated. Hence

$$D_1 \in -D_2(U_\varepsilon(0)) \pm \tfrac{1}{2}\varepsilon^{2-\bar{\zeta}}.$$

As in the first half of the proof we want to exclude the possibility that D_2 has numerically small eigenvalues, i.e., the set $R_c(\varepsilon)$ from (2.11) for some appropriate c. As above an occurrence of the event $R_c(\varepsilon)$ implies that

$$\begin{aligned} \operatorname{vol}\{-D_2(U_\varepsilon(0)) \pm \tfrac{1}{2}\varepsilon^{2-\bar{\zeta}}\} &\le \big(c\varepsilon^{2-\bar{\zeta}} + \tfrac{1}{2}\varepsilon^{2-\bar{\zeta}}\big)\big(\varepsilon^{1-\bar{\eta}} + \tfrac{1}{2}\varepsilon^{2-\bar{\zeta}}\big)^{p-1} \\ &\le c_1 \varepsilon^{p+1-\bar{\zeta}-(p-1)\bar{\eta}} = o(\varepsilon^p) \end{aligned} \tag{2.18}$$

as $\varepsilon \to 0$, where $c_1 > 0$ is some constant. As in calculation (2.16), this shows that the set $R_c(\varepsilon)$ may be ignored for any $c > 0$. For any $v \in \mathbb{R}^p$ with $\|v\| = 1$ and any $y \in S(\varepsilon)$, the inequality

$$|v^{\mathsf T} D_2(\beta_1) v - v^{\mathsf T} D_2 v| \le \varepsilon M(\varepsilon) \le \varepsilon^{1-\zeta} \tag{2.19}$$

shows that also the set on which $D_2(\beta_1)$ has any eigenvalue less than or equal to $c\varepsilon^{1-\zeta}$ may be ignored because it has probability $1 - o(\varepsilon^p)$. We still have the bound (2.9) for any $y \in S(\alpha)$ and therefore for any $y \in (L(\varepsilon) \cap S((1+\delta)\varepsilon)) \setminus R_c(\varepsilon)$ and any v with $\|v\| = 1$:

$$\begin{aligned} &\tilde\gamma(y;\beta_1) - \tilde\gamma(y;\beta_1 + \delta\varepsilon v) \\ &\quad = \gamma(y;\beta_1) - \gamma(y;\beta_1 + \delta\varepsilon v) \pm \tfrac{1}{3}(1+\delta)^3 \varepsilon^{3-\zeta} \\ &\quad = -\tfrac{1}{2}\delta^2\varepsilon^2 v^{\mathsf T} D_2(\beta_1) v \pm \left\{\tfrac{1}{3}(1+\delta)^3\varepsilon^{3-\zeta} + \tfrac{1}{6}(\delta\varepsilon)^3\varepsilon^{-\zeta}\right\} \\ &\quad > 0 \end{aligned} \tag{2.20}$$

if the constant c is chosen appropriately. Thus, it follows that $\tilde\gamma(y;\beta)$ has a local maximum in the set $U_{\delta\varepsilon}(\beta_1) \subseteq U_{\varepsilon(1+\delta)}(b)$ and the inequality opposite to (2.17) is derived by the argument analogous to the one leading to (2.17). We conclude that

$$g(b;\beta_0) = \tilde g(b;\beta_0). \tag{2.21}$$

On the set of negative definite D_2's the local maximum of $\tilde\gamma(y;\beta)$ is located at $b - D_2^{-1}D_1$, which has the conditional density

$$|D_2| f(d_1 | D_2; \beta_0)$$

at the point $b - D_2^{-1} d_1$, given D_2. From the arguments above it is seen that the set of singular D_2's, as a subset of $R_c(\varepsilon)$, does not contribute to the limit (2.7) and, therefore, the density $\tilde g(b;\beta_0)$ of the local maximum of $\tilde\gamma(y;\beta)$ at b equals the first expression in (2.3). The second expression in (2.3) is a simple recast of the first. □

3. Application to conditional inference. In this section we demonstrate the use of (2.2) to derive approximate formulas for the density of the estimator. In particular it will appear that Barndorff-Nielsen's formula for the conditional density for the maximum likelihood estimator given an ancillary statistic may be derived from (2.3). In the sequel we abandon any kind of rigour in the sketched proofs. The point is to show the potential of the theorem, not to provide new proofs of known results. In particular the technical problems related to the possibility of multiple local maxima are ignored. Throughout the section we consider only maximum likelihood estimation, i.e., the function $\gamma(y;\beta)$ is the logarithm of the density of y at β. We shall denote this function by $l(\beta;y)$ in the sequel. The likelihood considered is the likelihood from the model, even when we consider conditional inference, in which case an alternative would have been to maximize the conditional likelihood. In particular, it should be noticed that the functions $D_1(\beta)$ and $D_2(\beta)$ are the first two derivatives of the log-likelihood, unaffected by conditioning. However, we may want a formula for the conditional density of the (unconditional) maximum likelihood estimator given some exact or approximate ancillary statistic; then the theorem of the previous section applies to the conditional distributions, i.e., in the conditions and the resulting formula (2.3), all densities and expectations are conditioned on the ancillary.

Barndorff-Nielsen's formula [cf. Barndorff-Nielsen (1980, 1983)] for the conditional density of $\hat{\beta}$ at the point b given the event $A(y) = a$, where $A = A(y)$ is an exact or approximate ancillary statistic, is

$$g(b|a;\beta_0) \sim c(a)|\,j(b;y)|^{1/2} \exp\{l(\beta_0;y) - l(b;y)\}, \tag{3.1)s}$$

where $c(a)$ is a constant depending only on a, $j(\beta; y)$ equals $-D_2(\beta)$ and $y = y(b, a)$ is any data value for which $A(y) = a$ and $\hat{\beta}(y) = b$. Thus, for the formula to be well defined it is required that the joint statistic $(\hat{\beta}, A)$ be sufficient, in which case the formula does not depend on which of the possible data values y is chosen. The constant $c(a)$ is usually taken either as the general approximation $(2\pi)^{-p/2}$ or as the normalizing constant for which the integral of g with respect to b is 1. Some virtues of the formula are that it is exact for transformation models, it is equivalent to a saddlepoint approximation for full exponential family, it is an accurate approximation for subfamilies of exponential families and the formula applies to a large class of ancillary statistics. Several proofs of its validity as an asymptotic approximation for curved exponential families have been given [cf. Barndorff-Nielsen (1980), McCullagh (1987, Section 8.6) and Fraser (1988)], so the reader may well question the desirability of yet another. However, the main point of the following derivation, apart from its simplicity, is that it sheds some light over some features of the formula (3.1), including cases of exactness, its nature as an asymptotic approximation, its relation to the saddlepoint approximation and the appearance of the observed rather than the expected Fisher information in the formula. In the original proof of the general asymptotic nature of the approximation in Barndorff-Nielsen (1980), the observed information remained to some extent an optional choice compared to the expected information. This choice was, however, fully justified by a number of cases for which the formula turned out to be exact.

The discussion below is related, in particular, to the one in Fraser (1988). A difference is that Fraser, as McCullagh (1987), works with the score function at the point β_0; another is that the formula (2.3) applies also to cases of nonsufficiency for which other saddlepoint-type expansions may be derived from it [cf. Skovgaard (1985b)]; a third difference is that we focus our attention on the nature of (3.1) as an asymptotic expansion of the large deviation type.

To see how (3.1) follows from the theorem in Section 2, let us rewrite the last expression in (2.3) as a product of three factors,

$$\begin{aligned} g(b|a;\beta) &= \{f(0|a;b)\}\{f(0|a;\beta_0)/f(0|a;b)\} \\ &\quad \times \{E_{\beta_0}(|D_2|I_{\text{neg}}(D_2)|D_1 = 0, a)\}, \end{aligned} \tag{3.2}$$

where the conditioning on $A = a$ is included as opposed to (2.3). Now, as is always the case when (3.1) is applicable, we require $(\hat{\beta}, A)$ to be sufficient.

Then, for a fixed b, (D_1, A) will also be sufficient because of the one-to-one correspondence, at least locally, between these two pairs of statistics. From Neyman's factorization criterion it follows that D_2 is a function of (D_1, A) and hence that the conditional expectation in the third factor in (3.2) trivially equals $|D_2| = |j(b;y)|$. Thus, we may rewrite (3.2) as

$$(3.3)\qquad g(b|a;\beta) = \{f(0|a;b)\}\{f(0|a;\beta_0)/f(0|a;b)\}|j(b;y)|.$$

We now discuss the reduction of (3.3) to the approximation in (3.1). We consider the case when A is exactly ancillary and then the general case of an approximate ancillary. We also discuss the two cases of special interest, namely the full exponential families and the transformation models.

Full exponential families. In the cases of full exponential families the formula (3.1) is known to be equivalent to a saddlepoint expansion for the density of the score statistic. In this case there is no conditioning since the maximum likelihood estimator $\hat{\beta}$ is itself sufficient. Hence, recalling the sufficiency of the score function D_1 at any parameter value for this class of models, the second factor in (3.3) equals $\exp\{l(\beta_0;y) - l(b;y)\}$.

The mean of D_1 is zero and its variance is $-D_2$, which equals the observed as well as the expected information. Therefore the normal approximation to the P_b-density of D_1 at zero is

$$(3.4)\quad f(0;b) \sim (2\pi)^{-p/2}|\mathrm{var}_b(D_1)|^{-1/2} = (2\pi)^{-p/2}|j(b;y)|^{-1/2}.$$

Thus (3.1) follows from the approximation (3.3). Notice that the only approximation involved is (3.4), which is a normal approximation to a density at its mean. This approximation is identical to a saddlepoint approximation to the density of D_1 at zero and results in a relative error of $O(n^{-1})$ in (3.4) and consequently also in (3.1) for the case of n independent replications.

Exact ancillaries. If $A(y)$ is exactly ancillary, as is the case in transformation models where the maximal invariant statistic is used, the second factor in (3.3) equals $\exp\{l(\beta_0;y) - l(b;y)\}$. The first factor is the density at zero of the score statistic D_1 at the point b in terms of the conditional distribution induced by the parameter value b. The expected value in this distribution is zero, and consequently the normal approximation to the density at this point becomes

$$(3.5)\qquad f(0|a;b) \sim (2\pi)^{-p/2}|\mathrm{var}_b(D_1|a)|^{-1/2}.$$

Viewed together with the separation of the first two factors in (3.3), this approximation is entirely identical to a saddlepoint approximation as was the case for the exponential family.

It remains to obtain an approximation to the conditional variance of D_1 given A. As argued above, D_2 is a function of D_1 and A. Consider the case of n independent replications and assume that the ancillary statistic is in one-to-one correspondence with a function of the minimal sufficient statistic and that this function does not depend on the number of observations. This is the case if A is chosen, e.g., as the affine ancillary in a curved exponential family [cf. Barndorff-Nielsen (1980)]. Then, perhaps after a one-to-one transformation of A, we may write

$$n^{-1}D_2 = h(n^{-1}D_1, A),$$

where h is a function that is independent of n. Under mild conditions this function may be expanded as

$$(3.6)\qquad \frac{1}{n}D_2 \sim h(0, A) + \frac{1}{n}h'(0, A)D_1 + \frac{1}{2n^2}D_1^{\mathsf{T}}h''(0, A)D_1 + \cdots,$$

where, in a formal notation valid for each coordinate of D_2, we have used h' and h'' to denote the first two derivatives of h with respect to its first argument. If we take expectations on the right in (3.6) for fixed A, we see that the expectation of D_2 equals its value $D_2(0, A)$ as a function of $D_1 = 0$ and A, apart from an error term which is $O(n^{-1})$. But the expectation of $-D_2$ in the conditional distribution equals the conditional variance of D_1 if A is exactly ancillary, and therefore the conditional variance of D_1 may be approximated by $-D_2(0, a)$ which is identical to $j(b; y)$. On combination with (3.3) and (3.5), we arrive at (3.1) with $c = (2\pi)^{-p/2}$. Notice that the argument related to (3.6) shows why the observed rather than the expected Fisher information appears in the formula.

Transformation models. For the case of a transformation model the ancillary statistic is the maximal invariant which is exactly ancillary. From the discussion above it follows that the only remaining step to prove that (3.1) is exact is to prove the transformation invariance of the expression

$$|j(b; y)|^{1/2} f(0|a; b)$$

for $y = y(b, a)$, i.e., $b = \hat{\beta}(y)$ and $a = A(y)$. Then it follows that this expression is a function of a alone and (3.1) is therefore exact if $c(a)$ is chosen correctly. To avoid the theory of group actions we shall not verify this invariance; the arguments are given in Barndorff-Nielsen (1983). The reader may easily check the result in the case of a location model for independent identically distributed random variables, even if the model is reparametrized from the location parameter to a smooth function of this.

Approximate ancillaries. So far we have neglected the problem that A need not be exactly ancillary and as a consequence the only approximation involved was to the first factor in (3.3), based on the expansion (3.6). This approximation is of the large deviation type, i.e., it involves an error, here $O(n^{-1})$, which is added to a term that is bounded away from zero for b in a compact set. Thus, the resulting error is a relative error of this order of magnitude as $n \to \infty$. In particular, the relative error remains of this order of magnitude if b is kept fixed as $n \to \infty$, which is a sequence of large deviations in the standardized distribution of $\hat{\beta}$. The saddlepoint expansion has the same characteristics whereas, e.g., Edgeworth expansions provide bounds only on the additive errors to the density. The approximation derived above, for exact ancillaries, is not normalized. If the error involved is differentiable as a function of b, a renormalization obtained by dividing by the integral over some compact set will reduce the order of magnitude of the additive error to $O(n^{-3/2})$.

Compared to the case of an exact ancillary, three more approximations are generally required. The second factor in (3.3) is not exactly equal to the ratio between the marginal likelihoods. Instead it equals the ratio of the conditional likelihoods, and consequently the error introduced by use of the marginal likelihoods equals the ratio of the densities of A at a with respect to the parameters b and β_0. If A is a first order ancillary, e.g., as the Efron and Hinkley (1978) ancillary, this ratio is, by definition, $1 + O(n^{-1/2})$. A renormalization again improves the error of the resulting approximation to $O(n^{-1})$. For a second order ancillary the error of the approximation is $O(n^{-1})$, or $O(n^{-3/2})$ when renormalized. It should be noticed, however, that the approximation only applies to normal deviates of the ancillary A whenever this is only approximately ancillary since we have no control of the ratio between the two densities of A at a outside a set of normal deviations. If extreme accuracy is required, a remedy is to use the conditional likelihoods in (3.1) [cf. Barndorff-Nielsen (1983)].

The other two inaccuracies that appear when A is only approximately ancillary occur in the approximations to the conditional mean and variance of D_1 given $A = a$. If A is a first order asymptotic ancillary, then the variance of the conditional mean of D_1 given A is $O(1)$; see, e.g., Skovgaard (1985a). Hence the conditional mean, having mean zero over the distribution of A, is itself $O(1)$ and since the variance of D_1 is of order n, the error involved in the normal density approximation (3.5), due to the bias, is $O(n^{-1})$.

For the conditional variance of D_1 given A we notice that the argument based on the expansion (3.6) is still valid for a first order approximate ancillary, but the linear term in D_1 in (3.6) now contributes an amount $O(n^{-1/2})$ to the expectation, and an amount of the same order of magnitude is due to the fact that the variance of D_1 given A is no longer exactly equal to the conditional mean of $-D_2$. If the ancillary is of second order, both of these contributions to the error become $O(n^{-1})$.

It should be noticed that even for an asymptotic ancillary statistic that is only first order ancillary, the resulting expansion (3.1) is still of the large deviation type in the sense that it keeps a bounded relative error in a fixed interval around β_0. Therefore the order of magnitude of the additive error can be improved to $O(n^{-1})$ by renormalization. The error is also improved to this order of magnitude, without renormalization, if we restrict attention to normal deviates of $\hat{\beta}$, i.e., to values of b within a neighbourhood of size $O(n^{-1/2})$ around β_0. This is so because any first order ancillary is a local ancillary of second order [cf. Cox (1980) and Skovgaard (1985a)]. In this connection it is of interest to notice that Amari and Kumon (1983) have proved that in a (k, p) exponential family, a second order ancillary of the type in (3.6) exists, but in general no higher order ancillaries that are independent of the sample size as required in (3.6).

The general conclusion is that the formula (3.1), like the saddlepoint expansion, has the features of a large deviation type expansion, namely by keeping a uniform relative error over a fixed interval of parameter values. If the approximate ancillary is of first order, the error is $O(n^{-1/2})$; if it is of second order the error is $O(n^{-1})$. In any case the additive error is improved by one order of magnitude by renormalization. However, the formula applies only to normal deviates of the ancillary A unless this is exactly ancillary.

Acknowledgments. This manuscript was completed while the author was on leave at the University of British Columbia. Partial support was provided by the Natural Science and Engineering Research Council of Canada and the Danish Natural Science Research Council.

REFERENCES

AMARI, S. and KUMON, M. (1983). Differential geometry of Edgeworth expansions in curved exponential family. *Ann. Inst. Statist. Math.* **35** 1–24.

BARNDORFF-NIELSEN, O. E. (1980). Conditionality resolutions. *Biometrika* **67** 293–310.

BARNDORFF-NIELSEN, O. E. (1983). On a formula for the distribution of the maximum likelihood estimator. *Biometrika* **70** 343–356.

BHATTACHARYA, R. N. and GHOSH, J. K. (1978). On the validity of the formal Edgeworth expansion. *Ann. Statist.* **6** 434–451.

COX, D. R. (1980). Local ancillarity. *Biometrika* **67** 279–286.

EFRON, B. and HINKLEY, D. V. (1978). Assessing the accuracy of the maximum likelihood estimator: Observed versus expected Fisher information. *Biometrika* **65** 457–487.

FIELD, C. (1982). Small sample asymptotic expansions for multivariate *M*-estimates. *Ann. Statist.* **10** 672–689.

FRASER, D. A. S. (1988). Normed likelihood as saddlepoint approximation. Technical Report No. 1, Univ. Toronto.

HOUGAARD, P. (1985). Saddlepoint approximations for curved exponential families. *Statist. Prob. Lett.* **3** 161–166.

MCCULLAGH, P. (1987). *Tensor Methods in Statistics*. Chapman and Hall, London.

PAZMAN, A. (1984). Probability distribution of the multivariate nonlinear least squares estimates. *Kybernetika* **20** 209–230.

SKOVGAARD, I. M. (1985a). A second-order investigation of asymptotic ancillary. *Ann. Statist.* **13** 534–551.

SKOVGAARD, I. M. (1985b). Large deviation approximations for maximum likelihood estimators. *Probab. Math. Statist.* **6** 89–107.

DEPARTMENT OF MATHEMATICS
ROYAL VETERINARY AND AGRICULTURAL UNIVERSITY
THORVALDSENSVEJ 40
DK-1871 FREDERIKSBERG C
DENMARK

Chapter 8

Modelling relations between instrumental and sensory measurements in factorial experiments

Introduction by Per Bruun Brockhoff

DTU COMPUTE, Danish Technical University

This paper was presented by Ib at the 2nd Sensometrics Meeting in Edinburgh, 16-18 September 1994. Personally, I had the pleasure to be Ib's first PhD student and travel together with him to this conference as a PhD student. Sensometrics was the topic of my PhD work, Brockhoff (1994), initiated by Ib's engagement in the food science research taking place at the Royal Veterinary and Agricultural University of Denmark (KVL) at that time. For me, this trip was one of the memorable events of my life, as Ib and I had the chance to have many and long one-on-one talks about statistics and life in general in the free evenings. For a young PhD student, this openness from a supervisor, professionally as well as personally, made a lasting impression.

The paper is a nice example of how Ib has always managed to range from his strong contributions in mathematical statistics to various more applied contributions related to the many areas of relevance to KVL including agricultural science, genetics, food science and bio-sciences. Without doubt, the strong position and respect held by the statistics group at KVL had its roots in two key persons: Mats Rudemo and Ib Skovgaard. Without their combination of formal statistical professionalism, true interest in the integration of this into all the relevant bio, food- and agri-sciences and their open and strong personalities, statistics within these fields in Denmark would not have been the same.

Sensometrics, the modelling and analysis of sensory and consumer data (Brockhoff, 2011; Næs *et al.*, 2010) grew out of sensory science again historically anchored within food science. My PhD project was initiated by Ib in

1991 and in 1992 the first international meeting on sensometrics took place in Leiden in Holland. Since then the conferences have taken place every second year, and in 1998 Ib and I arranged the 4th Sensometrics meeting in Copenhagen together. Due to Ib's foresight I am now among a handful of people having attended all these meetings over the years. The sensory science field itself also matured during the same period of time. The first Pangborn Sensory Science Symposium was held in Järvenpää, Finland in 1992, and is now the main global conference in the field. And now also European, Asian and South American conference series exist (Eurosense, SenseAsia, SensIber). The journal *Food Quality and Preference* is the official journal of the Sensometrics Society.

One recurrent theme in sensory research is paradoxically to work towards situations where the somewhat expensive and often quite noisy human perceptual measurements are not needed and could be substituted by faster, cheaper and more precise chemical and/or instrumental measurements. Similarly, often scientific chemical explanations for various sensory properties, be it desirable or the opposite, are sought. Both of these purposes lead to experimental work in which sensory measurements as well as chemical/instrumental measurements are taken on the same (food) items to investigate the relations. In this paper, Ib is reanalyzing some data from a meat oxidation experiment Stapelfeldt *et al.* (1992) as an example of this. Sensory measurements of the co-called "Warmed over flavour" (WOF), an undesirable flavour related to oxidation and ocurring in re-heated meat, was taken together with chemical measurements of the amount of thiobarbituric reaction substances ("TBARS") hypothesized to be a possible explanation for the WOF. A low number of meat packages (3-6) for each of 5 different storage times was measured in an independent samples design. In the paper it is suggested to use the ultrastructural model of Dolby (1976) for this situation.

With bivariate data observed in (treatment) groups (X_{ij}, Y_{ij}), $j = 1, \dots, n_i, i = 1, \dots, I$, corresponding to a oneway (M)ANOVA setting, the ultrastructural model expresses a linear functional relation between the group means together with a linear structural relation within groups and can be expressed either in a measurement error/errors-in-variables form

$$\begin{aligned} X_{ij} &= V_{ij} + \varepsilon_{ij},\ \varepsilon_{ij} \sim N(0, \sigma^2), \\ Y_{ij} &= \alpha + \beta V_{ij} + e_{ij},\ e_{ij} \sim N(0, \omega^2), \\ V_{ij} &= \mu_i + \delta_{ij},\ \delta_{ij} \sim N(0, \tau^2), \end{aligned} \tag{8.1}$$

or as a bivariate normal model

$$\begin{pmatrix} X_{ij} \\ Y_{ij} \end{pmatrix} = N\left\{\begin{pmatrix} \mu_i \\ \alpha + \beta\mu_i \end{pmatrix}, \begin{pmatrix} \sigma^2 + \tau^2 & \beta\tau^2 \\ \beta\tau^2 & \omega^2 + \beta^2\tau^2 \end{pmatrix}\right\}. \quad (8.2)$$

Both these model versions are provided in Ib's paper and the relations to a number of other similar model approaches are clarified. In Dolby (1976) maximum likelihood parameter estimates are derived, but in Ib's paper the link to canonical variates (CVA) is made, and a test for linearity as well as the estimate of the linear relation is based on this. Williams (1967) is cited by Ib for the result that the slope can be estimated as minus the ratio of the coefficients of the second canonical variate. However, in Anderson (1984), Dolby (1976) together with other new results are said to be "rediscovered and extended" versions of similar results in Anderson (1951). Also Fisher (1938) is acknowledged by Anderson (1984) for "having considered a related problem". The CVA-approach can be understood to be different from maximum likelihood by two observations: the between groups lack-of-linear-fit is ignored, as is the fact the slope parameter enters the covariance matrix. But if the linearity assumption is validated and the information about the slope from the covariance matrix is weak, the CVA-based approach can be a good solution, as also pointed out by Ib.

In Carstensen (2004) Ib's paper is cited as part of a discussion of the ultrastuctural model for the purpose of modelling method comparison studies. The ultrastructural model is expressed there in univariate form, by letting $Z_{1ij} = X_{ij}$, $Z_{2ij} = Y_{ij}$, $\alpha_1 = 0, \beta_1 = 1$ and

$$Z_{mij} = \alpha_m + \beta_m(\mu_i + \delta_{ij}) + e_{mij},\ \delta_{ij} \sim N(0, \tau^2),\ e_{mij} \sim N(0, \sigma_m^2), \quad (8.3)$$

where now $e_{1ij} = \varepsilon_{ij}$, $e_{2ij} = e_{ij}$, $\sigma_1^2 = \sigma^2$ and $\sigma_2^2 = \omega^2$.

In this form it becomes clear how close the ultrastructural model is to another contribution of Ib's to the sensometrics field, that is now part of standard software used and still today impacts ongoing research. At one of our regular meetings, Ib felt that I needed some more boosting of and inspiration to my PhD work, so he brought a piece of paper, where he just rapidly had scribbled the following model to me:

$$Z_{mij} = \alpha_m + \beta_m\mu_i + e_{mij},\ e_{mij} \sim N(0, \sigma_m^2), \quad (8.4)$$

as a potential model for descriptive analysis data aka sensory profile data, where sensory properties are scored on a continuous line scale – typically today directly on a computer or tablet screen. Such data were and are still collected on a daily basis in many research institutions and in industry. Here the m now refers to the different assessors used in the evaluation, typically in the order of 10–20, the i to the different products evaluated (typically 5–20) and j to the (randomized) replications of each assessor-product combination. The model is the submodel of the ultrastructural model where the product-by-replicates interaction, using sensometrics terminology, is assumed to be non-existent, $\tau^2 = 0$. The model was introduced, treated and discussed for the sensometrics field in Brockhoff and Skovgaard (1994) and has now become known under the name *the assessor model*. The main feature of the model is that it models the individual differences in use of the scale range by the scale range parameters β_m. More generically put, it is a multiplicative model for the assessor-by-product interaction effects linearly related to the product main effects μ_i.

Very recently, also jointly with Ib, Brockhoff *et al.* (2015) presented mixed model extensions of this with the following *mixed assessor model (mam)* as the basic model structure

$$\begin{aligned} Z_{mij} = \alpha_m + b_m\mu_i + d_{mi} + e_{mij},\ e_{mij} \sim N(0,\sigma^2), \\ b_m \sim N(\beta,\sigma_b^2); \quad d_{mi}^2 \sim N(0,\sigma_d^2). \end{aligned} \tag{8.5}$$

This model is within what can also be termed a mixed multiplicative model Smith *et al.* (2003), i.e. a non-linear mixed model. From a sensory perspective this is now a model that distinguishes properly between the specific scale range differences and general perceptual disagreements, both of which are part of the assessor-by-product interaction effects. In this form, the model and related versions have now developed into practical data analysis procedures for use in the sensometrics field, see e.g. Peltier *et al.* (2013) and Kuznetsova *et al.* (2015b). The latter brings together recent developments from Kuznetsova *et al.* (2017) and Kuznetsova *et al.* (2015a) with the extended mixed assessor model framework. Further development of this is ongoing research, e.g. utilizing recent developments within automatic differentiation within the R-framework (Kristensen *et al.*, 2016) to a new R-package for multiplicative mixed models (Jensen *et al.*, 2016). The impact of the scribble on the little piece of paper by Ib has yet to reach its peak.

In the paper on the use of the ultrastructural model Ib also points out that quite often already in the early 1990s it was probably more common to have multivariate data settings of various kinds than just two specific measurements as examplified in the paper. Hence multivariate extensions of the relational analyses are of high importance, and obviously CVA-based approaches would work in higher dimensions as well, although challenged by error covariance low-rank situations where regularized versions would be needed. Also, for prediction purposes, often various regularized and/or latent variable regression approaches would be used. At KVL we also in the early 1990s experienced some of probably many pre-waves of the digital tsunami of big data hitting us today in terms of spectroscopic and other multivariate measurement data evolving, and a still today strong research group within the field of chemometrics grew out of the Food Science and Technology Department (Brereton, 2014). The sensometrics area was heavily influenced by the chemometrics field also and was and is an area where all different sorts of researchers, be it psychophysical modellers, inferential empirical statisticians and multivariate chemometricians meet with each other and with the research and industry using all this for their diverse purposes (Brockhoff, 2011).

Modelling relations in general, obviously today, now in the midst of the big data digital revolution, is of higher relevance than ever and doing this in situations where data comes in groups, sub-samples, hierarchies, multi-levels, you name it, continues to pose an analytical and interpretational challenge, in particular for people without a strong formal training in statistics. Surely, the number of situations in research, industry and society, where people without such a formal training meet loads of data with relational analysis wishes is growing faster every day. This contribution by Ib is a little reminder to all of us to think carefully about what we do or advise in such contexts, and to not forget the classical results and thoughts of some of the originators of the science of statistics.

ELSEVIER

0950-3293(95)00023-2

Food Quality and Preference **6** (1995) 239–244
Elsevier Science Limited
Printed in Great Britain
0950-3293/95 \$9.50+.00

MODELLING RELATIONS BETWEEN INSTRUMENTAL AND SENSORY MEASUREMENTS IN FACTORIAL EXPERIMENTS

Ib M. Skovgaard

Department of Mathematics and Physics, The Royal Veterinary and Agricultural University, Thorvaldsensvej 40, DK-1871 Frederiksberg C, Denmark

(Received 18 January 1995; accepted 3 April 1995)

ABSTRACT

*In connection with the sensory evaluation of foods, chemical or physical measurements of the relevant sensory properties are often sought. The data available for testing whether one particular instrumental measurement is linearly related to a sensory variable usually arise from a factorial experiment rather than from some random sub-population. Thus, to calibrate the two variables properly and to test whether a linear relationship is compatible with the data, we need a statistical model that reflects the design. The ultrastructural model suggested by Dolby (*Biometrika **63,** *39–50, 1976) is one such model. Furthermore, this model allows for random variation in both of the variables considered. In this paper it is shown how testing and estimation may be carried out using this model, although the methods suggested are not optimal. Finally, it is argued that the use of factorial designs in this connection is sensible, or even necessary, despite the complications that it leads to in terms of statistical modelling and analysis.*

Keywords: Calibration; canonical variates; functional relation; latent variables; structural relation; ultrastructural relation.

INTRODUCTION

Let X_i and Y_i denote a chemical and a sensory measurement, respectively, on the ith test sample of some food product. The two measurements are assumed to be related and may even measure the same 'quality parameter'. The measurements are results from some experiment designed to investigate this possibility and, if possible, lead to a calibration model that may be used to predict the sensory property from the chemical measurement. This situation is a simplification of a very common situation in which several chemical, physical, and sensory properties are measured on each sample instead of only two. Thus, instead of Y_i we may have several sensory characteristics described by certain words such as 'fruity taste' or 'sweetness' in a sensory profile analysis, while X_i may be replaced by a number of chemical measurements, for example a fluorescence spectrum.

The problem is to investigate whether the chemical measurement, X, can be used as a measurement of the sensory property, Y. In other words: do the two variables measure the same property? In this case a closely related problem is the estimatation of the relationship between the two variables.

At first sight this may look like a simple regression problem, but in most cases the pairs (X_i, Y_i) are not sampled at random from some population. Instead they arise from a planned factorial design, for example a simple one-way layout with some replications from each of a number of different treatments.

To fix ideas, we discuss the approach to the problem in terms of the following storage experiment, reported by Stapelfeldt *et al.* (1992), investigating oxidation of cooked sliced beef. Samples of meat were cooked and stored for a number of days as shown in Table 1.

Two measurements were made on each pack, denoted 'TBARS' and 'WOF', where 'TBARS' denotes a chemical measurement of the amount of thiobarbituric reactive substances, and 'WOF' denotes a sensory evaluation of 'warmed-over flavour', which is an undesirable flavour related to oxidation and occurring in reheated meat. The TBARS measurement is also related to oxidation products and was a priori assumed to be a good candidate for measuring WOF in an objective way. More experimental details may be found in Stapelfeldt *et al.* (1992).

Figure 1 shows the increase of TBARS, plotted on a logarithmic scale, and of WOF with storage time. Notice that the different storage times used in the design are the main source of variation for both variables.

The structure of the experiment corresponds to a one-way analysis of variance (ANOVA) which may be used to analyze each of the two variables separately, and to confirm that both TBARS and WOF increase with storage time. That was, however, known beforehand

TABLE 1. Allocation of Packs of Meat on Storage Time in the Storage Experiment

Storage time (days)	0	1	3	6	10
Number of units (packs)	3	5	6	4	6

and this analysis completely mistakes the reason for performing the experiment. The question may be phrased as follows: "Can TBARS be used as a measurement of WOF?" It is evident that this question cannot be answered affirmatively by a single experiment as this, and a better way to phrase the question may be: "Are these data compatible with the hypothesis that TBARS and WOF measure the same thing?"

Obviously, some calibration is necessary for the two types of measurement to become comparable. Closer inspection and previous experience suggest the logarithm of TBARS to be linearly related to the sensory evaluation, which itself is defined on a somewhat arbitrary scale, however. In any case we consider log(TBARS) as the *X*-variable and want to test whether it may be linearly related to WOF. Figure 2 shows that this may be quite reasonable.

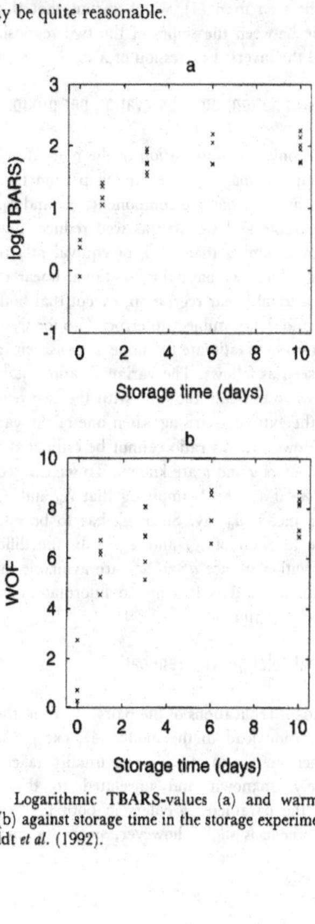

FIG. 1. Logarithmic TBARS-values (a) and warmed-over flavour (b) against storage time in the storage experiment from Stapelfeldt *et al.* (1992).

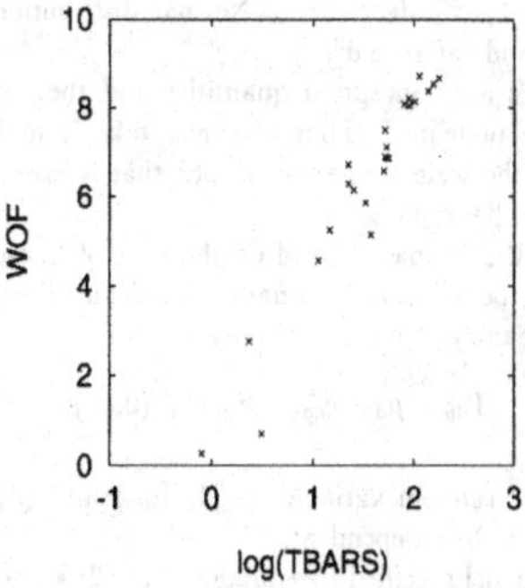

FIG. 2. Warmed-over flavour against logarithmic TBARS-values in the storage experiment from Stapelfeldt *et al.* (1992).

It is the claim here that no statistical test is currently available for this quite simple and common test problem. Furthermore, it is the purpose here to point out that the natural statistical model for this problem is the ultrastructural model suggested by Dolby (1976) and presented in the following section. This model, which is a specialization of the multivariate analysis of variance model, seems mostly to have been 'overlooked' in the statistical literature and practice, and it ought to play a much more prominent role.

The main problem, namely that of testing the adequacy of the ultrastructural model, is not solved in the present paper, although a test is suggested which may be satisfactory in many situations. This test, which is based on the use of canonical variates, is shown in Section 3, and the closely related estimation of the linear relation is discussed in Section 4. Section 5 discusses a few alternative statistical methods sometimes met in the food science literature. The question of how to design an experiment for calibration in the current setting is addressed in Section 6, and other problems and generalizations are finally discussed in Section 7.

THE ULTRASTRUCTURAL MODEL

Let X_{dp} and Y_{dp} denote the log(TBARS) and WOF measurements, respectively, for pack p on day d. Suppose that these two measurements were, in fact, measuring the same property of the particular pack of meat, apart from being measured on different scales. If V_{dp} denotes the 'true value' of the property in question we would then have

$$\log(\text{TBARS})_{dp} = \alpha + \beta V_{dp} + \epsilon_{dp},\ \epsilon_{dp} \sim \mathcal{N}(0,\sigma^2), \quad (1)$$

$$\text{WOF}_{dp} = a + bV_{dp} + e_{dp},\ \ e_{dp} \sim \mathcal{N}(0,\omega^2),$$

where $\mathcal{N}(0, \sigma^2)$ denotes the Normal distribution with mean 0 and variance σ^2.

The Vs are conceptual quantities and their scale is therefore undefined. Thus, we may take it to be the same as the scale as the X-variable, that is, we may let $\alpha=0$ and $\beta=1$.

Since the Vs may depend on the time of storage and also vary between packs, a natural model for these is the one-way analysis of variance model

$$V_{dp} = \mu_d + \delta_{dp}, \quad \delta_{dp} \sim \mathcal{N}(0, \tau^2). \tag{2}$$

All the random variables $\{\epsilon_{dp}\}$, $\{e_{dp}\}$ and $\{\delta_{dp}\}$ are assumed to be independent.

The model specified by equations (1)–(2) is the ultrastructural model vaguely suggested by Kendall and Stuart (1967, Section 29.35) and promoted further by Dolby (1976). It is a type of linear regression model with both variables being subject to random variation. Usually two types of such errors-in-variables models are considered: the structural and the functional relation, see, for example, Kendall and Stuart (1967, Chapter 29). The ultrastructural relation generalizes both of these, as is seen from the discussion later in this section.

Notice that apart from usual qualifications concerning normality, variance homogeneity, and so on, the ultrastructural model arises almost automatically from the claim that the two variables are measuring the same property on different scales. The experiment described here is a single-factor experiment, but might as well be any factorial experiment. This would just lead to a change of the model, (2), for the latent variables $\{V_d\}$.

Let us investigate the relation between the ultrastructural model and more well-known statistical models a little further. With the specification $\alpha=0$ and $\beta=1$ in equations (1) the model is

$$\begin{aligned} \log(\text{TBARS})_{dp} &= V_{dp} + \epsilon_{dp}, & \epsilon_{dp} &\sim \mathcal{N}(0, \sigma^2), \\ \text{WOF}_{dp} &= a + bV_{dp} + e_{dp}, & e_{dp} &\sim \mathcal{N}(0, \omega^2), \\ V_{dp} &= \mu_d + \delta_{dp}, & \delta_{dp} &\sim \mathcal{N}(0, \tau^2), \end{aligned} \tag{3}$$

with parameters a, b, $\{\mu_d\}$ and the variances σ^2, ω^2, τ^2.

Elimination of the latent variables $\{V_d\}$ leads to the bivariate Normal distribution of (X_{dp}, Y_{dp}) with parameters

$$\begin{aligned} E\begin{pmatrix} X_{dp} \\ Y_{dp} \end{pmatrix} &= \begin{pmatrix} \mu_d \\ a + b\mu_d \end{pmatrix}, \ \text{var}\begin{pmatrix} X_{dp} \\ Y_{dp} \end{pmatrix} \\ &= \begin{pmatrix} \sigma^2 + \tau^2 & b\tau^2 \\ b\tau^2 & \omega^2 + b^2\tau^2 \end{pmatrix}, \end{aligned} \tag{4}$$

where the different pairs (X_{dp}, Y_{dp}) are independent.

The statistical models discussed below may now be compared to the ultrastructural model using either of the specifications (3) or (4). The discussion is specific to the present case of only two variables. The multivariate generalizations are more complicated.

Structural relation: only one group.

If the μ_ds are replaced by a constant μ, for example because only one storage time is used in the experiment, we have the structural linear relation which describes the relation between X and Y in terms of their covariance matrix, while the means are constant. It is a known fact that the slope, b, cannot be estimated consistently in this model because of the over-parametrization of the covariance matrix, which has four parameters to describe three quantities. However, the covariance matrix restricts the slope, b, to the interval

$$\left(\frac{\text{cov}(X, Y)}{\text{var}X}, \frac{\text{var}Y}{\text{cov}(X, Y)} \right), \tag{5}$$

as may be seen from (4), which means that the slope should be between the slopes of the two regressions: Y on X and the inverted regression of X on Y.

Functional relation: one observation per group.

If there is only one observation of the pair (X, Y) from each group we may omit the index p from the equations. In that case the two components, μ_d and δ_d, of V_d are inseparable and we may as well reduce the model further by assuming that $\tau=0$, or equivalently that all δ_ds vanish. Then we have the functional linear relation which is a usual linear regression, except that both variables, X and Y, are subject to error. Also for this model is it impossible to estimate the slope, b, consistently. This may be seen as follows. The variance ratio σ^2/ω^2 determines how to estimate the line, with the two regression lines as the extremes, arising when one of the variances vanish. However, this ratio cannot be estimated even if the parameters a and b are known. To see this, consider the case $a=0$ and $b=1$, implying that X_d and Y_d have the same mean, μ_d say. Since μ_d has to be estimated from the location of X_d and Y_d, only the differences $Y_d - X_d$, with variance $\sigma^2 + \omega^2$, are available for variance estimation, thus leaving no information for estimation of the ratio.

Functional relation with replications.

Independent replications of the pairs (X, Y) in the functional relation lead to the model (4), except that the covariance matrix in this case is usually taken to be completely unknown and unrelated to the means, whereas the parameter b enters in both places in (4). The difference is slight, however, since the covariance matrix in (4) provides limited information on b, as discussed in connection to the structural relation.

Factor analysis.

The factor analysis model for pairs (X_i, Y_i) is defined as

$$X_i = \alpha + \beta V_i + \epsilon_i, \quad \epsilon_i \sim N(0, \sigma^2)$$

$$Y_i = a + bV_i + e_i, \quad e_i \sim N(0, \omega^2),$$

where $V_i \sim N(0, \tau^2)$ and all ϵs, es and Vs are independent. In this, bivariate, case the factor analysis model is identical to the structural relation.

Multivariate analysis of variance (manova).

The ultrastructural model is a specialization of the MANOVA-model. This model specifies an arbitrary and unknown 'within-group' covariance matrix, whereas the 'between-group' variation is specified in terms of arbitrary and unknown group means, $E(X_{dp}) = \mu_d$ and $E(Y_{dp}) = \nu_d$, say. Thus, the ultrastructural model assumes these pairs of means to be on a line,

$$\nu_d = a + b\mu_d, \tag{6}$$

for some a and b.

A TEST FOR LINEARITY

The ultrastructural relation implies that the pairwise group means $(E(X_{dp}), E(Y_{dp}))$ are on a line as noted in equation (6). This hypothesis is not entirely equivalent to the ultrastructural model, since the slope of this line further has to be in the interval (5) given by the within-group variances and covariance. This is a fairly weak requirement, especially when the interval is wide, which it is when the within-group correlation is relatively small, implying that the two regressions are far apart. From equations (3) it is seen that the correlation increases with the variation between the packs, i.e. with the variation between the Vs. Hence the within-group correlation is usually considerably smaller than the overall correlation. Therefore it may be reasonable to disregard the restriction on the slope and test the ultrastructural model simply by testing whether the group means are on a line in the bivariate analysis. This is a well-known hypothesis in multivariate analysis, and the test is usually performed in terms of canonical variates, see for example, Williams (1967). Each canonical variate represents a direction separating the groups in the two-dimensional space of response variables. Thus, if only one canonical variate is significant, the bivariate groups means are separated significantly only in a single direction,

TABLE 2. Test Statistics as χ^2-variates with Degrees of Freedom as Index, and Approximate P-values for the Hypotheses that the Group Means Lie in a q-dimensional Subspace

q	χ^2-statistic	P (approx.)
0	$\chi^2_8 = 499.4$	<0.0001
1	$\chi^2_3 = 5.17$	0.16

meaning that they are on a line except for variation which may be random.

In the storage experiment the tests for the significance of the first and the second canonical variate are shown in Table 2. The tests are performed as likelihood ratio tests using the χ^2-approximations, see for example, Chatfield and Collins (1980, Section 8.5).

The first line of the table shows the test for the hypothesis that the group means are the same, that is, whether there is no effect of storage time on the two variables measured. This is highly significant but of less relevance here. The second line shows the test for the hypothesis, (6), that the group means are on a line. This is far from being significant, and the conclusion is therefore that these data are compatible with the ultrastructural model, which is again based on the hypothesis that log(TBARS) and WOF measure the same property. It should be emphasized that this is by no means a proof of the hypothesis, of course, or even a strong support of it. To obtain such support, one or several larger experiment, varying many factors, would be necessary.

ESTIMATING THE LINEAR RELATIONSHIP

Having found that the two variables may be linearly related as described by the ultrastructural model, the next step is to estimate the linear function. We may specify the purpose of this estimation as a calibration resulting in a function of X designed to predict Y. The usual approach to linear calibration is to regress either X on Y or Y on X, denoted classical and inverse calibration, respectively, see, for example, Osborne (1991). It is, however, unrealistic to assume that the allocation of levels of the design factors in our experiment represents the 'random' variation in future applications of the calibrated equation. Therefore it is better to estimate the underlying 'true' functional relation $y = a + bx$ which is less likely to be affected by the design.

This estimation may be carried out by a straightforward extension of the method used to test the linearity of the group means from the previous section. We are looking for a function of (X, Y) which is constant, that is,

$$Y - (a + bX) = 0,$$

except for random variation. In view of equations (3) this is equivalent to the search for coefficients (a, b) satisfying the relation (6) between the group means. From Williams (1967), for example, we know the solution to this problem: the direction $(-b, 1)$ should be the same as that of the canonical variate corresponding to the smallest canonical correlation in the bivariate one-way analysis of variance. To see this, notice that the hypothesis tested in the previous section is that this canonical correlation is zero, implying that relation (6) is satisfied when $(-b,1)$ is chosen proportionally to the coefficients of the corresponding canonical variate.

In the storage experiment the coefficients of the second canonical variate are (6.41, −1.68), resulting in the estimate

$$b = 6.41/1.68 = 3.8$$

in good agreement with Figure 2, of course.

ALTERNATIVE STATISTICAL METHODS

Quite often the problem treated here is attacked by use of other statistical methods, a few of which are mentioned below.

The correlation coefficient between log(TBARS) and WOF might be calculated, ignoring the design, and a correlation close to a plus or minus one might then be taken as an indication that there is a close linear relation between the two variables. Although this is not completely false, the argument against it may be found in several textbooks: in a designed experiment the (co)-variance and hence the correlation may be greatly influenced by the choice of variation implicitly made by design. In the storage experiment the (overall) correlation coefficient is 0.97, whereas the within-group correlation is 0.69. This latter correlation may be thought of as the correlation from a homogeneous population of packs stored for the same period, and this is, not surprisingly, considerably smaller than the overall correlation.

When several variables are measured, principal component analysis and related methods using the correlation matrix are frequently used. They are, however, subject to the same criticism, since they are also based on overall correlations.

In the paper by Stapelfeldt *et al.* (1992) a method of covariance was used. An analysis of variance first showed a clearly significant dependence of WOF on storage time. Next, when log(TBARS) was included as a covariate in this model, the significance of storage time disappeared. In other words a linear function of log(T-BARS) was able to explain all the variation in WOF that was due to storage time. This method is closely related to the one suggested in the present paper, but it ignores that also log(TBARS) may be subject to error. In the storage experiment reported here and in the paper mentioned above this error was negligible, however, compared to that of the WOF measurement.

DESIGN CONSIDERATIONS

An important question is how an experiment ought to be designed to investigate the question whether two variables can be calibrated to measure the same property.

In the theoretical literature on calibration, linear relations are almost invariably assumed to be estimated either from a random sample from some population or from an experiment in which the X-variable is controlled.

In food science these two design methods are rarely applicable. The reason is that foods are manufactured products, and to obtain a random sample from a well defined population one would have to use the same treatment for all units, at least in an experimental setting. In that case, however, the variation would usually be too small to compare the variables in question on a reasonable scale, and the variation would be dominated by measurement and sub-sample variation instead of a realistic variation between products. Neither is it possible by design to control the value of a sensory property, and rarely a chemical quantity, to any useful degree of accuracy, because of the biological variation in the raw material.

The conclusion is that the only possible way is to perform an experiment in which treatments are intentionally varied. It is interesting that the ultrastructural model shows that, when both variables are subject to error, this is also necessary for the linear relation to be estimable, because it would reduce to the ordinary structural relation in the case of a single treatment group. The random sample calibration would correspond to the structural relation, and the controlled calibration to the functional relation, and in neither of these is the estimation possible.

What kind of factorial experiment to use is another question not addressed here. But to infer anything like a linear relation, many factors have to be varied in a single or in several experiments.

DISCUSSION

The interpretation of a calibration result, as discussed here, needs particular care. A usual reservation is that it is not possible, on purely empirical grounds at least, to infer a linear relationship between two variables, although it is possible to reject it. In the present context this precaution must be taken even more seriously. The problem is that many factors may influence either or both of the variables. Therefore a designed experiment, varying only one or a few factors, may result in similar patterns of variation of the two variables, even if these are physically unrelated, and a factor not included in the experiment might have different effects on them. Furthermore, a linear relationship may not always be good for prediction purposes. As an extreme counter-example, one of the variables may be constant, in which case it cannot predict the other. Separate analyses of variance of the two variables may rule out this possibility if treatment effects are strongly significant for both.

For example, in the storage experiment discussed here, TBARS and WOF both increase with time. Therefore it might well happen that the relation between them could be described as linear within a limited period of storage, even if they were unrelated in a physical or chemical sense. It should be stressed that the storage experiment is used here only for illustration of the method and should, in its real context, be seen as a series of experiments investigating the relation under many different conditions. In fact, other aspects were also investigated in the experiment.

On the other hand, it should also be noted that there is no other empirical way to find out whether two variables are related, than to vary as many factors as is realistic, and then to see whether they follow some simple relationship. For such an investigation the ultrastructural model seems to be the simplest appropriate model.

It is still an open problem how to test the ultrastructural model and how to estimate the parameters within the model. The methods shown here are surely not optimal because the within-group information about the relationship is not used, but since this information is limited, they may be satisfactory in many cases. Conventional likelihood methods are suspected to perform poorly because they are known to fail in the functional relation, which is a limiting case of the ultrastructural relation. This problem should be taken seriously, since experiments of the present kind often have only few replications, which is sensible because it is more important to have more factors and levels represented. The design is not the problem, but the statistical method is.

The possible relation between two variables is the simplest possible calibration problem, but in the context of sensory analysis, multivariate investigations are much more common. The complex path from chemical composition to sensory perception suggests that one would not know in advance which chemical variable to compare with. If many chemical variables are measured, some multivariate method of calibration must be used. Thus, a generalization of the statistical treatment of the ultrastructural model to a multivariate setting is needed. On top of these complications are the usual problems of over-fitting known from multivariate calibration, for which methods like principal component regression and partial least squares were invented. Combination of such methods with the ultrastructural model is another challenge.

REFERENCES

Chatfield, C. & Collins, A. J. (1980). *Introduction to Multivariate Analysis*. Chapman and Hall, London.

Dolby, G. R. (1976). The ultrastructural relation: A synthesis of the functional and structural relations. *Biometrika* **63,** 39–50.

Kendall, M. G. & Stuart, A. (1967). *The Advanced Theory of Statistics*. Vol. **2,** 2nd edition. Griffin, London.

Osborne, C. (1991). Statistical calibration: A review. *Int. Statist. Rev.* **59,** 309–36.

Stapelfeldt, H., Bjørn, H., Skovgaard, I.M., Skibsted, L. H. & Bertelsen, G. (1992). Warmed-over flavour in cooked sliced beef. Chemical analysis in relation to sensory evaluation. *Z. Lebensm. Unters. Forsch.* **195,** 203–8.

Williams, E.J. (1967). The analysis of association among several variates (with discussion). *J. R. Statist. Soc. B* **29,** 199–242.

Chapter 9

An explicit large-deviation approximation to one-parameter tests

Introduction by Thomas A. Severini
Northwestern University

This paper is a fundamental contribution to both the theory and methodology of likelihood ratio tests.

Consider a model for continuously-distributed data with parameter β and log-likelihood function $\ell(\beta)$. Suppose that β is of the form $\beta = (\alpha, \psi)$ where $\beta \in \mathbb{R}^p$, ψ is the real-valued parameter of interest and α is a nuisance parameter. Then the likelihood ratio statistic for testing the hypothesis $\psi = \psi_0$ is given by $W = 2\{\ell(\hat{\beta}) - \ell(\tilde{\beta})\}$ where $\tilde{\beta} \equiv \tilde{\beta}_{\psi_0}$ is the maximum likelihood estimator (MLE) of β under the hypothesis. It is generally preferable to use the signed version of the likelihood ratio statistic,

$$R = \operatorname{sgn}(\hat{\psi} - \psi_0)\big[2\{\ell(\hat{\beta}) - \ell(\tilde{\beta})\}\big]^{\frac{1}{2}},$$

where $\hat{\psi}$ denotes the MLE of ψ.

Under standard regularity conditions, the asymptotic distribution of R is standard normal with error of order $O(1/\sqrt{n})$, where n denotes the sample size; see, e.g., Bickel and Ghosh (1990). The present paper is concerned with constructing a modifed version of R that improves the order of this normal approximation.

When β is a canonical parameter of a full-rank exponential family model a saddlepoint approximation may be used to construct a statistic that agrees with R to order $O_p(n^{-\frac{1}{2}})$ and which has a standard normal distribution with error of order $O(n^{-\frac{3}{2}})$ (Jensen, 1992).

In more general models, a modified signed likelihood ratio statistic may be based on the likelihood-ratio approximation to the conditional distribution of the MLE given $A = a$ where A is an ancillary statistic such that $(\hat{\beta}, A)$ is sufficient; see Barndorff-Nielsen (1983), Barndorff-Nielsen and Cox (1994) and Skovgaard (1990). This approximation is given by $p^*(\hat{\beta}|a;\beta) = c\exp\{\ell(\beta) - \ell(\hat{\beta})\}|j(\hat{\beta})|^{1/2}$ for a normalizing constant c; note that, in this expression, $\ell(\beta)$ is viewed as a function of $(\hat{\beta}, a)$, as well as of β.

Using the p^* approximation, it is shown in Barndorff-Nielsen (1986) that it is possible to construct a statistic of the form

$$R^* = R + \frac{1}{R}\log\left(\frac{U}{R}\right),$$

where U is statistic such that $U/R = 1+O_p(n^{-\frac{1}{2}})$, such that R^* is asympototically normally distributed with error $O(n^{-\frac{3}{2}})$. Furthermore, for curved exponential family models, the normal approximation to the distribution of R^* holds in the large-deviation range in the sense that the error in the normal approximation to $\Pr(R^* > r^*)$ is of order $O(n^{-1})$ for r^* of order $O(\sqrt{n})$ (Jensen, 1992; Barndorff-Nielsen and Wood, 1998); thus, we expect that the normal approximation to R^* will be accurate in the tails of the distribution. In addition to such theoretical results, there is considerable empirical evidence indicating the high accuracy of the normal approximation to R^*; see, e.g., Skovgaard (2001).

The argument used to derive R^* approximates the distribution of R by using a change-of-variable argument applied to p^*; the form of this approximation leads to the expression for R^*. The change-of-variable naturally involves Jacobian terms, which may be written in terms of derivatives of the $\ell(\beta)$ with respect to $\hat{\beta}$; this type of derivative is often described as a *sample space derivative.* Calculation of such derivatives requires the specification of an ancillary statistic A such that $(\hat{\beta}, A)$ is sufficient.

For certain classes of models, such an ancillary statistic is readily available. For instance, for a full-rank exponential family model $\hat{\beta}$ itself is sufficient and no ancillary is needed; in models with a transformation structure, the ancillary statistic may often be based on a maximal invariant statistic. However, in the general case, it may be difficult to find an appropriate ancillary statistic, making the calculation of R^* difficult or impossible.

This fact led to a number of approximations to R^*; see, e.g., Barndorff-Nielsen and Chamberlin (1991), DiCiccio and Martin (1993), and Barndorff-Nielsen and Chamberlin (1994). Although these approximations provide useful improvements to the unadjusted statistic R, results on the accuracy of the normal approximation to their distributions do not extend to arguments in the large-deviation range. Thus, these results suggest that it may be necessary to sacrifice large-deviation accuracy in order to obtain a generally applicable approximation to R^*. Note that maintaining large-deviation accuracy puts important constraints on the methods of approximation used; for instance, local approximations, such as those provided by Taylor's series expansions, are generally not sufficiently accurate in the large-deviation sense.

In the present paper, Ib constructs an approximation to R^*, which he denotes by $\tilde{R}$, that addresses important theoretical questions and provides a useful methodology for small-sample likelihood-based inference. In particular, $\tilde{R}$ does not require specification of an ancillary statistic; furthermore, calculation of $\tilde{R}$ is of roughly the same complexity as calculation of the expected information matrix, making it useful in a wide range of models. The asymptotic distribution of $\tilde{R}$ is standard normal with error of order $O(n^{-1})$, improving on the error in the normal approximation to the unadjusted statistic R. In curved exponential family models, the relative error of the approximation is of order $O(n^{-\frac{1}{2}})$ in the large-deviation range, indicating that the normal approximation to the distribution of $\tilde{R}$ is generally accurate in the tails of the distribution. Finally, many of these properties continue to hold conditionally on the observed value of an ancillary statistic, without requiring specification of that statistic.

The following is a brief description of the method used to derive $\tilde{R}$. Consider an exponential family model for data y with model function $\exp\{t(y)^T\theta - \kappa(\theta)\}$; here $t(y)$ is a k-dimensional function of y and θ is a k-dimensional parameter. Consider a one-dimensional curved submodel given by $\theta = \theta(\beta)$ for a real-valued parameter β. Unlike the models in the paper, this model does not include nuisance parameters, but it has the advantage of illustrating the main idea in a simple setting and it may be viewed as an introduction to the more general results presented there.

Consider n independent observations from this model and let $\bar{t}$ denote the mean of the observations of t; note that $\bar{t}$ is sufficient. Calculation of the statistic R^* requires the sample space derivative

$$\bar{t}' \equiv \frac{\partial}{\partial\hat{\beta}}\bar{t}. \tag{9.1}$$

As noted previously, calculating (9.1) requires $\bar{t}$ to be written as a function of $(\hat{\beta}, A)$ for an ancillary statistic A; here we consider approximating (9.1).

Let $\tau(\theta) = \mathrm{E}(\bar{t};\theta)$ and let $\hat{\tau} = \tau\{\theta(\hat{\beta})\}$. Suppose there exists an ancillary statistic A such that $A = 0$ corresponds to $\bar{t} = \hat{\tau}$ and that $A = O_p(n^{-\frac{1}{2}})$. Note that only the existence of such a statistic is required.

Let $\Sigma(\theta) = n\text{Cov}(\bar{t};\theta)$, $\hat{\Sigma} = \Sigma\{\theta(\hat{\beta})\}$ and define

$$D\hat{\theta} = \frac{d}{d\beta}\theta(\beta)|_{\beta=\hat{\beta}}.$$

Then $\hat{\Sigma}$ is a $k\times k$ matrix and $D\hat{\theta}$ is a $k\times 1$ vector; note that $\Sigma(\theta) = d\tau(\theta)/d\theta$.

An important point is that when $A = 0$ the above results show that

$$\bar{t}' = \frac{d}{d\theta}\tau(\theta)|_{\theta=\theta(\hat{\beta})}\theta'(\hat{\beta}) = \tau'(\theta(\hat{\beta}))D\hat{\theta} = \hat{\Sigma}D\hat{\theta},$$

so that $\bar{t}'$ is easily calculated. When $\bar{t}$ is not equal to $\hat{\tau}$, i.e., when A is not 0, then $\bar{t}'$ may not be calculated exactly without knowledge of A.

A key part of Ib's ingenious method of approximation is to consider the projection of $\bar{t}'$ onto the space spanned by the vector $\hat{\Sigma}D\hat{\theta}$; more generally, when $\dim(\beta) > 1$, the relevant projection is onto the space spanned by the columns of $\hat{\Sigma}D\hat{\theta}$. Note that this approach avoids any local approximations that may affect large-deviation accuracy. The orthogonal projection is with respect to the inner product corresponding to $\hat{\Sigma}^{-1}$, a natural choice in this setting. The projection matrix is given by

$$\hat{P} = \hat{\Sigma}D\hat{\theta}\{(D\hat{\theta})^T\hat{\Sigma}\hat{\Sigma}^{-1}\hat{\Sigma}D\hat{\theta}\}^{-1}(D\hat{\theta})^T\hat{\Sigma}\hat{\Sigma}^{-1} = \hat{\Sigma}D\hat{\theta}\hat{i}_1^{-1}(D\hat{\theta})^T.$$

Here $\hat{i}_1 = (D\hat{\theta})^T\hat{\Sigma}D\hat{\theta}$ is the expected information based on a single observation, evaluated at the MLE.

Note that the score function in the curved model is given by

$$D_1(\beta;\bar{t}) \equiv n\{(D\theta)(\beta)\}^T[\bar{t} - \tau\{\theta(\beta)\}];$$

therefore,

$$(D\hat{\theta})^T\bar{t}' = \frac{1}{n}\frac{\partial}{\partial\hat{\beta}}D_1(\beta;\bar{t})|_{\beta=\hat{\beta}}.$$

This expression may be simplified further by noting that, by differentiating the likelihood equation $D_1(\hat{\beta};\bar{t}) = 0$ with respect to $\hat{\beta}$,

$$\frac{\partial}{\partial\hat{\beta}}D_1(\beta;\bar{t})|_{\beta=\hat{\beta}} = n\hat{j}_1,$$

where $\hat{j}_1$ denotes the average observed information evaluated at the MLE. Hence, $(D\hat{\theta})^T\bar{t}' = \hat{j}_1$.

It follows that $\hat{P}(\bar{t}') = \hat{\Sigma}(D\hat{\theta})\hat{i}_1^{-1}\hat{j}_1$. That is, even though $\bar{t}'$ cannot be calculated without knowledge of A, we can calculate the projection $\hat{P}(\bar{t}')$ and this projection can be used to approximate $\bar{t}'$. This general approach may be used to approximate the sample space derivatives appearing in R^*, leading to the approximation $\tilde{R}$.

Bernoulli **2**(2), 1996, 145–165

An explicit large-deviation approximation to one-parameter tests

IB M. SKOVGAARD

Department of Mathematics and Physics, The Royal Veterinary and Agricultural University, Thorvaldsensvej 40, DK-1871 Frederiksberg C, Denmark

An approximation is derived for tests of one-dimensional hypotheses in a general regular parametric model, possibly with nuisance parameters. The test statistic is most conveniently represented as a modified log-likelihood ratio statistic, just as the R^*-statistic from Barndorff-Nielsen (1986). In fact, the statistic is identical to a version of R^*, except that a certain approximation is used for the sample space derivatives required for the calculation of R^*. With this approximation the relative error for large-deviation tail probabilities still tends uniformly to zero for curved exponential models. The rate may, however, be $O(n^{-1/2})$ rather than $O(n^{-1})$ as for R^*. For general regular models asymptotic properties are less clear but still good compared to other general methods. The expression for the statistic is quite explicit, involving only likelihood quantities of a complexity comparable to an information matrix. A numerical example confirms the highly accurate tail probabilities. A sketch of the proof is given. This includes large parts which, despite technical differences, may be considered an overview of Barndorff-Nielsen's derivation of the formulae for p^* and R^*.

Keywords: conditional inference, large-deviation expansions, modified log-likelihood ratio test, nuisance parameters, parametric inference

1. Introduction

The purpose of the present paper is to derive an explicit general approximation for testing a one-dimensional, possibly composite, hypothesis in a well-behaved parametric model. By a one-dimensional hypothesis is meant a hypothesis that a single coordinate of the parameter vector assumes a particular value.

The entire approach and the result is highly related to and based on the line of theory developed by Barndorff-Nielsen through the p^* and the R^* formulae (see, in particular, Barndorff-Nielsen 1980; 1986; 1991). Thus, the paper deals with likelihood inference, and what is described may be seen as an attempt to improve some of the classical, normal-based, asymptotic results, especially the chi-squared approximation to minus twice the log-likelihood ratio statistic. In the case of a one-dimensional hypothesis the signed square root of this statistic has a standard normal distribution under the hypothesis, and the simplest way of presenting the present result is as a modification of this statistic, quite analogously to R^*. This statistic, here denoted $\tilde{R}$, is given in formula (2) in combination with formula (1) in Section 3.

The standard normal approximation to the tail probabilities of the distribution of $\tilde{R}$ is of

1350-7265

the large-deviation type for curved exponential models, i.e., the relative error of the tail probability tends to zero uniformly in a region of large deviations. The rate at which this relative error tends to zero is at least as good as $O(n^{-1/2})$ under repeated sampling. For comparison, the relative error for R^* is $O(n^{-1})$.

The calculation of $\tilde{R}$ involves only likelihood quantities of a computational complexity comparable to that of the Fisher information matrix. These quantities are well defined for any well-behaved parametric model, and the result is therefore not confined to curved exponential families. However, the proof is presented for curved exponential families, and for more general models some asymptotic results then follow by approximation to curved exponential families (see Section 4).

The reason why the R^* formula cannot be applied to the problem addressed here is that this statistic involves some sample space derivatives (see Barndorff-Nielsen 1991, Sections 2 and 3). These are derivatives of likelihood quantities with respect to the maximum likelihood estimate, through the minimal sufficient statistic. When the maximum likelihood estimate is not sufficient an ancillary statistic must be specified for the calculation of such sample space derivatives. The non-uniqueness of ancillaries that may be used for this purpose and the difficulties involved in their specification make it difficult to calculate R^* in general.

Barndorff-Nielsen and Chamberlin (1991; 1994) solve the problem by approximating R^*, but at the price of losing the general large-deviation properties of the approximation. The same is true for DiCiccio and Martin (1993), although their method of approximation is quite different.

Jensen (1992) constructs a statistic similar to R^* for curved exponential families by specifying the ancillary statistic as a series of one-dimensional signed log-likelihood ratios from the full exponential model down to the curved model. By means of this construction he can prove remarkably good large-deviation properties of the approximation, but the result may be hard to compute and may depend on the series of one-dimensional hypotheses chosen.

The main idea behind the present approach is that an explicit general approximation can be made to the sample space derivatives required. This approximation is sufficiently good to keep large-deviation properties and sufficiently simple to make it possible to derive explicit results.

In fact, in the case of no nuisance parameters, the approximate sample space derivatives used here are identical to the sample space derivatives in Fraser and Reid (1988), who define the ancillary statistic in terms of the sample space derivatives of the log-likelihood. The present use of these derivatives is different though, namely as approximations to sample space derivatives arising from ancillaries that are directed versions of the log-likelihood ratio for testing the model against the full exponential family.

After the introduction of notation in Section 2, the main result is given in Section 3. Its properties, especially in terms of asymptotic errors, are described in Section 4. This part of the paper, possibly together with the numerical example in Section 5, should suffice for readers who are not interested in the methods and proofs used. The proof is outlined for curved exponential families, for which notation and basic concepts are introduced in Section 6. A conceptual and mathematical description of ancillaries and sample space derivatives is given in Section 7, while the crucial approximation to the sample space derivatives is discussed in Section 8. With this approximation substituted into the expression for R^* in Barndorff-Nielsen (1991) the result might be derived, but this would not reveal the accuracy of the result. Partly for this reason, an entire outline of the proof from the beginning is given in Section 9. Another reason for this is that some results in this section may be of independent interest, in particular a number of intermediate results that hold for any ancillary statistic, regardless of its distributional properties. Finally, the discussion in Section 10 mainly points out some further problems.

2. Set-up and notation

Let $\{f(y;\beta); \beta \in B \subseteq \mathbb{R}^p\}$ be a family of densities of the random variable Y, indexed by the p-dimensional parameter β, possibly restricted to some subset B. The domain of Y and the underlying measure are of no importance in the present connection, except that the asymptotic results require absolute continuity of the distribution of the canonical sufficient statistic with respect to Lebesgue measure.

The problem is to derive a test for a one-dimensional hypothesis. More specifically, let

$$\beta = (\beta_1, \ldots, \beta_{p-1}, \beta_p) = (\alpha, \psi),$$

where $\alpha = (\beta_1, \ldots, \beta_{p-1})$ and $\psi = \beta_p$; we wish to test the hypothesis

$$H_0 : \psi = \psi_0.$$

Let $\hat{\beta}(y) = \hat{\beta} = (\hat{\alpha}, \hat{\psi})$ denote the maximum likelihood estimate of the full parameter vector, and let $\tilde{\beta}(y) = \tilde{\beta} = (\tilde{\alpha}, \psi_0)$ denote the maximum likelihood estimate under the hypothesis.

The log-likelihood function is denoted

$$\ell(\beta) = \ell(\beta; y) = \log f(y;\beta),$$

where the first version is used when y or the value of some sufficient statistic is understood. The kth derivative of the log-likelihood function is denoted D_k, i.e.,

$$D_k(\beta) = D_k(\beta; y) = \frac{\partial^k}{\partial \beta^k} \ell(\beta; y),$$

which is a k-sided array with p^k entries. In particular, D_1 is the score function.

The cumulants of the log-likelihood derivatives are denoted by χ's, with indices corresponding to the derivatives in question. Thus, for example,

$$\chi_k(\beta) = E_\beta\{D_k(\beta; Y)\}, \qquad \chi_{km}(\beta) = \mathrm{cov}_\beta\{D_k(\beta; Y), D_m(\beta; Y)\},$$

denote means and variances. For the more common information quantities we use the special notation

$$i(\beta) = \chi_{11}(\beta) = -\chi_2(\beta), \qquad j(\beta; y) = -D_2(\beta; y).$$

In particular, $i(\beta)$ is the (expected) Fisher information.

We use abbreviations such as $\hat{\ell} = \ell(\hat{\beta})$, $\hat{i} = i(\hat{\beta})$, $\hat{j} = j(\hat{\beta}; y)$, and $\tilde{j} = j(\tilde{\beta}; y)$, and frequently omit the argument y. Note that $\hat{j}$ is the observed Fisher information.

Finally, we need two less familiar quantities, $\hat{q}$ and $\hat{S}$, defined below. These are based on covariances of likelihood differences and derivatives. More specifically, let

$$\chi_{10}(\beta_1, \beta_2; \beta) = \mathrm{cov}_\beta\{D_1(\beta_1; Y), \ell(\beta_1; Y) - \ell(\beta_2; Y)\},$$

and

$$\chi_{11}(\beta_1, \beta_2; \beta) = \mathrm{cov}_\beta\{D_1(\beta_1; Y), D_1(\beta_2; Y)\},$$

and define

$$\hat{q} = \chi_{10}(\hat{\beta}, \tilde{\beta}; \hat{\beta}), \qquad \hat{S} = \chi_{11}(\hat{\beta}, \tilde{\beta}; \hat{\beta}).$$

Note that $\hat{q}$ is a p-vector, while $\hat{i}, \hat{j}$ and $\hat{S}$ are $p \times p$ matrices.

The determinant of a matrix, M say, is denoted $|M|$, and its transpose is denoted M^{T}.

3. The approximate test

Standard asymptotic theory would suggest an asymptotic chi-squared test based on twice the log-likelihood ratio

$$R^2 = 2\{\ell(\hat{\beta}) - \ell(\tilde{\beta})\}.$$

Alternatively, one-sided tests may be calculated from the standard normal distribution of the directed log-likelihood ratio test statistic R, equipped with the sign of $\hat{\psi} - \psi_0$.

Far better asymptotic performance is obtained by use of the modified signed log-likelihood ratio introduced by Barndorff-Nielsen (1986; 1991) and given by

$$R^* = R - \frac{1}{R}\log(R/U),$$

where U is a quantity which unfortunately is difficult to calculate since it requires specification of an ancillary statistic – or at least a local specification of the change of the log-likelihood difference with a change of the estimate. When the estimate is not sufficient this 'sample space derivative' is only defined in the conditional distribution given a supplementary statistic which together with the estimate is sufficient. When this supplementary statistic is chosen in a certain way the standard normal approximation to R^* becomes accurate to third order, i.e., with an error of order $O(n^{-3/2})$ in repeated sampling of n observations in a well-behaved model (see Barndorff-Nielsen 1986; 1991).

The point of the present paper is to provide an approximation, $\tilde{U}$, to U, which is easily calculated and sufficiently accurate to maintain the high-quality asymptotic behaviour, although one order of magnitude may be lost compared to R^*. This approximation is

$$\tilde{U} = [\hat{S}^{-1}\hat{q}]_p |\hat{j}|^{1/2} |\tilde{j}_{\alpha\alpha}|^{-1/2} |\hat{i}|^{-1} |\hat{S}|, \tag{1}$$

where $[\cdots]_p$ denotes the pth coordinate of the vector, and $\tilde{j}_{\alpha\alpha}$ denotes the upper left $(p-1) \times (p-1)$ submatrix of $\tilde{j}$. Notice that $\tilde{j}_{\alpha\alpha}$ is simply the observed Fisher information for the parameter α under the hypothesis.

Insertion in the expression for R^* now defines the statistic

$$\tilde{R} = R - \frac{1}{R}\log(R/\tilde{U}) \qquad (2)$$

and the claim is, as for R^*, that a standard normal distribution provides an accurate approximation to its distribution under the hypothesis. This statement will be made more precise in the following section.

It may be noted that in terms of asymptotic approximation an entirely equivalent result may be obtained by use of a different type of Laplace approximation to a tail integral, using a method from Bleistein (1966), also known from Lugannani and Rice (1980). This gives the right tail probability as

$$1 - \Phi(R) + \frac{\phi(R)}{R}(R/\tilde{U} - 1), \qquad (3)$$

instead of

$$1 - \Phi(\tilde{R}).$$

The asymptotic equivalence of the two expressions is proved in Jensen (1992, Lemma 2.1). Numerical examples seem to indicate, however, that the R^* version is preferable (see, for example, Pierce and Peters 1992).

The quantities $\hat{q}$ and $\hat{S}$ are usually of the same computational complexity as $\hat{i}$, since they are also covariances of ordinary likelihood quantities. Alternative expressions for $\hat{q}$ and $\hat{S}$ are in terms of derivatives of the Kullback–Leibler distance

$$KL(\beta, \beta_1) = E_\beta\{\log f(Y;\beta) - \log f(Y;\beta_1)\},$$

from which we obtain

$$\chi_{10}(\beta, \beta_1; \beta) = \frac{\partial}{\partial \beta} KL(\beta; \beta_1)$$

and

$$\chi_{11}(\beta, \beta_1; \beta) = -\frac{\partial}{\partial \beta}\frac{\partial}{\partial \beta_1} KL(\beta; \beta_1).$$

The first of these derivatives also appears in Sweeting (1995, Section 5).

4. Properties of the approximation

Since the expression for $\tilde{R}$ only involves likelihood quantities it is well defined for all sufficiently regular parametric models and does not require an embedding in an exponential family. Furthermore, it is trivially invariant under sufficient transformations of the data.

The expression is also invariant under relevant smooth one-to-one transformations of the parameter, i.e., under transformations of the parameter of interest, ψ, and under transformations of β preserving ψ. There is no loss of generality in the formulation of the hypothesis as a hypothesis concerning a single coordinate since we can always make the one-dimensional parameter of interest a coordinate.

Concerning the asymptotic properties, we confine ourselves to the case of n independent replications, although the approximation may well also be reasonable in other cases. The results are stated in terms of right tail probabilities, but hold analogously for the left tail. The asymptotic results are only valid for absolutely continuous distributions.

Assume first that the model is a 'curved exponential family', i.e., a smooth submodel of an exponential family. Let $\bar{t}$ denote the mean of the canonical sufficient statistic and let $\bar{W} = \bar{W}(\bar{t})$ denote n^{-1} times minus twice the log-likelihood ratio test statistic for testing the model against the full exponential family. Scaling by the divisor n ensures that $\bar{W}$ depends only on $\bar{t}$, not on n. The conditional result below holds given any ancillary statistic of the form $A = A(\bar{t})$ which is such that $\bar{W}$ is a function of A, combined with the requirement that $(\hat{\beta}, A)$ is sufficient. Thus A is any 'directed log-likelihood ratio' as opposed to the ancillaries used for R^* which are directed *modified* log-likelihood ratios. Any statistic A of the kind used in the present paper will generally be a first-order ancillary statistic in the sense that its standardized distribution is free of β under the model apart from a term of order $n^{-1/2}$.

Now the result for the conditional tail probability, given any of the ancillary statistics of the type mentioned above, is that, under the hypothesis,

$$1 - \Phi(\tilde{r}) = \text{pr}_\beta\{\tilde{R} \geq \tilde{r} | A\}\{1 + O(n^{-1}) + O(\|\hat{\psi} - \psi_0\| \, \|A\|)\}$$

as $n \to \infty$ uniformly over $(\hat{\psi}, A)$ in some fixed neighbourhood of $(\psi_0, 0)$, where $A = 0$ is chosen to correspond to $\bar{W} = 0$. Since $(\hat{\psi}, A)$ converges to its mean at rate $n^{-1/2}$, this is a large-deviation region. Notice, however, that both of the normal deviations of A and $\hat{\psi} - \psi_0$ are of order $n^{-1/2}$, but either or both of them may become of order $O(1)$ in a large-deviation region. Thus the result states that the error is of order n^{-1} in a normal-deviation region, whereas the relative error is of order $n^{-1/2}$ in a large-deviation region of either the ancillary or the estimate, but not both.

Unconditionally this implies that the relative error is $O(n^{-1/2})$ uniformly in a fixed set of values of $\hat{\psi}$, i.e., in a large-deviation region. For comparison the relative error for R^* is $O(n^{-1})$. For the numerical quality of the approximation it is presumably more important, however, that the large-deviation property is kept, since this results in a far better tail behaviour of the approximation than the central type of expansions that form the basis of standard asymptotic theory.

For models that are not submodels of exponential models it is more difficult to derive conditional results since it is hard to come up with useful ancillary statistics in a general form. However, for analytic models (see Skovgaard 1990), it follows that unconditionally we have

$$1 - \Phi(\tilde{r}) = \text{pr}_\beta\{\tilde{R} \geq \tilde{r}\}\{1 + O(n^{-1} + n^{-1/2}\|\hat{\psi} - \psi_0\| + n^{-K}\text{pr}_\beta\{\tilde{R} \geq \tilde{r}\}^{-\epsilon})\}$$

for any $\epsilon > 0$ and $K > 0$, uniformly in a fixed set of $\hat{\psi}$ around ψ_0. With $\epsilon = 0$ the uniform relative error would have been kept, so the result is that this is almost the case. Since any analytic model can be approximated locally to any order by a curved exponential family, this result follows almost immediately from this general approximation theory in Skovgaard (1990, Section 2.7), but the proof will not be given here. That the normal deviation error is of order $O(n^{-1})$ is a trivial consequence, because then $\|\hat{\psi} - \psi_0\| = O(n^{-1/2})$.

As a final property it is worth noting that $\tilde{R}$ agrees exactly with R^* when no ancillary variable is necessary, which essentially is when the model is a full exponential family. However, the hypothesis may be curved, so for the Behrens–Fisher example given in Jensen (1992) the two statistics are identical.

5. An example

As a numerical example we consider a one-way analysis of variance with random effects. Thus, let $i = 1, \ldots, m$ enumerate the groups and $j = 1, \ldots, n_i$ the observations within groups, and let X_{ij} be normally distributed with mean μ and variance ω^2. Observations from different groups are assumed to be independent, while the within-group correlation is ρ. We allow for negative correlations such that the lower bound for ρ becomes

$$\rho > -(n_{\max} - 1)^{-1},$$

where $n_{\max}$ is the largest group size. Notice that the variance of the largest group mean tends to zero as ρ approaches its lower bound.

Usually this random effects model is written in the form

$$X_{ij} = \mu + B_i + \epsilon_{ij}, \tag{4}$$

where the B_i's and the ϵ_{ij}'s are all independent with standard deviations σ and σ_B, respectively. This formulation covers only cases with non-negative within-group correlations, however.

We wish to test the hypothesis $\mu = 0$ and consider three test statistics. First, R denotes the unmodified directed square root of the log-likelihood ratio statistic. Second, the modified statistic $\tilde{R}$ from (2) is considered. Third, we compute an approximate F-test statistic based on the approximation suggested by Satterthwaite (1946). This test is based on the ratio of the between-group mean square to its estimated expectation which is a linear combination of the same mean square and the residual mean square. The distribution of the ratio is then approximated by an F-distribution for which the number of degrees of freedom for the denominator is chosen to make the variance of the denominator correct. This third statistic, denoted 'Satterthwaite' in Table 1, may not be as attractive as the likelihood ratio based statistics for variance components in general, but it has the advantage of admitting Satterthwaite's approximation to the null distribution – an approximation known to be quite accurate.

The computation of $\tilde{R}$ is straightforward and uninteresting, so it will not be shown here. It involves nothing beyond matrix inversions of the size of the information matrix, i.e. 3×3 matrices, and calculation of cumulants of order 4 or less in a normal distribution.

The distributions are, like the entire model, symmetric about zero, so only two-sided tail probabilities will be considered. The approximate F-statistic is two-sided by construction although it might easily be reformulated as a one-sided t-test.

Only one set of simulations will be shown. Others have been done which indicate the same accuracy of the approximations. The example has 5 groups of sizes 1, 2, 3, 4, and 5,

Table 1. Numbers of exceedances of nominal two-sided significance levels in 100 000 simulations of the one-way analysis of variance with random effects; the nominal significance levels are given indirectly as the 'expected' numbers of exceedances

Expected	50 000	20 000	10 000	2000	200	20	2
R	56 635	27 771	16 399	4780	917	146	27
$\tilde{R}$	50 061	20 229	10 128	2101	224	24	3
Satterthwaite	50 470	19 948	9875	1904	181	13	1

respectively, and 100 000 samples were generated from the model (4) with $\mu = 0$, $\sigma = 1.0$, and $\sigma_B = 0.04$. The pseudo-random number generator RAN2 from Press *et al.* (1986, Chapter 7) was used. The numbers of exceedances of nominal two-sided tail probabilities are seen in Table 1, together with the nominal expected numbers according to the respective approximate distribution.

The usual asymptotic approximation to the uncorrected log-likelihood ratio statistic, R, rejects the hypothesis far too often in the tails. For the assessment of $\tilde{R}$ this shows that although the example is a fairly 'nice' one, standard likelihood asymptotic approximations are not automatically of high quality. In contrast, the adjusted log-likelihood ratio test, $\tilde{R}$, behaves well, even in the extreme tail. This type of behaviour is typical of approximations with large-deviation properties as opposed to central approximations such as for R.

The approximation based on Satterthwaite's method is of the same quality as $\tilde{R}$. This method is known to give quite accurate approximations, but is limited to the approximate F-tests, which may not be the most desirable tests in variance component models in general. One reason is that there are many ways of constructing such tests for more complicated models.

There was one case, not counted in the table, for which the two likelihood methods broke down because the numerical algorithm searching for the solution to the likelihood equation did not converge. In all other cases such a solution was found. However, strictly speaking, the likelihood methods were inapplicable in all cases, because the likelihood function tends to infinity when μ is held equal to the mean of the largest group while the within-group correlation tends to its lower bound. This is just one reason why the method of maximum likelihood should not be used for variance component models; the more structured approach of restricted maximum likelihood should be used instead. It is an important challenge to extend the type of methods described in the present paper to such structured inference.

6. Curved exponential families

Consider an exponential family with densities

$$f(y;\theta) = \exp\{\langle\theta, t\rangle - \kappa(\theta)\}$$

with respect to some underlying measure, where the parameter vector θ and the canonical sufficient statistic $t = t(y)$ both belong to $\mathbb{R}^k$, and $\langle\cdot,\cdot\rangle$ denotes the inner product in $\mathbb{R}^k$.

We use the special notation

$$\tau(\theta) = \mathrm{E}_\theta\{t\}, \qquad \Sigma(\theta) = \mathrm{var}_\theta\{t\}$$

for the first two derivatives of the cumulant generating function κ. The representation is assumed to be minimal, such that $\Sigma(\theta)$ has full rank.

A curved exponential model is given by

$$\theta = \theta(\beta)$$

where $\beta \in \mathbb{R}^p$. The derivatives of the mapping θ at β are denoted $D\theta(\beta)$, $D^2\theta(\beta)$, and so on, and $D\theta$ is assumed to have rank p.

Assume that n independent observations are obtained from a distribution within this model. The mean, $\bar{t}$, of the n observations of t is sufficient, and we may then write $\bar{t}$ in place of y in all relevant statistics. The score statistic is

$$D_1(\beta;\bar{t}) = n\langle D\theta(\beta), \bar{t} - \tau(\theta(\beta))\rangle$$

where the inner product is taken over $\mathbb{R}^k$, such that D_1 becomes a p-vector. Similarly, further differentiation yields

$$D_k(\beta;\bar{t}) = n\{\langle D^k\theta(\beta), \bar{t} - \tau(\theta(\beta))\rangle + \chi_k(\theta(\beta))\}. \tag{5}$$

The important thing to notice is that all these log-likelihood derivatives deviate from their means by linear functions of the canonical sufficient statistic, $\bar{t}$.

The maximum likelihood estimate, $\hat{\beta}$, solves the likelihood equation

$$\langle D\hat{\theta}, \bar{t} - \hat{\tau}\rangle = 0$$

with obvious abbreviations.

Some of the quantities entering the expression for $\tilde{U}$ from (1) are

$$\hat{q} = n(D\hat{\theta})^{\mathrm{T}}\hat{\Sigma}(\hat{\theta} - \tilde{\theta}),$$

where $\tilde{\theta} = \theta(\tilde{\beta})$ is the estimate under the hypothesis, and

$$\hat{S} = n(D\hat{\theta})^{\mathrm{T}}\hat{\Sigma}(D\tilde{\theta}).$$

Notice that the computation of these quantities is of the same complexity as the computation of the information $n(D\theta)^{\mathrm{T}}\Sigma(D\theta)$ and essentially requires only the covariance matrix of t and the first derivative of the mapping θ.

To make it easier to see the power of the n appearing in the various expressions, we use $\bar{\ell}$ to denote the function of $\bar{t}$ that is 'free of n', i.e.,

$$\bar{\ell}(\beta;\bar{t}) = \langle\theta(\beta), \bar{t}\rangle - \kappa(\theta(\beta)) = n^{-1}\ell(\beta;\bar{t}).$$

For the same reason we define the information quantities

$$i_1(\beta) = n^{-1}i(\beta), \qquad j_1(\beta;\bar{t}) = n^{-1}j(\beta;\bar{t}),$$

which depend on β and $\bar{t}$ only, not on n.

Note that all likelihood derivatives, D_k, and their cumulants are proportional to n, when viewed as functions of β and $\bar{t}$.

There is no loss of generality in considering n independent replications since all expressions are equally valid for the special case with $n = 1$, and hence $\hat{i}_1 = \hat{i}$, and so on. Asymptotic results are, however, only proved for n independent replications.

7. Ancillaries and sample space derivatives

We continue to consider a curved exponential family, but shall from time to time, when explicitly stated, revert to more general models.

Notice first that the likelihood equation may be rewritten as the orthogonality

$$\bar{t} - \hat{\tau} \perp_{\hat{\Sigma}^{-1}} \hat{\Sigma}(D\hat{\theta}) \tag{6}$$

of the two vectors in the $\bar{t}$-space with respect to the variable inner product $\hat{\Sigma}^{-1}$. Thus the observations of $\bar{t}$ that lead to the same estimate $\hat{\beta}$ are located in a $(k-p)$-dimensional linear subspace, and the observation deviates from the estimated mean value by the vector $\bar{t} - \hat{\tau}$ in this subspace. The orthogonality, (6), of this subspace to the tangent space, spanned by $\hat{\Sigma}(D\hat{\theta})$, is the reason why it is convenient to work with the variable inner product, as will be seen in Section 9.

In the development below, which goes through the p^* formula, it is necessary to be able to write the sufficient statistic as a function of $(\hat{\beta}, A)$, where A is some supplementary statistic of dimension $k - p$. In the present paper we require that $A = A(\bar{t})$ is a smooth function of $\bar{t}$, and that $\bar{t}$ and $(\hat{\beta}, A)$ are in one-to-one correspondence. We shall refer to any such statistic as a *supplementary* statistic and reserve the word 'ancillary' for a statistic for which further properties are at least desired.

For fixed A, the sufficient statistic $\bar{t} = \bar{t}(\hat{\beta}, a)$ moves along a p-dimensional level surface of A which may be thought of as 'parallel' to the model space $\{\bar{t} = \tau(\theta(\beta))\}$ parametrized by β.

The fact that $\bar{t}$ is a function of $\hat{\beta}$ and A means that derivatives of $\bar{t}$, and consequently of likelihood quantities, may be defined with respect to $\hat{\beta}$ for fixed A. From now on we reserve a prime to denote such a derivative, i.e.,

$$t' = \frac{\partial \bar{t}}{\partial \hat{\beta}}, \qquad \ell' = \ell'(\beta; \bar{t}) = \frac{\partial}{\partial \hat{\beta}} \ell(\beta; \bar{t}(\hat{\beta}, A)).$$

These derivatives are usually referred to as *sample space derivatives*. Notice that, for example, $\hat{D}_1' = D_1'(\hat{\beta})$ means $D_1(\beta, \bar{t}(\hat{\beta}, A))$ differentiated with respect to $\hat{\beta}$ before $\hat{\beta}$ is substituted for β. Furthermore, we use the convention that a sample space derivative corresponds to the 'last index' of an array, for example to the columns of the $p \times p$ matrix D_1'. From (5) it is seen that for curved exponential families we have

$$D_k'(\beta) = n\langle D^k \theta(\beta), \bar{t}' \rangle. \tag{7}$$

As ancillary statistic, $A = A(\bar{t})$, for curved exponential families we shall consider statistics with the first or both of the following two properties:

(A1) The model subspace $\{\bar{t} = \hat{\tau}\}$ is a level surface of A. In this case we let $A = 0$ represent this subspace.

(A2) The log-likelihood ratio test statistic of the model against the full exponential family is a function of A, i.e., A is a directed log-likelihood ratio.

The asymptotic result of the paper, as stated in Section 4, requires an ancillary statistic with both of the properties.

Let $\bar{\theta}$ denote the maximum likelihood estimate of θ in the full exponential family, i.e., $\bar{\theta} = \tau^{-1}(\bar{t})$. Then minus twice the log-likelihood ratio test statistic for the model against the full exponential family is

$$W = 2n(\bar{\ell}(\bar{\theta}) - \bar{\ell}(\hat{\theta})).$$

Since the model subspace corresponds to the set $\{W = 0\}$, which is of the same dimension as any level subspace for A, Property (A2) implies Property (A1).

8. Approximation of sample space derivatives

The most important point of the present paper is to show that the sample space derivatives needed for calculation of the tail probability corresponding to the R^* formula from Barndorff-Nielsen (1986; 1991) may be approximated sufficiently well to obtain a large-deviation approximation.

To do this we first need to define a tangent space projection related to the orthogonality in (6). Thus, let $\hat{P}$ denote the orthogonal projection

$$\hat{P} = \hat{\Sigma}(D\hat{\theta})\hat{i}_1^{-1}(D\hat{\theta})^{\mathrm{T}} \tag{8}$$

on $\hat{\Sigma}(D\hat{\theta})$ with respect to the variable metric, or inner product, $\hat{\Sigma}^{-1}$. This is the projection on the subspace for $\bar{t}$ tangent to the model space at $\hat{\tau}$.

The approximation of the sample space derivative used in the present paper is now simply given by

$$\bar{t}\,' \approx \hat{P}(\bar{t}\,'), \tag{9}$$

the point being that this has the unique explicit expression

$$\hat{P}(\bar{t}\,') = \hat{\Sigma}(D\hat{\theta})\hat{i}_1^{-1}(D\hat{\theta})^{\mathrm{T}}\bar{t}\,' = \hat{\Sigma}(D\hat{\theta})\hat{i}_1^{-1}n^{-1}\hat{D}_1' = \hat{\Sigma}(D\hat{\theta})\hat{i}_1^{-1}\hat{j}_1, \tag{10}$$

since $(D\hat{\theta})^{\mathrm{T}}\bar{t}\,' = \langle D\hat{\theta}, \bar{t}\,' \rangle = n^{-1}\hat{D}_1'$. This result holds for any supplementary statistic.

For general, non-exponential, models the projection may be defined as a regression of the log-likelihood derivatives on the score statistic, i.e.,

$$P(D_k) = \chi_k + \chi_{k1}\chi_{11}^{-1}D_1, \tag{11}$$

which is then used at $\hat{\beta}$. This gives

$$\hat{P}(\hat{D}_k') = \hat{\chi}_{k1}\hat{\chi}_{11}^{-1}\hat{D}_1' = \hat{\chi}_{k1}\hat{i}^{-1}\hat{j}, \tag{12}$$

because of the well-known relation

$$\hat{D}_1' = \hat{j}$$

which is immediately obtained by differentiation of the equation $D_1(\hat{\beta}; y(\hat{\beta}, A)) = 0$ with respect to $\hat{\beta}$. This relation may be found, for example, in Barndorff-Nielsen and Cox (1994, Section 5.2).

It may easily be checked that the general definition of the projection agrees with that for curved exponential families. At the same time, it gives a more statistical interpretation of the projection.

Our first, quite simple, result concerning approximation (9) is the following.

Lemma 1. *For any supplementary statistic, A, satisfying condition* (A1), *approximation* (9) *is exact on the model subspace, which may be characterized by $\hat{D}_k = \hat{\chi}_k$ for all k.*

This result follows trivially from (10) because on the model subspace $\hat{i}_1 = \hat{j}_1$ and

$$\bar{t}' = \frac{\mathrm{d}}{\mathrm{d}\hat{\beta}}\tau(\theta(\hat{\beta})) = \hat{\Sigma}(D\hat{\theta}).$$

It turns out that the sample space derivatives needed are $\ell'(\hat{\beta}) - \ell'(\tilde{\beta})$ and $\tilde{D}_1'$, where the data point argument in all functions is $\bar{t} = \bar{t}(\hat{\beta}, A)$. One of the main points in the present paper is that these sample space derivatives can be calculated explicitly in general when approximation (9) is used, and that this approximation is sufficiently accurate.

Lemma 2. *If $\hat{P}(\hat{D}_k')$ is substituted for $\hat{D}_k'$ for all k, we obtain*

$$\ell'(\hat{\beta}) - \ell'(\tilde{\beta}) = \hat{q}^{\mathrm{T}}\hat{\imath}^{-1}\hat{\jmath} \tag{13}$$

and

$$\tilde{D}_1' = \hat{S}^{\mathrm{T}}\hat{\imath}^{-1}\hat{\jmath}. \tag{14}$$

The relative error due to the substitution is $O(\|\hat{\beta} - \tilde{\beta}\|\,\|A\|)$ in both cases, for any supplementary statistic with Property (A1).

Proof. To see this, consider first the log-likelihood sample space derivative in (13). Expand the log-likelihood difference in an infinite Taylor series about $\hat{\beta}$ as

$$\ell(\tilde{\beta}) - \ell(\hat{\beta}) = \hat{D}_1(\tilde{\beta} - \hat{\beta}) + \tfrac{1}{2}\hat{D}_2(\tilde{\beta} - \hat{\beta})^2 + \cdots,$$

where a suitable notation should be adopted to make this multivariate Taylor series expansion formally correct. Now differentiate the series with respect to $\hat{\beta}$ to obtain

$$\ell'(\tilde{\beta}) - \ell'(\hat{\beta}) = \hat{D}_1'(\tilde{\beta} - \hat{\beta}) + \tfrac{1}{2}\hat{D}_2'(\tilde{\beta} - \hat{\beta})^2 + \cdots,$$

because the log-likelihood derivatives with respect to the parameters vanish at the maximum values considered. Substitution of $\hat{P}(\hat{D}_k')$ from equation (12) for $\hat{D}_k'$ now leads to an infinite sum which, except for the sign reversal, may be identified with the right-hand side of expression (13). One way of doing this is to check that the expansions of the two expressions agree. Since $\hat{D}_1' = \hat{P}(\hat{D}_1')$ is an exact relation, the error from the approximation $\hat{D}_k' \approx \hat{P}(\hat{D}_k')$ in the infinite sum is $O(\|\hat{\beta} - \tilde{\beta}\|^2\|\hat{D}_k' - \hat{P}(\hat{D}_k')\|)$, which is known from Lemma 1 to be $O(n\|\hat{\beta} - \tilde{\beta}\|^2\|A\|)$, because of the smoothness of $\hat{D}_k' - \hat{P}(\hat{D}_k')$ which vanishes when $A = 0$. Since the leading term, $\hat{D}_1'(\tilde{\beta} - \hat{\beta}) = \hat{\jmath}(\tilde{\beta} - \hat{\beta})$, is of order $n\|\hat{\beta} - \tilde{\beta}\|$, the result for the first sample space derivative follows. The result for $\tilde{D}_1'$ is obtained in a similar way. □

9. Derivation of the result

In this section we sketch the proof of the result for n independent replications from a curved exponential family for which the distribution of the canonical sufficient statistic is absolutely continuous. The line of proof summarizes the development from Barndorff-Nielsen (1980; 1986; 1991), with some technical differences. The notation from the previous sections is used, in particular the n-free functions $\bar{\ell}$, i_1 and j_1 from the end of Section 6. The development assumes that the ancillary, A, has Properties (A1) and (A2) from Section 7, but several intermediate results of some general interest hold for any supplementary statistic, or assume only Property (A1). The assumptions will be explicitly stated in these cases.

9.1. APPROXIMATE CONDITIONAL DENSITY OF $\hat{\beta}$

We use f_β generically to denote the β-density of any of the statistics considered with respect to Lebesgue measure on the space in question. We start by transforming the density, $f_\beta(\bar{t})$, of $\bar{t}$ to the density of $(\hat{\beta}, A)$. It turns out to be convenient to use the variable metric $\hat{\Sigma}^{-1}$ to calculate the Jacobian of the transformation, which we may do if we multiply the density of $\bar{t}$ by $|\hat{\Sigma}|^{1/2}$, since the Riemannian measure corresponding to this metric has density $|\hat{\Sigma}|^{-1/2}$ with respect to Lebesgue measure. Determinants with respect to the variable metric are denoted $|\cdot|^*$.

The point is that we know that $\partial\bar{t}/\partial A$ belongs to the space which, in the variable metric, is orthogonal to $\hat{\Sigma}(D\hat{\theta})$, since $\hat{\beta}$ is constant when only A is changed.

Recall the definition of $\hat{P}$ from (8). The Jacobian of the transformation from $\bar{t}$ to $(\hat{\beta}, A)$ may be written

$$\left|\frac{\partial \bar{t}}{\partial(\hat{\beta}, A)}\right|^* = |\hat{P}(\bar{t}')|^* \left|\frac{\partial \bar{t}}{\partial A}\right|^*,$$

where the determinants on the right are generalized determinants, i.e.,

$$|M|^* = |M^{\mathrm{T}}\hat{\Sigma}^{-1}M|^{1/2} \tag{15}$$

for $M = \hat{P}(\bar{t}')$ or $M = \partial\bar{t}/\partial A$. Generalized determinants and related computations may be found in Tjur (1974, Section 11).

From equations (10) and (15) we see that

$$|\hat{P}(\bar{t}')|^* = |\hat{j}_1 \hat{i}_1^{-1} \hat{j}_1|^{1/2} = |\hat{j}_1|\,|\hat{i}_1|^{-1/2}.$$

Thus the density of $(\hat{\beta}, A)$ becomes

$$f_\beta(\hat{\beta}, A) = |\hat{\Sigma}|^{1/2} f_\beta(\bar{t}) | \hat{j}_1 | |\hat{i}_1|^{-1/2} \left| \left(\frac{\partial \bar{t}}{\partial A}\right)^{\mathrm{T}} \hat{\Sigma}^{-1} \left(\frac{\partial \bar{t}}{\partial A}\right) \right|^{1/2} \tag{16}$$

for any supplementary statistic A.

We now wish to isolate factors that mainly depend on A. The precise meaning of this is as follows.

Definition 1. *A function $h(\hat{\beta}, A)$ is called an A-function if it is constant on the model space given by $\{\bar{t} = \hat{\tau}\}$, or equivalently by $\{A = 0\}$ if A satisfies Property (A1).*

With ancillary statistics that are directed log-likelihood statistics, we can now show the following result.

Lemma 3. *For any supplementary statistic satisfying Property (A2),*

$$|\partial \bar{t}/\partial A|^* \text{ is an } A\text{-function.} \tag{17}$$

Proof. First note that at $A = 0$ we have

$$\hat{\Sigma}^{-1}\bar{t}' = \hat{\Sigma}^{-1}\hat{P}(\bar{t}') = (D\hat{\theta})\hat{i}_1^{-1}\hat{j}_1 = D\hat{\theta},$$

according to Lemma 1 and the fact that $\hat{i} = \hat{j}$ when $\bar{t} = \hat{\tau}$.

Since $W = 2n(\bar{\ell}(\bar{\theta}) - \bar{\ell}(\hat{\theta}))$ is assumed to be constant on level surfaces of A its derivative with respect to $\hat{\beta}$ is zero, i.e.,

$$\langle \bar{\theta} - \hat{\theta}, \bar{t}' \rangle = 0.$$

The second derivative of this equation with respect to A at 0 is the suitably symmetrized version of

$$2\left\langle \hat{\Sigma}^{-1} \frac{\partial \bar{t}}{\partial A}, \frac{\partial \bar{t}'}{\partial A} \right\rangle - \hat{\Lambda}\left(\hat{\Sigma}^{-1} \frac{\partial \bar{t}}{\partial A}, \hat{\Sigma}^{-1} \frac{\partial \bar{t}}{\partial A}, D\hat{\theta} \right) = 0, \tag{18}$$

where $\Lambda(\theta) = (\partial/\partial\theta)\Sigma(\theta)$ and we have used the notation $\hat{\Lambda}(\cdot, \cdot, \cdot)$ to denote a matrix-like multiplication of the three arguments on the three sides of the symmetric p^3-dimensional array $\hat{\Lambda}$.

This equation turns out to be what is needed to show that the derivative of $|\partial \bar{t}/\partial A|^*$ is zero on $A = 0$. To see this we simply calculate the derivative

$$\frac{\partial}{\partial \hat{\beta}} \left\{ \left(\frac{\partial \bar{t}}{\partial A}\right)^{\mathrm{T}} \hat{\Sigma}^{-1} \left(\frac{\partial \bar{t}}{\partial A}\right) \right\} = 2\left(\frac{\partial \bar{t}'}{\partial A}\right)^{\mathrm{T}} \hat{\Sigma}^{-1} \left(\frac{\partial \bar{t}}{\partial A}\right) - \hat{\Lambda}\left(\hat{\Sigma}^{-1} \frac{\partial \bar{t}}{\partial A}, \hat{\Sigma}^{-1} \frac{\partial \bar{t}}{\partial A}, D\hat{\theta} \right),$$

which vanishes according to (18). This proves (17). □

The final step in the rewriting of the density $f_\beta(\hat{\beta}, A)$ from (16) uses the standard saddlepoint approximation

$$f_{\bar{\theta}}(\bar{t}) = c_{k,n} |\Sigma(\bar{\theta})|^{-1/2} (1 + O(n^{-1})),$$

where $c_{k,n} = \{n/(2\pi)\}^{k/2}$, and the expansion holds uniformly for $\bar{t}$ in some bounded set. Some simple manipulations then give

$$f_\beta(\hat{\beta}, A) = c_{k,n} b(\bar{t}) |\hat{j}_1|^{1/2} e^{n\{\bar{\ell}(\beta) - \bar{\ell}(\hat{\beta})\}} (1 + O(n^{-1})), \tag{19}$$

and

$$b(\bar{t}) = (|\hat{\Sigma}|^{1/2} |\Sigma(\bar{\theta})|^{-1/2})(|\hat{j}_1|^{1/2} |\hat{i}_1|^{-1/2}) \left|\frac{\partial \bar{t}}{\partial A}\right|^* e^{n\{\bar{\ell}(\hat{\beta}) - \bar{\ell}(\bar{\theta})\}},$$

where we notationally allow ℓ to be a function of θ as well as of β. Note that the two factors in parentheses are A-functions, since $\bar{\theta} = \hat{\theta}$ when $\bar{t} = \hat{\tau}$. Also $|\partial \bar{t}/\partial A|^*$ has been shown to be an A-function when Property (A2) holds, and the exponential depends on $(\hat{\beta}, A)$ only through A. Thus, $b(\bar{t})$ is a function of A multiplied by an A-function. It is noteworthy that n only appears in the factor which is exactly independent of $\hat{\beta}$ – a direct consequence of Property (A2) of the supplementary statistic, A.

The marginal density of A is obtained by a Laplace-type integration over $\hat{\beta}$ using the fact that the exponent in (19) is maximal at $\hat{\beta} = \beta$. This involves some quantities defined at the point $\bar{t}_0 = \bar{t}(\beta, A)$. For fixed A, we have

$$\frac{\partial^2}{\partial \hat{\beta}^2} (\bar{\ell}(\hat{\beta}) - \bar{\ell}(\beta)) = \langle \bar{t}_0', D\theta(\beta) \rangle = n^{-1} D_1'(\beta; \bar{t}_0) = j_1(\beta; \bar{t}_0)$$

at $\hat{\beta} = \beta$. Thus, a Laplace approximation to the integral of the density in equation (19) over $\hat{\beta}$ gives

$$f_\beta(A) = c_{k-p,n} b(\bar{t}_0)(1 + O(n^{-1}))$$

uniformly in A in some bounded neighbourhood of zero. An inspection of this result shows that A, satisfying Property (A2), is a first-order ancillary statistic in the sense of having a limiting standardized distribution which is independent of β.

Division of $f_\beta(\hat{\beta}, A)$ by $f_\beta(A)$ now gives the conditional density approximation

$$\begin{aligned} f_\beta(\hat{\beta} | A) &= c_{p,n} |\hat{j}_1|^{1/2} e^{n\{\bar{\ell}(\beta) - \bar{\ell}(\hat{\beta})\}} \{b(\bar{t})/b(\bar{t}_0)\}(1 + O(n^{-1})) \\ &= c_{p,n} |\hat{j}_1|^{1/2} e^{n\{\bar{\ell}(\beta) - \bar{\ell}(\hat{\beta})\}} (1 + O(n^{-1}) + O(\|\hat{\beta} - \beta\| \, \|A\|)), \end{aligned} \tag{20}$$

where the omission of the factor $b(\bar{t})/b(\bar{t}_0)$ induces the relative error of order $O(\|\hat{\beta} - \beta\| \, \|A\|)$ because the factor is 1 if either $\hat{\beta} = \beta$ or $A = 0$. This is why the A-functions are collected in the factor $b(\bar{t})$. Notice that the factor omitted does not depend on n, because the ancillary has been chosen such that the exponentials in $b(\bar{t})$ and $b(\bar{t}_0)$ cancel exactly.

Formula (20) is identical to the simple version of the p^* formula from Barndorff-Nielsen (1980; 1983), using the general approximative constant $c_{p,n}$ instead of renormalizing the density. Also, several of the arguments used above, especially the omission of some factors in $b(\bar{t})$, may be recognized from the original proof in the 1980 paper.

9.2. TRANSFORMATIONS AND JACOBIANS

Having derived the density of $\hat{\beta}$ given A, we next transform it in two steps to the conditional density of $(\tilde{\alpha}, R)$. In Section 9.3 we get rid of $\tilde{\alpha}$, essentially by means of marginalization.

Consider the equation for the maximum likelihood estimate under the hypothesis,

$$D_1(\tilde{\beta}) - \hat{\lambda} e_p = 0, \tag{21}$$

where $\lambda \in \mathbb{R}$ is a Lagrange multiplier, and e_p is the vector $(0, \ldots, 0, 1)^{\mathrm{T}}$. The value of λ at the maximum is denoted $\hat{\lambda}$.

The following lemma gives the Jacobian for the transformation from $\hat{\beta}$ to $(\tilde{\alpha}, \hat{\lambda})$. It would be more correct to include A as a third component in the transformation, but since it is kept fixed throughout we omit it from the notation.

Lemma 4. *For any supplementary statistic, A, the Jacobian of the transformation from $\hat{\beta}$ to $(\tilde{\alpha}, \hat{\lambda})$ is*

$$\left| \frac{\partial \hat{\beta}}{\partial(\tilde{\alpha}, \hat{\lambda})} \right| = |\tilde{D}_1'|^{-1} |\tilde{j}_{\alpha\alpha}|, \tag{22}$$

and the partial derivatives are

$$\frac{\partial \hat{\beta}}{\partial \tilde{\alpha}} = (\tilde{D}_1')^{-1} (\tilde{j})_{\cdot\alpha}, \tag{23}$$

where $(\tilde{j})_{\cdot\alpha}$ denotes the first $p-1$ columns of $\tilde{j}$, and

$$\frac{\partial \hat{\beta}}{\partial \hat{\lambda}} = (\tilde{D}_1')^{-1} e_p. \tag{24}$$

Proof. Differentiation of equation (21) with respect to $\tilde{\alpha}$ for fixed $\hat{\lambda}$ gives

$$\tilde{D}_1' \left(\frac{\partial \hat{\beta}}{\partial \tilde{\alpha}} \right) + \tilde{D}_2 D\beta(\tilde{\alpha}) = 0,$$

where $\beta(\alpha)$ is the mapping $\alpha \mapsto (\alpha, \psi_0)$. This gives equation (23) since postmultiplication by the matrix $D\beta(\tilde{\alpha})$ has the effect of picking out the first $p-1$ columns of the previous matrix.

Differentiation of equation (21) with respect to $\hat{\lambda}$ for fixed $\tilde{\alpha}$ gives

$$\tilde{D}_1' \left(\frac{\partial \hat{\beta}}{\partial \hat{\lambda}} \right) - e_p = 0,$$

from which equation (24) follows. The determinant is now easily computed. □

Up to this point all the approximations and derivations can be made for multivariate hypotheses with only trivial modifications. The statistic R is, however, one-dimensional by construction and the analogue of the following transformation for multivariate hypotheses is not obvious.

We wish to transform $(\tilde{\alpha}, \hat{\lambda})$ to $(\tilde{\alpha}, R)$. Since $\tilde{\alpha}$ is unchanged we only need to work out $\partial\hat{\lambda}/\partial R$.

Lemma 5. *For any supplementary statistic, A, the relevant partial derivative of the transformation from $(\tilde{\alpha}, \hat{\lambda})$ to $(\tilde{\alpha}, R)$ is*

$$\frac{\partial\hat{\lambda}}{\partial R} = R/[(\ell'(\hat{\beta}) - \ell'(\tilde{\beta}))(\tilde{D}_1')^{-1}]_p, \tag{25}$$

where $[\cdot]_p$ denotes the last coordinate of the vector. At $\hat{\beta} = \tilde{\beta}$ the limiting value replaces the expression.

Proof. The equation defining R, apart from the sign, is

$$\tfrac{1}{2}R^2 = \ell(\hat{\beta}) - \ell(\tilde{\beta}),$$

which may be differentiated with respect to R for fixed $\tilde{\alpha}$ to give

$$R = (\ell'(\hat{\beta}) - \ell'(\tilde{\beta}))\left(\frac{\partial\hat{\beta}}{\partial\hat{\lambda}}\right)\left(\frac{\partial\hat{\lambda}}{\partial R}\right).$$

Substitution of the partial derivative from (24) now immediately gives the result. □

A combination of the results above shows that for any supplementary statistic, A, we have the relation

$$\begin{aligned} f_\beta(\tilde{\alpha}, R|A) &= f_\beta(\hat{\beta}|A)\left|\frac{\partial\hat{\beta}}{\partial(\tilde{\alpha}, \hat{\lambda})}\right|\left|\frac{\partial\hat{\lambda}}{\partial R}\right| \\ &= f_\beta(\hat{\beta}|A)|\tilde{D}_1'|^{-1}|\tilde{j}_{\alpha\alpha}|R/[(\ell'(\hat{\beta}) - \ell'(\tilde{\beta}))(\tilde{D}_1')^{-1}]_p, \end{aligned} \tag{26}$$

which is an exact transformation result on the domain where the transformation from $(\hat{\beta}, A)$ to $(\tilde{\alpha}, R, A)$ is one-to-one. This may be useful in, for example, transformation models where the p^* formula is known to apply, but the ancillary is different from here. For the special type of ancillaries used here the expressions for the sample space derivatives from Lemma 2 may be inserted to give a complete approximation to the conditional density of $(\tilde{\alpha}, R)$.

9.3. ELIMINATION OF NUISANCE PARAMETERS

As a final step, we need to eliminate the nuisance parameter α and its estimate $\tilde{\alpha}$. The two obvious ways of doing this are to condition on $\tilde{\alpha}$ and to marginalize from $(\tilde{\alpha}, R)$ to R. Both lead to the same result, to the order considered here, but they also both lead to the same technical difficulty which has to do with the fact that the parameter α does not disappear from the marginal or conditional density approximation for R.

Let us consider the marginalization and return to the modifications needed to resolve the difficulty mentioned above. Starting from the density (26) and the approximation (20) we approximate the integral over $\tilde{\alpha}$ by a Laplace approximation. The exponent

$$n\{\bar{\ell}(\beta) - \bar{\ell}(\hat{\beta})\} = n\{\bar{\ell}(\beta) - \bar{\ell}(\tilde{\beta})\} - \tfrac{1}{2}R^2,$$

considered as a function of $\tilde{\alpha}$ for fixed R, is maximal at $\tilde{\beta} = \beta$, or equivalently at $\tilde{\alpha} = \alpha$. Differentiation with respect to $\tilde{\alpha}$ yields

$$\frac{\partial}{\partial \tilde{\alpha}}\{\ell(\beta) - \ell(\tilde{\beta})\} = \{\ell'(\beta) - \ell'(\tilde{\beta})\}\left(\frac{\partial \tilde{\beta}}{\partial \tilde{\alpha}}\right),$$

and by use of equation (23) we see that minus the second derivative at the point $\tilde{\alpha} = \alpha$ is

$$(D\beta(\tilde{\alpha}))^{\mathrm{T}} \tilde{D}_1' \left(\frac{\partial \tilde{\beta}}{\partial \tilde{\alpha}}\right) = (j(\beta; \bar{t}_\alpha))_{\alpha\alpha},$$

where $\bar{t}_\alpha$ denotes the data point given by $\tilde{\alpha} = \alpha$, R and A. Also this result holds for any supplementary statistic, A.

Using the approximation (20) and the notation $\bar{R} = R/\sqrt{n}$, which is a function of $\bar{t}$ only, the Laplace integration of (26) now gives

$$f_\beta(\bar{R}|A) = \sqrt{\frac{n}{2\pi}} \mathrm{e}^{-(n/2)\bar{R}^2} g_{\tilde{\alpha}}(\bar{R}|A)(1 + O(n^{-1}) + O(\|\hat{\beta} - \tilde{\beta}\| \, \|A\|)), \tag{27}$$

where

$$\begin{aligned} g_{\tilde{\alpha}}(\bar{R}|A) &= |\hat{j}|^{1/2} |\tilde{D}_1'|^{-1} |\tilde{j}_{\alpha\alpha}|^{1/2} R / [(\ell'(\hat{\beta}) - \ell'(\tilde{\beta}))(\tilde{D}_1')^{-1}]_p \\ &= \{|\hat{j}|^{-1/2} |\tilde{j}_{\alpha\alpha}|^{1/2} |\hat{i}| \, |\hat{S}|^{-1} R / [\hat{S}^{-1}\hat{q}]_p\}(1 + O(\|\hat{\beta} - \tilde{\beta}\| \, \|A\|)), \end{aligned} \tag{28}$$

and with the modification that everything should be evaluated at the data point $\bar{t}_\alpha$. This is exactly the technical difficulty referred to above, for two reasons: first, the nuisance parameter α enters the approximation; and second, the data point $\bar{t}_\alpha$ is awkward since its determination requires specification of the ancillary.

One method of overcoming this difficulty is used by Barndorff-Nielsen (1991) and by Jensen (1992). First, we simply substitute $\tilde{\alpha}$ for α in the approximation for the density. Then we continue to derive R^*, or $\tilde{R}$, from this approximation. Having derived the expression for $\tilde{R}$, we then go back to the density of $(\tilde{\alpha}, R)$ and transform this to a density of $(\tilde{\alpha}, \tilde{R})$ and again integrate out $\tilde{\alpha}$ by Laplace's method. The resulting distribution is then shown to agree with a standard normal distribution to the order considered.

The problem with the substitution of $\tilde{\alpha}$ for α in the density approximation above is that the right-hand side then depends on $\bar{t}$ through $\tilde{\alpha}$ as well as through R and A, which is unfortunate since it is supposed to approximate the density of R given A. Thus, the approximation becomes random and difficult to formalize, which is why it is more convenient to revert to the transformation to $(\tilde{\alpha}, \tilde{R})$. We shall not go through these computations which have nothing new to say, but instead continue the derivation formally from the point where formula (27) has been proved with the data point $\bar{t}$ appearing instead of $\bar{t}_\alpha$ in this formula as well as in expression (28).

Note that $g_{\tilde{\alpha}}(0\,|\,A) = 1$ and that the n's cancel in $g_{\tilde{\alpha}}(\bar{R}\,|\,A)$ so that n only appears in (27) where it is explicitly written. For such density approximations, tail probabilities may be approximated either by the Lugannani–Rice method, which gives

$$\mathrm{pr}_\beta\{\bar{R} \geq \bar{r}\,|\,A\} \approx 1 - \Phi(\sqrt{n}\bar{r}) + \frac{\phi(\sqrt{n}\bar{r})}{\sqrt{n}\bar{r}}\{g_{\tilde{\alpha}}(\bar{r}\,|\,A) - 1\},$$

or by the R^* method, giving

$$\mathrm{pr}_\beta\{\bar{R} \geq \bar{r}\,|\,A\} \approx 1 - \Phi\left(\sqrt{n}\bar{r} - \frac{1}{\sqrt{n}\bar{r}} \log g_{\tilde{\alpha}}(\bar{r}\,|\,A)\right)$$

with a relative error of order $O(n^{-1} + \|\hat{\beta} - \tilde{\beta}\|\,\|A\|)$ in both cases. As mentioned, the unpleasant fact that $\tilde{\alpha}$ appears on the right in these expressions is avoided by a rigorous reformulation as in Barndorff-Nielsen (1991) or Jensen (1992).

Formula (2) for $\bar{R}$, or the equivalent Lugannani–Rice type approximation (3), with expression (1) for $\bar{U}$, now follows by substitution of expression (28) for $g_{\tilde{\alpha}}(\bar{R}\,|\,A)$. The relative error is as stated in Section 4, because $\|\hat{\beta} - \tilde{\beta}\| = O(\|\hat{\psi} - \psi_0\|)$.

10. Discussion

The positive side of the result of the present paper is that an explicit expression has been obtained which may be of sufficient accuracy for general use, and which may therefore serve as a replacement for the usual normal-based approximations in a number of situations. There are, however, several questions, limitations and open problems, some of which are discussed below.

First of all, the present development and result deal exclusively with one-dimensional hypotheses. Generalizations to multivariate hypotheses are definitely within reach, although it is not obvious which is the best way to proceed. One way to obtain large-deviation properties for multivariate hypotheses is to use a directional approach (see Fraser and Massam 1985; Skovgaard 1988). This approach would be based on a conditioning on the direction of the score statistic from the estimate under the hypothesis, thereby effectively reducing the problem to one dimension. Other statistics than the score statistic might be considered, however, but the convenient transformation to $\hat{\lambda}$ in Section 9, which is essentially the score statistic, makes this an obvious choice. In contrast, the maximum likelihood estimate, $\hat{\psi}$, of the parameter of interest would not lead to a parametrization-invariant result.

The elimination of nuisance parameters is included here, but it is far from obvious whether the approach is sufficiently effective to deal with a large number of nuisance parameters. Presumably this will not always be the case. There are also some technical problems in connection with the elimination of these parameters, as pointed out in Section 9, and it would be nice to have a more convincing technique for this. There do not seem to be important differences between results obtained from marginalization and conditioning on the estimates, but conditioning seems more appealing because it leads to a complete elimination when the nuisance parameters are canonical parameters in an exponential family. A study such as Pierce and Peters (1992), investigating the effects of the various correction factors, might throw some light on these problems.

As discussed in relation to the example in Section 5, the method of maximum likelihood is not always reasonable, and, in fact, breaks down in the example. For variance component models most statisticians would prefer restricted maximum likelihood, which uses a marginal likelihood to estimate the variance parameters. This method does not suggest a reasonable general way of testing hypotheses about the means, however. For transformation models in general, partition of the likelihood function into marginal and conditional parts seems intuitively correct, and an adaptation of modified likelihood methods to such structured inferences would be of great practical value.

A more technical question has to do with the properties of the approximation based on $\tilde{R}$. The asymptotic behaviour stated here has been proved, although the proof was not given in all its details, but it has not been established whether the properties might be even better. In Barndorff-Nielsen and Wood (1995) it is shown that the difference between the R^*-statistic obtained from conditioning on the modified and on the unmodified directed log-likelihood is negligible to the order considered. This suggests that $\tilde{R}$ is equally valid as an approximation to the R^* obtained from the modified ancillary, as from the unmodified as used here. However, conditional statements based on the two ancillaries are not the same. The consequences of this result in the present connection are not quite clear, however.

Acknowledgement

The author wishes to thank three referees for several helpful comments on the subject material as well as its presentation. This paper was presented at the 'Conference on likelihood, asymptotic and neo-Fisherian inference' in Brixen, June 1995, at which many lectures and discussions were related to the present research topic. It is a pleasure to thank the organizers for promoting the research in this way. Discussions with Professor Don Pierce have been particularly stimulating and have affected the final version of this paper.

References

Barndorff-Nielsen, O.E. (1980) Conditionality resolutions. *Biometrika*, **67**, 293–310.

Barndorff-Nielsen, O.E. (1983) On a formula for the distribution of the maximum likelihood estimator. *Biometrika*, **70**, 343–356.

Barndorff-Nielsen, O.E. (1986) Inference on full or partial parameters, based on the standardized signed log likelihood ratio. *Biometrika*, **73**, 307–322.

Barndorff-Nielsen, O.E. (1991) Modified signed log likelihood ratio. *Biometrika*, **78**, 557–563.

Barndorff-Nielsen, O.E. and Chamberlin, S.R. (1991) An ancillary invariant modification of the signed log likelihood ratio. *Scand. J. Statist.*, **18**, 341–352.

An explicit approximation to one-parameter tests 165

Barndorff-Nielsen, O.E. and Chamberlin, S.R. (1994) Stable and invariant adjusted directed likelihoods. *Biometrika*, **81**, 485–499.

Barndorff-Nielsen, O.E. and Cox, D.R. (1994) *Inference and Asymptotics*, London: Chapman & Hall.

Barndorff-Nielsen, O.E. and Wood, A.T.A. (1995) On large deviations and choice of ancillary for p^* and the modified directed likelihood. Research Report No. 299, Department of Theoretical Statistics, University of Aarhus.

Bleistein, N. (1966) Uniform asymptotic expansions of integrals with stationary point near algebraic singularity. *Comm. Pure Appl. Math.*, **19**, 353–370.

DiCiccio, T.J. and Martin, M.A. (1993) Simple modifications for signed roots of likelihood ratio statistics. *J. Roy. Statist. Soc. Ser. B*, **55**, 305–316.

Fraser, D.A.S. and Massam, H. (1985) Conical tests: observed levels of significance and confidence regions. *Statist. Hefte*, **26**, 1–17.

Fraser, D.A.S. and Reid, N. (1988) On conditional inference for a real parameter: a differential approach on the sample space. *Biometrika*, **75**, 251–264.

Jensen, J.L. (1992) The modified signed likelihood statistic and saddlepoint approximations. *Biometrika*, **79**, 693–703.

Lugannani, R. and Rice, S.O. (1980) Saddlepoint approximation for the distribution of the sum of independent random variables. *Adv. Appl. Probab.*, **12**, 475–490.

Pierce, D.A. and Peters, D. (1992) Practical use of higher order asymptotics for multiparameter exponential families (with discussion). *J. Roy. Statist. Soc. Ser. B*, **54**, 701–737.

Press, W.H., Flannery, B.P., Teukolsky, S.A. and Vetterling, W.T. (1986) *Numerical Recipes. The Art of Scientific Computing*. Cambridge: Cambridge University Press.

Satterthwaite, F.E. (1946) An approximate distribution of estimates of variance components. *Biometrics Bull.*, **2**, 110–114.

Skovgaard, I.M. (1988) Saddlepoint expansions for directional test probabilities. *J. Roy. Statist. Soc. Ser. B*, **50**, 269–280.

Skovgaard, I.M. (1990) *Analytic Statistical Models*. Lecture Notes, Monograph Ser. 15. Hayward, CA: Institute of Mathematical Statistics.

Sweeting, T.J. (1995) A framework for Bayesian and likelihood approximations in statistics. *Biometrika*, **82**, 1–23.

Tjur, T. (1974) Conditional probability distributions. Lecture Notes 2. Institute of Mathematical Statistics, University of Copenhagen.

Received July 1995 and revised January 1996

Chapter 10

Modified residuals in non-linear regression

Introduction by Anthony Davison
Ecole Polytechnique Fédérale de Lausanne

This hidden gem is typical of Ib's work: starting from first principles it develops an elegant, accurate and easily implemented approximation. It deserves to be much better known and appreciated.

Although the examination of residuals mounts to the antiquity of what we now call regression analysis, the need for tools for model criticism became more pressing with the arrival of statistical computing and rudimentary computer graphics. One consequence of this was a boom of research on *regression diagnostics* in the 1970s and 1980s, partly stimulated by Cook's 1977 formal definition of the influence of an observation on a regression model. Residuals of different sorts crop up everywhere when using these diagnostics, both in the construction of formal test statistics and in the accompanying plotting procedures. Initially research was focused on the classical linear model, but the rapid spread of generalised linear models and normal nonlinear regression models, rapidly led to the development of diagnostics for these broader classes. There is a menagerie of definitions for generalised linear models: Anscombe residuals, deviance residuals, Pearson residuals, Williams residuals, all standardized or not, but fortunately the situation is simpler for Gaussian nonlinear regression, with which this paper is concerned.

In a pioneering paper, Cox and Snell (1968) had proposed a general definition of residuals for potentially nonlinear regression models with general error distributions, and had obtained corrected means and covariances that account for their mutual dependence. Such corrections can be substantial in small samples, and it is particularly important to account for them when the residuals are used to construct test statistics (Anscombe, 1961; Cox and Snell, 1971). Later authors (e.g., Pierce and Schafer, 1986; McCullagh, 1987; Davison and Snell, 1991) made related suggestions, usually amounting to centering and possibly scaling quantities that would have zero mean and unit variance, were it not for the dependence among them introduced by the fitting procedure. The approach taken in this paper is more elegant and should be more accurate.

The starting-point is the principle that inference on the fit of a model should be based on an ancillary statistic, whose distribution will by definition be invariant to the parameter values. In the classical linear model, in which the $n \times 1$ response vector y has a Gaussian distribution with mean vector $X\beta$ and covariance matrix $\sigma^2 I_n$, where X is a $n \times p$ matrix of rank $p < n$ consisting of explanatory variables, β is a $p \times 1$ parameter vector and σ^2 the response variance, this suggests that model-checking should be based on the raw residuals

$$y - X\hat{\beta} = (I_n - H)y,$$

where $\hat{\beta}$ is the maximum likelihood, least squares, estimate of β and $H = X(X^{\mathrm{T}}X)^{-1}X^{\mathrm{T}}$ is the so-called hat matrix. However the raw residuals may have very different variances, so it is better to use the Studentized or the deletion residuals,

$$r_j = \frac{y_j - x_j^{\mathrm{T}}\hat{\beta}}{s(1 - h_{jj})^{1/2}}, \quad t_j = r_j \left(\frac{n - p - 1}{n - p - r_j^2} \right)^{1/2}, \quad j = 1, \ldots, n,$$

where y_j and x_j^{T} are the jth rows of y and X, s^2 is the unbiased estimate of σ^2, and h_{jj} is the jth diagonal element of H. An advantage of the deletion residuals is that, if the model assumptions are met, they have exact t_{n-p-1} distributions, and thus can be transformed to exact normality, though of course they are dependent. In the corresponding nonlinear model, the mean $x_j^{\mathrm{T}}\beta$ of y_j is replaced by a nonlinear function $\mu_j(\beta)$, and the raw residual $y_j - \mu_j(\hat{\beta})$ does not have an exact normal distribution, so although the same transformation can be applied, the transformed deletion residuals z_j do not satisfy exact normality. Nevertheless it is desirable that if either $\mu(\beta)$ approaches linearity in β, or if $n \to \infty$ with p fixed, the corresponding exact results should appear, and these are the desiderata that the paper sets out to meet.

The basic idea is to correct the z_j to allow for nonlinearity, and the paper does this by a sequence of ingenious steps that modify the corresponding exact computations for the linear model. In the first step an essentially exact approximation to the joint density of the score statistic and an ancillary is integrated to provide the marginal density of the latter, and this marginal density is approximated using Laplace's method. The resulting expression, which applies in wide generality to estimators defined via estimating functions, is then specialised to the nonlinear normal model by taking as the *approximate ancillary* the unit vector $A(\beta)$ given by

$$A(\beta) = \frac{\{I_n - H(\beta)\}\{y - \mu(\beta)\}}{\|\{I_n - H(\beta)\}\{y - \mu(\beta)\}\|}, \quad H(\beta) = X(\beta)\{X(\beta)^{\mathsf{T}}X(\beta)\}^{-1}X(\beta)^{\mathsf{T}},$$

with $X(\beta)$ the $n \times p$ matrix of derivatives with (j, r) element $\partial\mu_j(\beta)/\partial\beta_r$. For the normal linear model, $H(\beta) \equiv H$ is constant and the marginal distribution of $A(\beta) \equiv A$ is uniform on the unit sphere in the column space of $I - H$, so neither the unit vector nor its distribution depend on β. In the nonlinear case, the unit vector and its distribution depend on β, though to an extent that depends on the nonlinearity of the model, and may be hoped to be mild in many cases.

The second step of the development is to compare the Laplace approximation in the nonlinear case with that in the linear case, leading in the notation of the paper to an approximation,

$$f_\omega(\hat{A} = a) \doteq \left\{\frac{|\jmath(\omega; a)|}{|\imath(\omega)|}\right\}^{1/2} \times f_{\text{lin}}(\hat{A} = a),$$

to the density of the approximate ancillary $\hat{A} = A(\hat{\omega})$, where $\hat{\omega}$ is the maximum likelihood estimator of the parameter $\omega = (\beta, \sigma)$, $\imath(\omega)$ is the expected information matrix, $\jmath(\omega; a)$ is the observed information matrix evaluated at $\hat{\omega}$, and f_{lin} is the uniform density of the ancillary statistic in the linear case.

The third step is to associate A_1 with a particular residual, setting $r_j = A_1(n-p)^{1/2}$ as in the linear case, and then to integrate out $A_2, \ldots, A_n$ to obtain an approximate marginal density for r_j, yielding

$$f_\omega(r_j = x) \propto \left\{1 - x^2/(n-p)\right\}^{(n-p-3)/2} \times \mathrm{E}_{\text{lin}}\left\{ |\imath(\omega)^{-1}\jmath(\omega; a)|^{1/2} \Big| A_1 \right\};$$

the first term on the right-hand side of this expression is the exact density of the Studentized residual in the linear case.

The fourth step is a delta method approximation to the expectation, which yields a quantity $\rho_j(\beta)$ given in the paper; this is identically unity for a linear model, and otherwise depends on the curvature of the model manifold. In the paper 1_p denotes the identity matrix of size p. The factor $\rho_j(\beta)$ is key, since it appears in the improved, 'Skovgaard', residual

$$z_j^* = z_j - z_j^{-1} \log \rho_j(\hat{\beta}). \quad (10.1)$$

Table 10.1 Residuals for nonlinear least squares fit to carrot top data (Nelder, 1961).

x	y	Residual					ρ
		Raw	Studentized, r	Deletion, t	Normal, z	Skovgaard, z^*	
−2.15	1.272	0.859	1.655	2.202	1.855	1.844	1.020
−1.50	1.832	−0.217	−0.253	−0.223	−0.215	−0.208	1.001
−0.85	2.255	−1.475	−1.659	−2.213	−1.862	−1.856	1.012
−0.08	2.828	−0.616	−0.704	−0.657	−0.624	−0.620	1.003
0.52	3.199	0.068	0.078	0.068	0.066	0.065	1.000
1.10	3.520	0.897	1.007	1.009	0.941	0.936	1.004
2.28	3.912	0.949	1.027	1.035	0.964	0.956	1.008
3.23	4.128	0.879	0.954	0.942	0.882	0.876	1.005
4.00	4.239	0.575	0.645	0.595	0.568	0.566	1.001
4.65	4.206	−0.905	−1.070	−1.092	−1.014	−1.009	1.004
5.00	4.239	−1.014	−1.245	−1.351	−1.230	−1.222	1.010

Expression (10.1) results from the fifth and final step of the argument, which is based on the observation that in the integral approximation

$$\left(\frac{n}{2\pi}\right)^{1/2}\int_{\infty}^{\tilde{x}} a(x)e^{-nh(x)}\left\{1+O(n^{-1})\right\}\,dx \doteq \Phi\{r^*(\tilde{x})\}\left\{1+O(n^{-1})\right\},$$

where Φ is the standard normal integral and

$$r^*(x)=n^{1/2}r(x)-\frac{1}{n^{1/2}r(x)}\log\left\{\frac{a(x)r(x)}{h'(x)}\right\},\quad r(x)=\pm\{2h(x)\}^{1/2},$$

changing $a(\cdot)$ alters only the second term of $r^*(x)$. This fact can be used to assess the sensitivity of corresponding Bayesian computations to the choice of prior, but here the clever twist is to note that if $h(x)$ corresponds to a density for which the integral is exact with $a(x)\equiv 1$, then we can expect unusually good approximations also when $a(x)\doteq 1$ or equivalently here $\rho_j(\hat{\beta})\doteq 1$, as will be the case for nearly-linear models.

The paper contains no numerical work, so to illustrate the correction, we consider data shown in the first two columns of Table 10.1 on the growth of carrot tops in a field experiment. A simple three-parameter nonlinear regression model with means $\beta_1+\beta_2\exp(\beta_3 x_j)$ is readily fitted to these data, yielding estimates $\hat{\beta}_1=4.68_{0.12}$, $\hat{\beta}_2=-1.76_{0.15}$, $\hat{\beta}_3=-0.318_{0.029}$, where the standard errors are given as subscripts, and $s=0.0844$. Here $n=11$ and with $p=3$ parameters there is a strong effect of deletion for several of the observations, which is reduced by transforming the residuals to the normal scale, giving the z_j. The pattern of signs of the residuals suggest that the model may not fit very well, and indeed a different model is fitted in the original paper.

The quantities $\rho_j(\hat{\beta})$ are readily computed, and in this case are very close to unity, so the Skovgaard residuals z_j^* differ little from the z_j, perhaps because only one parameter enters the model non-linearly. All the ρ_j exceed unity, so $|z_j^*| < |z_j|$, perhaps due to the slight lack of fit. The main correction stems from the transformation to the normal scale, rather than from the higher-order effect of including the $\rho_j(\hat{\beta})$, though these are so easily programmed that they could be used essentially automatically.

This is a beautiful little paper. More study of the Skovgaard residual, and especially more comprehensive numerical work, seems long overdue.

Acknowledgement

The work was supported by the Swiss National Science Foundation.

Modified residuals in non-linear regression

Ib Skovgaard,
The Royal Veterinary and Agricultural University, Denmark

Published in
Proceedings of a Symposium in Honour of Ole E. Barndorff-Nielsen
Memoirs, 16, Univ. Aarhus, Aarhus, 2000 96 – 101.

Introduction

In linear normal models theory of residuals and their distribution is well established. In non-linear models one usually relies on asymptotic approximations, essentially approximating the non-linear model with its linear counterpart which is tangential to the non-linear model at the estimate. The resulting asymptotic approximations are usually of first order, although in some sense better because they also approach exactness as the non-linear model approaches the linear. However, approximations for residuals have yet only to a limited extent drawn upon the more recent high precision asymptotic approximations, which have been developed for estimators and tests, in particular through the foundational work by Barndorff-Nielsen; see Barndorff-Nielsen (1980, 1986, 1991). For reviews of and contributions to residuals and their distributions see Davison & Snell (1991) and Pierce & Schafer (1986).

The aim here is to obtain approximations for residuals in non-linear regression with properties in line with Barndorff-Nielsen's p^*-formula and r^*-statistic. Thus, we want a relative error tending to zero in some large deviation region as the number of observations tend to infinity, despite the limited information on the single residual; in the terminology of Pierce & Schafer (1986) we thus consider n-asymptotics. Furthermore we want to keep the property of exactness for the linear model so that the approximation becomes "double asymptotic" in the sense that it holds either for the number of observations tending to infinity or for the model approaching linearity. Finally, it turns out that these properties can be obtained without using deletion residuals, so that only estimation with the full data set is necessary. After having obtained the estimate the expressions are explicit.

The main result

We consider a model of independent observations $y_1, \ldots, y_n$ which are normally distributed with means $\mu_1(\beta), \ldots, \mu_n(\beta)$ and variance σ^2. Here the unknown parameters are $\beta \in \mathbf{R}^p$ and $\sigma^2 > 0$. The μ_is are known smooth mappings, which are linear for the linear normal model. The full parameter vector is $\omega = (\beta, \sigma^2)$ and its maximum likelihood estimate is denoted $\hat{\omega} = (\hat{\beta}, \hat{\sigma}^2)$. We let $\ell(\omega) = \ell(\omega; \hat{\omega}, A)$ denote the log-likelihood function when $(\hat{\omega}, A)$ is a sufficient statistic. Similarly $j(\omega; \hat{\omega}, A)$ is the observed Fisher information and $i(\omega)$ is the expected Fisher information.

Let $r_i = (y_i - \hat{\mu}_i)/(s\sqrt{1 - \hat{h}_{ii}})$ denote the Studentized residual, where $s^2 = \hat{\sigma}^2 n/(n-p)$ is the usual estimate of the variance and $\hat{h}_{ii}$ is the ith diagonal element of the hat matrix, $\hat{H}$, for the approximating linear model, that is $\hat{H} = X(X^TX)^{-1}X^T$ where X is the $n \times p$ matrix of partial derivatives of μ with respect to β at $\hat{\beta}$. In the linear case the deletion residual,

$$t_i = r_i \sqrt{\frac{n-p-1}{n-p-r_i^2}},$$

is distributed exactly as a t-distribution with $n-p-1$ degrees of freedom. Let F denote the distribution function of this distribution and define the normal deviate for the residual as

$$z_i = \Phi^{-1}(F(t_i)),$$

which is normally distributed in the linear case, of course.

The modified residual for the non-linear model is now defined as

$$z_i^* = z_i - \frac{1}{z_i} \log \hat{\rho}_i,$$

where $\hat{\rho}_i$ is given further ahead. Intuitively it is the contribution to the determinant ratio, $(|\hat{\jmath}|/|\hat{\imath}|)^{1/2}$, between the observed and expected information matrices, stemming from the ith residual.

The remaining part of this note explains the derivation of the result, which is based on the almost exact distribution of the ancillary statistic together with a method of perturbing a known distribution (from the linear model) to an approximately similar distribution (from the non-linear model).

The ancillary and distributions for the linear case

For a linear model the estimate of the mean vector, $\hat{\mu} = \mu(\hat{\beta})$ is simply the projection, Py say, of the observation vector onto the space, L say, spanned by the derivatives of μ with respect to β. Furthermore $y - \hat{\mu} = y - Py = P^{\perp}y$, and the residual sum of squares, RSS $= n\hat{\sigma}^2$, is the squared length of this vector. The unit vector $A = P^{\perp}y/||P^{\perp}y||$ is an ancillary statistic and its conditional distribution given $\hat{\omega}$ as well as its marginal distribution is a uniform distribution on the unit sphere in $L^{\perp}$.

Consider now the residual corresponding to y_i, to be specific let us say y_1. Without altering the distribution we may rotate A so that we may write the Studentized residual, r_1, as

$$r_1 = A_1 \sqrt{n-p},$$

where A_1 is the first coordinate of A with respect to a unit coordinate system in $L^{\perp}$. The distribution of A_1 can be found directly by integration and it is useful for the sequel to note that its density is

$$f(A_1 = x) = \text{const} \cdot (1 - x^2)^{(n-p-3)/2}$$

from which the t-distribution of the deletion residual, t_1, may be derived. The expression above shows that the density is of Laplace type (see below) as n tends to infinity.

Distribution of the ancillary statistic

It seems intuitively reasonable that model checking in general should be based on the ancillary information in the data relative to the model we are using. In fact this is an obvious requirement if we want to end up with a statistic which we can assess with a distribution we know, implying that the distribution does not depend on the parameter. Thus we derive our approximation for the residual starting from the distribution of a more general ancillary statistic.

For each fixed ω let $D_1(\omega)$ denote the score statistic and let $A(\omega)$ denote a supplementary statistic such that $(D_1(\omega), A(\omega))$ is sufficient and has a smooth density with respect to a Lebesgue measure, possibly transferred to a differential geometric surface like a sphere in the present case. The ancillary statistic is of the form $\hat{A} = A(\hat{\omega})$. For such a statistic, not necessarily ancillary, the density of $(\hat{\omega}, \hat{A})$ may be shown to be

$$f_{\omega}(\hat{\omega} = v, \hat{A} = a) = |\hat{\jmath}|\, e^{\ell(\omega)-\ell(\hat{\omega})} f_v(D_1(v) = 0, A(v) = a),$$

where we use $f(X = x)$ as general notation for the density of X at x and the likelihood functions are calculated at the values (v, a) of the sufficient statistic $(\hat{\omega}, \hat{A})$. By $\hat{\jmath}$ and $\ell(\hat{\omega})$ we understand the observed Fisher information and the log-likelihood calculated at the parameter value $\hat{\omega} = \hat{\omega}(v, a)$ corresponding to the sufficient statistic. This result which is essentially exact was given without proof or regularity conditions in Skovgaard (2000). Some smoothness and moment conditions are required and by "essentially exact" is meant that it is exact if we can ignore the problem of multiple or no solutions to the maximum likelihood equation; in reality this usually implies an exponentially small error as for the related result in Skovgaard (1985) and in Jensen & Wood (1998). A precise version of the result is available from the author and will be published elsewhere.

The next step is to marginalize to the distribution of $\hat{A}$ using a Laplace approximation to the integral over $\hat{\omega}$. Thus, in the integral

$$\int |\hat{\jmath}(v,a)|\, e^{\ell(\omega;v,a)-\ell(\hat{\omega};v,a)} f_v(D_1(v)=0, A(v)=a)\, dv$$

the exponent has a maximum at $v=\omega$ and its second differential at that point is $j(\omega;\omega,a)$. Thus a Laplace approximation gives

$$f_\omega(\hat{A}=a) \sim c_{p+1}^{-1} |\hat{\jmath}(\omega,a)|^{1/2} f_\omega(D_1(\omega)=0, A(\omega)=a),$$

where $c_k=(2\pi)^{-k/2}$. The relative error of this approximation is $O(n^{-1})$ under reasonable regularity conditions.

As the (approximately) ancillary statistic, $A(\omega)$, for the non-linear regression we use essentially the same as in the linear case. Since the space, L_β, spanned by the derivatives of μ with respect to β depends on β we need, however, to transform the unit vector

$$P_\beta^\perp(y-\mu(\beta))/\|P_\beta^\perp(y-\mu(\beta))\|$$

from the $(n-p)$-dimensional subspace $L_\beta^\perp$ to $\mathbf{R}^{n-p}$. This is just a matter of defining a smoothly changing coordinate system in $L_\beta^\perp$ which for the purpose of investigating the residual corresponding to y_1 should keep its first basis vector in the direction corresponding to this residual, such that $r_1 = A_1\sqrt{n-p}$ as in the linear case. This allows us to assume that the first Studentized residual is a function of $\hat{A}$. A similar construction can be assumed to derive the distribution of the other residuals. The resulting statistic $\hat{A}$ can be shown to be a first order ancillary statistic under usual regularity conditions, in particular that all eigenvalues of the information matrix grow at the rate of n.

Now one of the main points of the present development is to match the distribution from the non-linear model to that from the linear model. So consider the linear model which is tangential to the non-linear model at ω. In the linear case the observed and expected Fisher information matrices are identical so the derivation of the ancillary distribution above gives $|i(\omega)|$ instead of $|\hat{\jmath}(\omega,a)|$, but otherwise the same. Thus we see that for the non-linear case

$$f_\omega(\hat{A}=a) \sim \left(\frac{|\hat{\jmath}(\omega,a)|}{|i(\omega)|}\right)^{1/2} f_{\text{lin}}(\hat{A}=a),$$

where f_{lin} denotes the density from the linear case, that is the uniform distribution on the unit circle.

To obtain the distribution of A_1 for the non-linear case we have to marginalize with respect to the remaining coordinates of A, that is to integrate round a sphere of radius $\sqrt{1-A_1^2}$. This gives the density from the linear case multiplied by the factor

$$E_{\text{lin}}\left\{(|\hat{\jmath}(\omega,a)|/|i(\omega)|)^{1/2} \,\middle|\, A_1\right\}.$$

The delta method, effectively plugging in zeroes for the remaining coordinates of A because of symmetry, gives

$$\rho_1(\beta) = \left|1_p - M(\beta)^{-1}N_1(\beta)sr_1/\sqrt{1-h_{11}}\right|^{1/2}$$

as an approximation to this factor, where $M(\beta)$ is the $p \times p$ matrix with (s,t) coordinate $(\partial\mu/\partial\beta_s)(\partial\mu/\partial\beta_t)$, and $N_1(\beta)$ is a $p \times p$ matrix with (s,t) coordinate

$$\sum_{i=1}^{n} \frac{\partial^2\mu_i}{\partial\beta_s\partial\beta_t} q_{1i}$$

with q_{ij} as the coordinate of the orthogonal projection matrix $P_\beta^\perp = 1_p - H(\beta)$, where $H(\beta)$ is the hat matrix at β.

Modifying a tail probability approximation

Consider a sequence of densities on $\mathbf{R}$ admitting a Laplace-type approximation of the form

$$f_n(x) = \sqrt{\frac{n}{2\pi}}\, a(x)e^{-nh(x)}(1+O(n^{-1}))$$

where a and h are smooth functions, and h has a unique minimum at x_0, say, at which point $h(0) = 0$, $h'(0) = 0$ and $a(x_0) = \sqrt{h''(x_0)}$. For reviews of such approximations see Jing & Robinson (1994) and Skovgaard (2000). One of the features of this class of approximations is that it admits highly accurate tail probability approximations, either by use of the Lugannani & Rice formula (Lugannani & Rice, 1980), or by use of the r^*-type normal deviate; see Barndorff-Nielsen (1986, 1991). The latter gives the right tail probability from x as the normal tail probability $1 - \Phi(r^*(x))$, where

$$r^*(x) = \sqrt{n}\, r(x) - \frac{1}{\sqrt{n}\, r(x)} \log\left(\frac{a(x)r(x)}{h'(x)}\right)$$

and $r(x) = \pm\sqrt{2h(x)}$, the sign matching that of $x - x_0$.

Now suppose that g_n is another sequence of densities admitting the same kind of Laplace approximations except that the function $a(x)$ is replaced by another function, $b(x)$ say. Then the tail probability from x for the sequence g_n may be approximated as above, and we see that the two normal deviates, r_f^* and r_g^* say, satisfy the relation

$$r_g^*(x) = r_f^*(x) - \frac{1}{\sqrt{n}\, r(x)} \log\left(\frac{b(x)}{a(x)}\right).$$

A point here is that if we know r_f^* we may obtain the analogue for g_n without using $h'(x)$ which may be awkward to calculate. But more essentially, instead of modifying r_f^* we may use the exact normal deviate from the sequence f_n if these distributions can be calculated exactly. This is just the case in our application where we know the distribution of the residual for the linear case and want to modify it to the non-linear case, and the approach has the advantage of giving the correct limit as the non-linear model approaches the linear.

Thus, to obtain the normal deviate for a residual from the non-linear regression we do as follows. First calculate the tail probability using the t-distribution from the approximating (tangential) linear model, then convert it to a normal deviate, r^*. To arrive at the final normal deviate for the residual subtract

$$\frac{1}{r^*(x)} \log\left(\frac{b(x)}{a(x)}\right)$$

where $b(x)/a(x)$ is the modifying factor for the non-linear residual density relative to the linear case. The $\sqrt{n}$ is inherent in the r^*, and it does not change the order of approximation to use r^* instead of r in the correction term. (In fact symmetry of the residual distribution assures that this replacement keeps the singularity at $r = 0$ removable, although still numerically unpleasant).

Thus $b(x)/a(x)$ is a modifying factor which in our application for the density of a residual from a non-linear model relative to its linear approximating model is $\rho_1(\beta)$ given in the previous section. As a final approximation we need in practice to calculate it at the estimate $\hat{\omega}$. Because A is ancillary to first order this imposes a relative error of order $O(n^{-1/2}\|\hat{\beta} - \beta\|)$. Expressed in terms of the unmodified residual rather than r_1, we get

$$\hat{\rho}_1 = \left| 1_p - M(\hat{\beta})^{-1} N_1(\hat{\beta})(y_1 - \hat{\mu}_1)/(1 - h_{11}) \right|^{1/2}.$$

Some further remarks

Parts of the development given here are not linked to the class of non-linear normal models. This is true, in particular, for the argument leading to the distribution of the ancillary statistic, but also for the method of adjusting a tail probability using the modifying factor.

The result obtained for the marginal distribution of the supplementary statistic $\hat{A}$ did not require the statistic to be ancillary. But using the result also for ω at $\hat{\omega}$ a simple division gives the conditional density

$$f_\omega(\hat{\omega} = v|\hat{A} = a) \sim c_{p+1}|\hat{\jmath}|^{1/2}e^{\ell(\omega)-\ell(\hat{\omega})}f_v(\hat{A} = a)/f_\omega(\hat{A} = a)$$

so the p^*-formula drops out if A is ancillary. This is a more direct and easier version of the argument given in Skovgaard (1990), and it links the p^*-formula solely to the property if A being ancillary, beside regularity conditions.

References

Barndorff-Nielsen, O.E. (1980). Conditionality resolutions. *Biometrika* **67**, 293–310.

Barndorff-Nielsen, O.E. (1986). Inference on full or partial parameters, based on the standardized signed log likelihood ratio. *Biometrika* **73**, 307–322.

Barndorff-Nielsen, O.E. (1991). Modified signed log likelihood ratio. *Biometrika* **78**, 577–563.

Davison, A.C. & Snell, E.J. (1991). Residuals and diagnostics. In **Statistical theory and modelling. In honour of Sir David Cox**. Eds: D.V. Hinkley, N. Reid and E.J. Snell, Chapman & Hall, London, pp 83–106.

Jensen, J.L. & Wood, A.T.A. (1998). Large deviation and other results for minimum contrast estimators. *Ann. Inst. Statist. Math.*, 673–695.

Jing, B. & Robinson, J. (1994). Saddlepoint approximations for marginal and conditional probabilities of transformed variables. *Ann. Statist.* **22**, 1115–1132.

Lugannani, R. & Rice, S.O. (1980). Saddlepoint approximation for the distribution of the sum of independent random variables. *Adv. in Appl. Probab.* **12**, 475–490.

Pierce, D.A. & Schafer, D.W. (1986). Residuals in generalized linear models. *J. Amer. Statist. Assoc.* **81**, 977–986.

Skovgaard, I.M. (1985). Large deviation approximations for maximum likelihood estimators. *Probab. Math. Statist.* **6**, 89–107.

Chapter 11

Likelihood asymptotics

Introduction by Ruggero Bellio and Donald A. Pierce

University of Udine, Oregon State University

This paper is based on a series of lectures at the 17th Nordic Conference on Mathematical Statistics, held in Helsingør, Denmark, in July 1998. One of us (R.B.) was fortunate enough to be among the audience, and could appreciate the brilliant overview of likelihood asymptotic theory illustrated by Ib in those lectures.

The paper is primarily a review of important developments in the theory and methods of statistics, largely initiated in the 1980s by O.E. Barndorff-Nielsen. Ib played a major role in these developments; see, for example, Skovgaard (1985b) and Skovgaard (1996), to which we refer below.

The likelihood asymptotics in mind are mainly what are also called *neo-Fisherian* inference, see Pace and Salvan (1997), which complements the Neyman–Pearson theory of optimality, that was in terms of power maximization, by relying totally on sufficiency. For this approach to be broadly applicable, it is necessary to condition on a notional approximate ancillary statistic a intended to render the maximum likelihood estimator (MLE) as conditionally approximately sufficient while sacrificing little information. Though not a necessary restriction, the reader can for clarity consider this ancillary to be the ratio of observed to expected Fisher information, as studied by Efron and Hinkley (1978) and by Skovgaard (1985b).

This conditional sufficiency approach leads fairly simply to the likelihood ratio approximation to the conditional distribution of the MLE, given below. This is commonly referred to as the p^* formula. As can happen, what was a saga of discovery, involving many workers around 1980, becomes much simpler in retrospect.

With this approach to conditional sufficiency, clarified in Skovgaard (1985b), it is possible to use a relatively simple argument proposed in Durbin (1980) to see that the ancillary-conditional distribution of the MLE of the full model parameter ω is to excellent approximation given by the likelihood ratio approximation

$$p^*(\hat{\omega} \mid a; \omega) = \frac{|\jmath(\hat{\omega})|^{1/2}}{(2\,\pi)^{p/2}} \frac{L(\omega; y)}{L(\hat{\omega}; y)} = p(\hat{\omega} \mid a; \omega) \left\{1 + O(n^{-1})\right\}.$$

It is in principle useful, then, as proposed by Barndorff-Nielsen (1986), to transform from this highly accurate ancillary-conditional distribution of the MLE to that of the directed signed root r of the likelihood ratio statistic for testing an hypothesis on a scalar interest parameter $\psi(\omega)$. This raises the quantity that is central to theory, Barndorff-Nielsen's r^*, which is a modification of r that has a standard normal distribution to third order. It is useful, as in some developments, to emphasize instead the more inferentially direct matter that highly accurate tail probabilities for the conditional distribution of r can be expressed in terms of $\Phi(r^*)$. As emphasized by the title of Skovgaard (1996), inferences based on r^* have protection for large deviations, which is more important than whether the approximations are second or third order in powers of $n^{-1/2}$.

In §§5.4-5.5 of this paper is illustrated a notable strength of how Ib generally tends to work. He shows how to argue this result for r^* without relying on the Barndorff-Nielsen argument, starting from the beginning. His approach involves the computation of tail probabilities based on Laplace-type approximations to densities, a class of functions which include the p^* formula. The more standard way is indicated in §6.6 of Barndorff-Nielsen and Cox (1994), and in §7.3 and 7.4 of Severini (2000).

The Barndorff-Nielsen r^* theory remained largely only theoretical, until the breakthrough in Skovgaard (1996). What was lacking in the Barndorff-Nielsen development of the distribution of r was a way to compute the Jacobians required for transformation of the conditional distribution of $\hat{\omega}$ to the corresponding distribution of r, which must be partial derivatives holding fixed the ancillary. Part of the difficulty in this was that the ancillary to be conditioned upon is largely a notional concept. The Skovgaard approximation to these Jacobians was developed specifically to meet these difficulties. The details of Skovgaard approximation are given in §§7 and 8 of Skovgaard (1996). The following is aimed at clarifying that line of thought, since the Ib argument involves quite sophisticated thinking of the geometry of second-order ancillarity; see also the introduction to Skovgaard (1996) by T.A. Severini in this volume.

The approach to this was suggested in Skovgaard (1985b). There he indicated in relation to Figure 1 of that paper that for asymptotic ancillaries the spaces of fixed ancillary are asymptotically parallel to one another in terms of orthogonality in the information-based inner product. This is clearest in the context of curved exponential families but generalizes widely. It seems clear that Ib had in mind asymptotic ancillaries with the character discussed in his 1985 paper that satisfy the "parallel subspaces" for fixed ancillary spaces considered in that paper, as opposed to more specialized ones. The point of that Figure 1, and the further development in that paper, is to indicate why the term $\hat{\imath}^{-1}\,\hat{\jmath}$ in (11.1) and (11.2) below quantifies the effect of the ancillary conditioning.

The type of partial derivatives, holding the ancillary fixed, are given by the left sides of the relations below, with the Ib approximations being given by the right sides

$$\frac{\partial^2 l(\tilde{\omega};\hat{\omega},a)}{\partial\omega\partial\hat{\omega}} \doteq \mathrm{cov}_{\hat{\omega}}\left\{D_1(\tilde{\omega}),D_1(\hat{\omega})\right\}\hat{\imath}^{-1}\hat{\jmath}\,, \tag{11.1}$$

$$\frac{\partial\{l(\hat{\omega};\hat{\omega},a)-l(\tilde{\omega};\hat{\omega},a)\}}{\partial\hat{\omega}} \doteq \mathrm{cov}_{\hat{\omega}}\left\{l(\hat{\omega})-l(\tilde{\omega}),D_1(\hat{\omega})\right\}\hat{\imath}^{-1}\hat{\jmath}\,. \tag{11.2}$$

Here ν is the nuisance parameter with MLE $\hat{\nu}$ and constrained MLE $\hat{\nu}_\psi$, and $\tilde{\omega}=(\psi,\hat{\nu}_\psi)$ is the constrained MLE of ω. Note that (11.1) and (11.2) involve derivatives with respect to parameter estimates, which are referred to as sample-space derivatives. As noted above, the sample-space derivative must be computed while holding an ancillary statistic fixed, which will be approximate and largely notional.

To clarify his approximations we focus on the first of these, eq. (11.1). The quantities $D_1(\cdot)$ are ordinary score statistics, and the parameter estimates where they are evaluated are considered as fixed values when computing the covariances. Note that when $\tilde{\omega}=\hat{\omega}$, the left side of (11.1) is equal to $\hat{\jmath}$. So the aim is to approximate it for nearby values of $\tilde{\omega}$, that is, conceptually in terms of corrections based on $(\tilde{\omega}-\hat{\omega})$. The key to understanding the basis for the approximation is to see that (11.1) is an orthogonal projection. Terms preceding the final $\hat{\jmath}$ are, in some disguise, the familiar ones for the orthogonal projection in a space with information-based inner product. This is discussed in §5.6 of the paper, but perhaps more clearly near (9), (10) of Skovgaard (1996). Grasping this "projection onto D_1" terminology requires seeing that this means onto the space of D_1-values as the data vary.

In §3 of the current paper there is particularly useful attention to the questions: "What has been achieved?", "Is it necessary?", "How good is it?", and "How complicated is it to use?". This is noteworthy, demonstrating the attention of Ib not only mathematical correctness and elegance, but also to practical usefulness of the methodology. While we surely share such attention, we note that part of the underlying issues there is that obtaining another decimal point or two in p-values is not one of the most important gains from these developments. Cox, in his discussion of Pierce and Peters (1992), has with characteristic insight itemized six desiderata for higher-order results, with distributional accuracy falling last in this list. First on his list is that the methods be "reasonably readily computed" for general models, and the `R` package mentioned below bears on that matter. It seems useful to note that, particularly outside the U.S.A., there is considerable dissatisfaction with the power-maximization approach of the Neyman–Pearson theory to isolating optimal inferences. See, for example, the discussion section of Reid (2005). The neo-Fisherian aspect of higher-order asymptotics provides a useful alternative, or complement, to that Neyman–Pearson approach, and for this purpose the developments reviewed by Ib are important.

In regard to Ib's comments on how complicated is it to use, we point to our R (R Core Team, 2016) package `LikelihoodAsy`, available on the `CRAN` repository. The software accompanies Pierce and Bellio (2016) which aims at providing a simpler view of the modern developments, intended for a wider audience, and to introduce that `R` package. The software can be used to compute the r^* statistic for a given scalar function $\psi(\omega)$ of the model parameters, requiring the user to supply only `R` code for the likelihood. Additionally required is a function to simulate a data set from the model, in order to specify the extra-likelihood model specification required for going beyond first-order approximations. Using the latter model specification, the covariances (11.1) and (11.2) can be readily approximated by Monte Carlo computation.

In addition to the review of likelihood asymptotic theory, this paper gives a novel development for the extension to testing an hypothesis on a multidimensional parameter. The resulting test statistic is called w^*, the name indicating that it is a direct extension of the r^* statistic. Indeed, although w^* is derived by replicating the same steps followed for r^*, the multidimensional setting presents some additional challenges, starting from the fact that "it is not even obvious what corresponds to a tail probability in the multidimensional case", to use Ib's words. A delicate step is the transformation of $\hat{\omega}$ to a vector of squared length w, a task for which there may be several alternatives. It is not surprising that the w^* method does not retain the same extreme accuracy of its scalar counterpart, as noted by Ib. A recent general proposal to improve on that is Fraser *et al.* (2016), extending previous work for linear exponential families (Davison *et al.*, 2014). These two papers propose a directional test that overcomes the omnibus nature of w and w^*.

Yet the multidimensional test based on w^* does have some remarkable properties shared with r^*, including protection for large deviations, parameterization invariance and the fact that its computation is as simple as that of r^*, as it involves standard likelihood quantities. The calculation of w^* also requires some sample space derivatives, but the paper provides general approximations very similar to those commented on above, so that also this statistic is generally computable. Furthermore, Monte Carlo evidence given in the paper and in several other papers published in the subsequent years showed that the statistic always provides a remarkable improvement over the likelihood ratio test, with accuracy comparable to that of the Bartlett corrected version of likelihood ratio test. For some examples see, among several others, Ferrari and Pinheiro (2011) and Sharma *et al.* (2014). The fact that there have been applications of w^* for statistical models such as nonlinear regression, beta regression, negative binomial regression, extreme values regression, multivariate regression models, measurement error models, testify to the generality and usefulness of this proposal.

Likelihood Asymptotics*

Ib M. SKOVGAARD

The Royal Veterinary and Agricultural University

ABSTRACT. The paper gives an overview of modern likelihood asymptotics with emphasis on results and applicability. Only parametric inference in well-behaved models is considered and the theory discussed leads to highly accurate asymptotic tests for general smooth hypotheses. The tests are refinements of the usual asymptotic likelihood ratio tests, and for one-dimensional hypotheses the test statistic is known as r^*, introduced by Barndorff-Nielsen. Examples illustrate the applicability and accuracy as well as the complexity of the required computations. Modern likelihood asymptotics has developed by merging two lines of research: asymptotic ancillarity is the basis of the statistical development, and saddlepoint approximations or Laplace-type approximations have simultaneously developed as the technical foundation. The main results and techniques of these two lines will be reviewed, and a generalization to multi-dimensional tests is developed. In the final part of the paper further problems and ideas are presented. Among these are linear models with non-normal error, non-parametric linear models obtained by estimation of the residual density in combination with the present results, and the generalization of the results to restricted maximum likelihood and similar structured models.

Key words: conditional inference, large deviation expansions, modified log likelihood ratio test, nusiance parameters, parametric inference

I. Introduction

Briefly, the present paper deals with improvements of classical asymptotic results for likelihood inference. The "improvements" refer to better asymptotic and numerical approximations, and the results aim at parametric tests and confidence intervals. We are concerned with general results for well-behaved parametric models and have in mind common types of models such as non-linear regression or mixed effects models. Thus, although the results may be considered as theoretical refinements, they should be feasible to implement and play a role in practical statistical work.

The main results presented are test probabilities obtained by Barndorff-Nielsen's statistic, r^*, and its generalization for testing hypotheses involving several parameters. The Bartlett correction of the usual likelihood ratio test is a good alternative, but is usually more difficult to calculate and has somewhat different asymptotic properties: higher order of accuracy but weaker properties for large deviations. The development leading to the results presented has primarily taken place over the last 20 years and has built on large deviation type asymptotic approximations like the saddlepoint approximations. In statistical terms this has been possible because of further development of Fisher's ideas about the use of ancillary information when the maximum likelihood estimate (MLE) is not sufficient. Pierce (1975) and Efron & Hinkley (1978) re-initiated this approach of approximate ancillarity, and Barndorff-Nielsen derived the remarkable p^*-formula and r^*-statistic using approximations of large deviation type, see, in particular, Barndorff-Nielsen (1980, 1986, 1991). General references of particular relevance to the current topic are Barndorff-Nielsen & Cox (1994), Jensen (1995) and Pace & Salvan (1997). The present review concentrates on Barndorff-Nielsen's methods with the further approximations to sample space derivatives derived in Skovgaard (1996), but it should be noted that alternatives

*This paper was presented as an Invited Lecture at the 17th Nordic Conference on Mathematical Statistics, Helsingør, Denmark, June 1998.

exist, in particular those due to Jensen (1992, 1997) and to Fraser, Reid and co-authors; see Fraser *et al.* (1999) and section 3.6.

Before going on a few words will explain why we concentrate on tests. It is commonly accepted that confidence sets and tests are two sides of the same coin: the confidence set consists of those parameter values that are not rejected by a test at the corresponding level. But what about estimates? We probably would not like to have our estimate outside a confidence set, and perhaps we should require it to be inside. The consequence is that an estimate is the intersection of confidence sets at all levels, in other words a zero-level confidence set as suggested in Skovgaard (1989). Aside from the logic in this, it is a convenient attitude because parametrization depending concepts such as unbiasedness and minimal variance vanish.

The organization of the paper is as follows. The problem and basic notation is introduced in section 2. In section 3 we review the main results and achievements from a practical viewpoint. Examples follow in section 4, and an outline of the derivation of the results is given in section 5. Regularity conditions, strict proofs and non-conceptual arguments are largely ignored in an attempt to make the important arguments stand out, but it should be noted that rigorous versions do exist for smooth and sufficiently regular absolutely continuous models. Finally in section 6 we discuss problems and ideas of further development. The appendix contains a brief review of exponential family notation together with some technical points on the profile score and on relative errors in large deviation regions.

2. Basic notation and setup

Let $\{f(y;\,\omega);\ \omega \in \Omega \subseteq \mathbf{R}^p\}$ be a family of densities of the random variable Y, indexed by the p-dimensional parameter ω, possibly restricted to some subset Ω. The domain of Y and the underlying measure are of no importance in the present connection, except that absolute continuity is required for rigorous derivation of the results. The random variable Y contains the entire data set. The theory has been developed for independent identically distributed (i.i.d.) random variables, but it is applied in many other situations as well, in the belief that similar results may hold more generally. Therefore there is no notational restriction to the i.i.d. situation, but it should be kept in mind that the results may not be suited for "strongly non i.i.d." situations like stochastic processes or situations with increasing number of nuisance parameters.

Let $\omega = (\nu, \psi)$ where $\nu = (\omega_1, \ldots, \omega_{p-q})$ and $\psi = (\omega_{p-q+1}, \ldots, \omega_p)$. We wish to test the hypothesis

$$H_0\colon \psi = \psi_0,$$

so ψ is the parameter of interest and ν is a nuisance parameter. We let $\hat{\omega}(y) = \hat{\omega} = (\hat{\nu}, \hat{\psi})$ denote the maximum likelihood estimate of the full parameter vector, while $\tilde{\omega}(y) = \tilde{\omega} = (\tilde{\nu}, \psi_0)$ denotes the maximum likelihood estimate under the hypothesis.

The log-likelihood function is denoted

$$l(\omega) = l(\omega;\, y) = \log f(y;\, \omega),$$

where the first version is used when y or the value of some sufficient statistic is understood. The kth derivative of the log-likelihood function is denoted D_k, i.e.

$$D_k(\omega) = D_k(\omega;\, y) = \frac{\partial^k}{\partial\omega^k}\, l(\omega;\, y)$$

which is a k-sided array with p^k entries. In particular, D_1 is the score vector.

Special notation is used for the (expected) Fisher information, $i(\omega) = \operatorname{var}_\omega D_1(\omega) = -ED_2(\omega)$, and for $j(\omega;\, y) = -D_2(\omega;\, y)$. We use abbreviations like $\hat{l} = l(\hat{\omega})$, $\hat{i} = i(\hat{\omega})$,

$\hat{j} = j(\hat{\omega}; y)$, and $\tilde{j} = j(\tilde{\omega}; y)$ and omit the argument y frequently. Note that $\hat{j}$ is the observed Fisher information.

The likelihood ratio test uses the statistic

$$w = 2(\hat{l} - \tilde{l}) = 2(l(\hat{\omega}) - l(\tilde{\omega})),$$

sometimes referred to as the deviance. Of importance is also its signed square root

$$r = \text{sign}(\hat{\psi} - \psi_0)\sqrt{w}$$

when ψ is one-dimensional.

Of conceptual importance are the so-called sample space derivatives of the log-likelihood and its derivatives. Assume that a supplementary statistic a exists such that $(\hat{\omega}, a)$ is sufficient. Then we may (and do) replace y in the various likelihood quantities by $(\hat{\omega}, a)$. Thus, the log-likelihood is $l(\omega; \hat{\omega}, a)$ and its derivative (vector) with respect to $\hat{\omega}$, called the sample space derivative is denoted $l'(\omega)$ with the statistic $(\hat{\omega}, a)$ understood. Except in section 5.4 where there is no risk of confusion, we reserve the "prime" to denote sample space derivatives, while other derivatives are denoted differently. Thus, $\tilde{D}_1'$ is the sample space derivative (with respect to $\hat{\omega}$ with a kept fixed) of the score vector at $\tilde{\omega}$, for example.

For approximation of the sample space derivatives we need two unusual likelihood quantities, namely

$$\hat{q} = \widehat{\text{cov}}(\hat{D}_1, \hat{l} - \tilde{l}), \quad \hat{S} = \widehat{\text{cov}}(\hat{D}_1, \tilde{D}_1),$$

using the convention that covariances are computed before the estimates are inserted. For example the strict way of writing the definition of $\hat{S}$ starts from the matrix

$$\text{cov}_{\omega_1}\{D_1(\omega_1; Y), D_1(\omega_2; Y)\}$$

and then inserts $\hat{\omega}$ for ω_1 and $\tilde{\omega}$ for ω_2 in the result. Note that $\hat{q}$ is a p-vector and $\hat{S}$ is a $p \times p$ matrix.

Vectors are column vectors unless transposed. The determinant of a matrix, M say, is denoted $|M|$, and its transpose is denoted M^{T}. Square brackets are used around vectors and matrices in coordinate expressions, like $[\tilde{D}_1']_{ij} = \partial[D_1]_i/\partial\hat{\omega}_j$, with simple exceptions like ω_j. A submatrix is indicated similarly, for example $[j(\omega)]_{\nu\nu}$ denotes the rows and columns of the matrix with coordinates corresponding to the ν-part of ω.

In parts of section 5 we use f to denote the density of the statistic in question, the context will show which.

A brief summary of exponential family notation is given in the appendix.

3. Review of results

Classical asymptotic theory for independent replications goes back to Fisher, see in particular Fisher (1922, 1925). Under usual regularity conditions the maximum likelihood estimator, $\hat{\omega}$, is consistent and $n^{1/2}(\hat{\omega} - \omega)$ converges in P_ω-distribution to the normal $N(0, ni(\omega)^{-1})$. The n appears in the variance here to cancel the factor n inherent in the information. Correspondingly, $\hat{\psi}$ is consistent and $n^{1/2}(\hat{\psi} - \psi_0)$ converges in distribution to the normal $N(0, n[i(\omega)^{-1}]_{\psi\psi})$ under the hypothesis, where $[\cdots]_{\psi\psi}$ denotes the square submatrix corresponding to the ψ-coordinates.

More in line with the present approach is Wilks' result, Wilks (1938), that minus twice the log-likelihood ratio, w, converges in distribution to $\chi^2(q)$ under the hypothesis. When ψ is one-dimensional the corresponding result for its signed square root, r, is that it converges to a standard normal distribution. Wilks' test is invariant to reparametrizations of the model and the hypothesis, an advantage shared by the asymptotically equivalent score test,

$$\tilde{D}_1^{\mathrm{T}} \tilde{i}^{-1} \tilde{D}_1.$$

The parametrization invariance is not shared by a third asymptotically equivalent test statistic, Wald's statistic, $(\hat{\psi} - \psi_0)^{\mathrm{T}}([\tilde{i}^{-1}]_{\psi\psi})^{-1}(\hat{\psi} - \psi_0)$, see Wald (1941, 1943).

Several asymptotically equivalent versions are available because of the convergence in probability of $n^{-1}i(\hat{\omega})$, $n^{-1}i(\tilde{\omega})$, $n^{-1}j(\hat{\omega})$ and $n^{-1}j(\tilde{\omega})$ to $n^{-1}i(\omega)$, so $\hat{i}$, $\tilde{i}$, i, $\hat{j}$, $\tilde{j}$ and j may be used interchangeably in the test statistics above, at least in terms of the asymptotic distribution under the hypothesis. Other arguments may suggest choices between these statistics, however, and in particular first-order asymptotic theory from stochastic processes documents strong advantages of using the observed rather than the expected Fisher information, for a review see Barndorff-Nielsen & Sørensen (1994).

3.1. Modern likelihood methods: what has been achieved?

The two main results for testing use Barndorff-Nielsen's r^*-statistic and the Bartlett correction, respectively. We first review the use of the r^*-statistic, and return to the Bartlett correction later. Following the r^*-statistic we present a generalization to multi-dimensional tests.

For testing the hypothesis $\psi = \psi_0$ for one-dimensional interest parameter ψ, Barndorff-Nielsen (1986, 1991) has derived the test statistic, in the sequel referred to as "Barndorff-Nielsen's (test) statistic",

$$r^* = r - \frac{1}{r} \log \gamma, \tag{1}$$

where the general expression for the quantity γ, usually written r/u, is

$$\gamma = |\hat{j}|^{1/2} |\tilde{D}_1'|^{-1} |\tilde{j}_{\nu\nu}|^{1/2} \frac{r}{[(\hat{l}' - \tilde{l}')^{\mathrm{T}} (\tilde{D}_1')^{-1}]_{\psi}}, \tag{2}$$

which involves the sample space derivatives $\tilde{D}_1'$ and $\hat{l}' - \tilde{l}'$. The statistic r^* is a "modified directed deviance" in the sense that it adds a correction to the directed deviance, r. Under the hypothesis $\psi = \psi_0$ the distribution of r^* is standard normal with high degree of accuracy conditionally on approximately ancillary statistics and hence also unconditionally. The quality of the approximation is much better than in the classical first-order results. This has been documented by asymptotic results as well as by numerical studies.

But the problem is to compute, or even define, γ in general, because the sample space derivatives are only well defined when an (approximately) ancillary statistic is specified which together with $\hat{\omega}$ is sufficient. And even if such an ancillary statistic is specified the computation may be difficult. Therefore it is desirable to have alternative expressions for these awkward quantities. Skovgaard (1996) gave the explicit approximations

$$(\hat{l}' - \tilde{l}')^{\mathrm{T}} \approx \hat{q}^{\mathrm{T}} \hat{i}^{-1} \hat{j}, \quad \tilde{D}_1' \approx \hat{S}^{\mathrm{T}} \hat{i}^{-1} \hat{j}. \tag{3}$$

This leads to the general approximation

$$\tilde{\gamma} = |\hat{j}|^{-1/2} |\hat{i}|\, |\hat{S}|^{-1} |\tilde{j}_{\nu\nu}|^{1/2} \frac{r}{[\hat{S}^{-1}\hat{q}]_{\psi}}, \tag{4}$$

which is exact when the model is a full exponential family. The approximation (4) to γ turns Barndorff-Nielsen's statistic, r^*, from (1) into an intrinsic expression in the sense that it is defined purely in model terms and does not depend on embedding into an exponential family, for example. When γ in (2) is not available we use $\tilde{\gamma}$ in (1) to obtain the test statistic and hence the test probability for the hypothesis.

In some cases we may compute γ explicitly, however; first of all in full exponential models and in simple transformation models, i.e. models generated from a single probability distribution by a group of transformations on the sample space.

For a full exponential family the maximum likelihood estimator is sufficient, hence no ancillary is necessary. This implies that the sample space derivatives may be computed explicitly, and we get

$$\gamma = \frac{r\{(t-\tilde{\tau})^{\mathrm{T}}\tilde{\Sigma}^{-1}(t-\tilde{\tau})\}^{1/2}}{(\hat{\theta}-\tilde{\theta})^{\mathrm{T}}(t-\tilde{\tau})}\left(\frac{|\tilde{\Sigma}|\,|\tilde{j}_{\nu\nu}|}{|\hat{\Sigma}|\,|\tilde{i}_{\nu\nu}|}\right)^{1/2}, \tag{5}$$

where θ is the canonical parameter, t is the corresponding canonical sufficient statistic, τ is the mean of t and Σ is its variance matrix; see the appendix for more specification. This expression is valid also when $\omega = (\nu, \psi)$ is a non-linear function of θ.

For transformation models we restrict ourselves to the simpler case of location models. Thus we consider a model for independent replications of observations with density

$$f(y; \omega) = f_0(y-\omega), \tag{6}$$

where y and ω may be p-dimensional, and f_0 is a known density. Then $\hat{l}' - \tilde{l}' = \tilde{D}_1$ and $\tilde{D}_1' = \tilde{j}$, so we get

$$\gamma = \frac{|\hat{j}|^{1/2}r}{|\tilde{j}_{\nu\nu}|^{1/2}\tilde{\lambda}}, \tag{7}$$

where $\tilde{\lambda}$ is the profile score, i.e. the derivative of $l(\tilde{\omega}, y)$ with respect to ψ_0.

The statistic r^* in (1) has been designed to test a one-dimensional hypothesis, possibly in the presence of nuisance parameters, but often it is desired to test an hypothesis involving several parameters, for example when testing for treatment effects in a linear or non-linear mixed model. A generalization of Barndorff-Nielsen's statistic to the multiparameter test situation with similar properties is as follows. As test statistic we use either

$$w^{**} = w - 2\log\gamma, \tag{8}$$

or the asymptotically equivalent version

$$w^* = w\left(1 - \frac{1}{w}\log\gamma\right)^2, \tag{9}$$

where $w = r^2$ is minus twice the log-likelihood ratio, and

$$\gamma = |\hat{j}|^{1/2}|\tilde{D}_1'|^{-1}|\tilde{j}_{\nu\nu}|^{1/2}|\tilde{\tilde{j}}_{\nu\nu}|^{-1/2}|\tilde{\tilde{j}}|^{1/2}\frac{\{\tilde{D}_1^{\mathrm{T}}\tilde{\tilde{j}}^{-1}\tilde{D}_1\}^{q/2}}{w^{q/2-1}(\hat{l}'-\tilde{l}')^{\mathrm{T}}(\tilde{D}_1')^{-1}\tilde{D}_1}, \tag{10}$$

where $\tilde{\tilde{j}} = j(\tilde{\omega}; \tilde{\omega}, A)$. The test statistics w^{**} and w^* are approximately distributed as $\chi^2(q)$ with high degree of accuracy, where $q = \dim\psi$. The version w^{**} is the one that arises naturally in the theoretical development and it also seems to be numerically better in some cases, but the version w^* has the advantage of being non-negative and of reducing to the square of Barndorff-Nielsen's statistic when $q = 1$.

Beside the sample space derivatives another awkward quantity, $\tilde{\tilde{j}}$, has appeared. Its computation also requires specification of the ancillary, and even so is hard to compute. With the same degree of accuracy as in the approximation to the sample space derivatives we may, however, approximate this by

$$\tilde{\tilde{j}} \approx \tilde{i}\,\hat{S}^{-1}\hat{j}\hat{i}^{-1}\hat{S}, \tag{11}$$

which may be used independently of the approximations (3) to the sample space derivatives. With both of the approximations we get

$$\tilde{\gamma} = |\tilde{i}|^{1/2}|\hat{i}|^{1/2}|\hat{S}|^{-1}|\tilde{j}_{\nu\nu}|^{1/2}|[\tilde{i}\hat{S}^{-1}\hat{j}\hat{i}^{-1}\hat{S}]_{\nu\nu}|^{-1/2}\frac{\{\tilde{D}_1^{\mathrm{T}}\hat{S}^{-1}\hat{i}\hat{j}^{-1}\hat{S}\tilde{i}^{-1}\tilde{D}_1\}^{q/2}}{w^{q/2-1}\tilde{D}_1^T\hat{S}^{-1}\hat{q}}. \tag{12}$$

As in the one-dimensional case we may calculate γ explicitly if the model is a full exponential model. In this case we get

$$\gamma = \frac{\{(t-\tilde{\tau})^{\mathrm{T}}\tilde{\Sigma}^{-1}(t-\tilde{\tau})\}^{q/2}}{w^{q/2-1}(\hat{\theta}-\tilde{\theta})^{\mathrm{T}}(t-\tilde{\tau})}\left(\frac{|\tilde{\Sigma}|\,|\tilde{j}_{\nu\nu}|}{|\hat{\Sigma}|\,|\tilde{i}_{\nu\nu}|}\right)^{1/2}, \tag{13}$$

which also agrees with γ from (5) when $q = 1$, as it should.

In the other reference case, the location model (6), (7) generalizes to

$$\gamma = \frac{\{\tilde{D}_1^{\mathrm{T}}\hat{j}^{-1}\tilde{D}_1\}^{q/2}}{w^{q/2-1}\tilde{D}_1^T\tilde{j}^{-1}\tilde{D}_1}\frac{|\tilde{j}_{\nu\nu}|^{1/2}|\hat{j}|}{|\hat{j}_{\nu\nu}|^{1/2}|\tilde{j}|}, \tag{14}$$

where we have used $\tilde{\tilde{j}} = \tilde{j}$ together with the relations $\hat{l}' - \tilde{l}' = \tilde{D}_1$ and $\tilde{D}_1' = \tilde{j}$.

A completely different asymptotic approach to the same problem leads to the so-called Bartlett correction to the likelihood ratio test, originating from Bartlett (1937) and Lawley (1956). The basic idea is that instead of using the chi-squared approximation to minus twice the log-likelihood ratio, w, we use the approximation

$$w\frac{q}{Ew} \sim \chi^2(q). \tag{15}$$

Moreover, since Ew is not explicitly available in general, we may use a general asymptotic approximation in terms of cumulants (or moments) of log-likelihood derivatives. The expression may be found, for example, in McCullagh & Cox (1986), Barndorff-Nielsen & Cox (1994, sect. 5.3) or Pace & Salvan (1997, sect. 9.4), but for comparison it is included here also. A noteworthy point is that only simple hypotheses need to be considered, because of the trivial relation $E(\hat{l} - \tilde{l}) = E(\hat{l} - l) - E(\tilde{l} - l)$. For the expression we need some special notation. Let $\kappa_{i,j,k}$ and $\kappa_{i,j,k,l}$ denote the third and fourth joint cumulants of the coordinates $[D_1]_i$, $[D_1]_j$, and so on, of the score vector. Similarly, let $[D_2]_{ij}$ and $[D_3]_{ijk}$ denote the (i, j)th and the (i, j, k)th coordinate of the second and third derivative of the log-likelihood, respectively; $\kappa_{i,jk}$ denotes the covariance between $[D_1]_i$ and $[D_2]_{jk}$, $\kappa_{i,j,kl}$ the joint third cumulant between $[D_1]_i$, $[D_1]_j$ and $[D_2]_{kl}$, and so on. Finally, $[i]^{ij}$ is the (i, j)th coordinate of the inverse of the information matrix, and everything is calculated at the "true" parameter ω. Then, for a simple test based on n independent replications,

$$\begin{aligned} Ew \sim p &+ [i]^{ij}[i]^{kl}\left(-\frac{1}{4}\kappa_{i,j,k,l} + \frac{1}{2}\kappa_{ik,jl} - \frac{1}{4}\kappa_{ij,kl} - \frac{1}{2}\kappa_{ij,k,l}\right) \\ &+ [i]^{ij}[i]^{kl}[i]^{mn}\left(-\frac{1}{2}\kappa_{i,km}\kappa_{j,ln} + \frac{1}{2}\kappa_{i,j,k}\kappa_{l,mn} + \frac{1}{4}\kappa_{i,jk}\kappa_{l,mn}\right. \\ &\left.+\frac{1}{6}\kappa_{i,k,m}\kappa_{j,l,n} + \frac{1}{4}\kappa_{i,j,k}\kappa_{l,m,n}\right) + O(n^{-2}), \end{aligned} \tag{16}$$

where summation over all 4 or 6 indices in each term is understood, and the error indicated is valid for absolutely continuous models under regularity conditions. If the model is a linear exponential family model all cumulants involving D_2 or D_3 vanish, so only terms based on

$\kappa_{i,j,k,l}$ and $\kappa_{i,j,k}$ remain. Properties of the Bartlett correction method will be discussed later, but it should be noted right away that the use of the correction (15) is more profound than it seems, because it actually leads to an asymptotic approximation with error $O(n^{-2})$, a property not shared by the similarly adjusted versions of other asymptotically equivalent chi-squared tests such as the score test or Wald's test. The proof of this property in Lawley (1956) was based on heavy computations with many terms cancelling at the end. With modern approaches related to the current paper it is much easier.

When the exact mean is available it is usually preferable to its approximation, but even in cases where the mean may not exist the general asymptotic approximation may give a valid distributional approximation—a phenomenon commonly known in asymptotic theory, having to do with convergence in mean being stronger than convergence in distribution.

3.2. *Is it necessary?*

The answer to this question relies primarily on the quality of the first-order approximations and might better be answered in connection to the numerical examples in section 4. There is a vast experience with the use of first-order asymptotics and the general perception may be that it works well for fairly simple models with a reasonable number of observations, but the behaviour is more or less unclear in complicated situations. But even in simple situations like test of independence in a two-way contingency table, non-linear regression, inference for variance parameters in variance component models, or logistic regression, one has to be careful with the use of first-order asymptotics. A surprisingly poor behaviour may be observed for multi-parameter tests, even in well-behaved models. As an example which we return to in section 4, consider a test of interaction between two factors in an ordinary two-way analysis of variance. If the data table has 5 rows, 4 columns and 2 observations per group, the F-test for interaction is based on (12, 20) degrees of freedom. The observation $F = 2.28$ corresponds to a test probability $p = 0.05$, but the likelihood ratio test gives $w = 34.46$ resulting in the test probability $p = 0.00057$ when the $\chi^2(12)$ approximation is used.

There is no guarantee, of course, that these problems are solved by using the present alternatives, but numerical and analytic results indicate improvements that are much better than small adjustments.

Another reason to look for better results is for implementation in statistical software packages doing standard analyses. For such programs confidence intervals for parameters are routinely computed by use of Wald's test. This may give rather imprecise results, especially for skew distributions, and implementation of the present methods is an alternative. The intermediate solution of implementing confidence intervals based on the likelihood ratio test is not commonly used, perhaps because of the computational burden. But if the belief of getting vast improvements is strong enough this (moderate) obstacle might be overcome.

An alternative is to use simulations, possibly in the form of bootstrap simulations, but even with modern computers it may well be too time consuming, in particular for routine applications like confidence intervals for individual parameters. Besides it is not a trivial matter to use simulation for the test of a composite hypothesis, or for confidence sets for components of a parameter.

3.3. *How good is it?*

From a numerical point of view: really good—at least for one-dimensional tests! This is the experience from investigations of r^* and related approximations in the literature, see, for example, Barndorff-Nielsen (1986, 1991), Barndorff-Nielsen & Cox (1989, 1994), Fraser

(1990), DiCiccio *et al.* (1990), Fraser & Reid (1993), Fraser *et al.* (1999), Jensen (1992, 1995), Pace & Salvan (1997), Pierce & Peters (1992, 1994) and Skovgaard (1996). Rather comprehensive numerical studies by R. Bellio and A. Brazzale (personal communication) confirm the excellent approximations by r^* and relatives for a one-dimensional interest parameter. Some results can be found in Bellio & Brazzale (1999) and in particular in Bellio (1999). The behaviour generally matches that of the saddlepoint approximation with a small relative error even in the extreme tail of the distribution. It should be kept in mind, however, that most of the examples reported have dealt with fairly simple and well-behaved models, and tests based on r^* have been univariate. It is not obvious that equally good results will be obtained for multivariate cases. In Skovgaard (1988) a multidimensional numerical example resulted in poor behaviour of related approximations in cases with very little information. For the analysis of variance example from section 3.2 the statistic w^* from (9) gives the p-value 0.0323, the exact value being 0.05. While this may be acceptable for practical use it does not match the quality from univariate examples.

These considerations refer to the distributional approximations, not to the statistical properties of the tests which are largely unexplored, except that all the tests considered in the present paper are equivalent to first order.

The numerical experience with the Bartlett correction is also very positive, although there is some variation in the results depending on the way the expectation of w is calculated or approximated. Generally it seems to work best when the exact mean is used (if available); in particular this possibility should be kept in mind for many tests in multivariate analysis. In the analysis of variance example the Bartlett corrected likelihood ratio test gives a test probability of 0.0219 when the general approximation to the mean is used, while use of the exact mean gives the impressive result of $p = 0.0485$. One should take care how to approximate the mean; for the same type of model Skovgaard (1988) used a direct approximation based on a stochastic expansion with much poorer results. There are many small numerical studies in the literature, in particular for Bartlett's test of homogeneity of variances, but also for other test problems, see, for example, Porteous (1985), Møller (1986) and Jensen (1993). Some numerical examples of r^*, w^* and the Bartlett correction are seen in section 4.

Theoretical properties of the approximations concern parametrization invariance, asymptotic properties and conditional properties. The tests using r^*, w^*, possibly with the general approximations (4) and (12), and the log likelihood ratio with or without Bartlett correction are all invariant under interest respecting reparametrizations, that is, one-to-one smooth reparametrizations of the form

$$(\nu, \psi) \mapsto (\alpha, \phi),$$

where ϕ is in one-to-one correspondence with ψ, see Barndorff-Nielsen & Cox (1994) for more thorough discussion of this point.

The asymptotic property of the distributional approximation to r^* is

$$P(r^* \geqslant x) = \{1 - \Phi(x)\}\{1 + O(n^{-1}\|\hat{\psi} - \psi_0\|)\}$$

uniformly in some large deviation region for a curved exponential family, see Barndorff-Nielsen & Wood (1998). This holds even conditionally on some approximately ancillary statistic while the error may go up by a factor $\sqrt{n}$ in the conditional distribution given other ancillaries within a reasonable class. It is seen from the expression that in normal deviation regions the (conditional) error is $O(n^{-3/2})$ for appropriate ancillaries and $O(n^{-1})$ for arbitrary ancillaries within a reasonable class. A thorough asymptotic study of the error for r^* and different ancillaries is given in Barndorff-Nielsen & Wood (1998).

The concept of conditional distributions that stay the same no matter which ancillary is used for conditioning is called stability, see McCullagh (1987) or Barndorff-Nielsen & Cox (1994), and it is a very satisfactory property from several points of view. Thus, there may be several good choices of ancillaries to choose from and one might not like inference to depend too much on the outcome of an ancillary statistic, and furthermore, an argument of relevance of inference may require restriction of the sample space to a conditional subset. All the statistics presented in section 3.1 have very good stability properties.

With the approximation (4) conditional properties are less clear because it was derived using a particular kind of ancillary statistic, but given this statistic and hence also unconditionally it has been proved to have relative error of order $O(n^{-1/2})$ in a large deviation region, implying a normal deviation error of order $O(n^{-1})$. Similar properties hold for the multivariate analogues.

The error of the chi-squared approximation to the log likelihood ratio statistic is $O(n^{-1})$, going down to $O(n^{-2})$ by use of the Bartlett correction. Detailed results on stability properties for these statistics are given in McCullagh (1987, ch. 8), the main result from the present viewpoint being that the error of the Bartlett corrected statistic is as low as $O(n^{-3/2})$ given any ancillary statistic.

An important distinction for these approximations is whether the relative errors are small for large deviations. For r^*, including the approximate versions, this is the case: at worst the relative error is $O(n^{-1/2})$, and even better in its best form. I am not aware of detailed accounts of the relative error of the usual chi-squared approximations to the log likelihood ratio, but using the Laplace integration techniques, see Woodroofe (1978) and section 5.4 below, it seems that with or without Bartlett correction the relative error is of order $O(1)$ for large deviations.

Finally we note that although most of the theoretical development of asymptotic distributions of r^* is for the case of a submodel of an exponential family, the approximation itself is not restricted to such curved exponential families. This is true, in particular, for the approximate versions using $\tilde{\gamma}$ from (4) and (12), but also for the general versions except that it may be difficult to find useful ancillaries to make the statistic computable. Asymptotic and especially conditional properties are, however, less clear outside exponential families and transformation models, although some indication is given in Skovgaard (1996).

3.4. How complicated is it to use?

As noted in section 3 it may be difficult or virtually impossible to compute r^* in its general form as given in (1) and (2), because of the unspecified sample space derivatives. However, the approximation (4) is defined explicitly and involves only quantities of complexity corresponding to the Fisher information matrix. Beside the log-likelihood ratio statistic, and the observed and expected information, the quantities $\hat{q}$ and $\hat{S}$ are required. While $\hat{i}$ is the variance of the score function, $\hat{S}$ is the covariance between the score functions calculated at two different parameter values, and $\hat{q}$ is similar except that a log-likelihood difference is substituted for one of the log-likelihood derivatives. For the multivariate versions (8), (9), (10) and (12), the same quantities appear, but the expressions are a bit longer. In most cases the main computational problem is the calculation of the estimates. The examples in section 4 will illustrate the complexity.

In comparison, the general approximation to the Bartlett correction factor requires calculation of cumulants or moments of order four of log-likelihood derivatives up to fourth order. The terms involve variances of second derivatives, mean of fourth derivative or fourth cumulant of the score function, covariances between first and second derivative and so on. While these terms are not in principle more complicated than information quantities, they are in practice of higher complexity and straight forward application of the summation formulae over indices may involve a high amount of computation simply because of the number of terms. For most well-behaved parametric models the computation of the Bartlett correction is feasible, but the computation of the approximate version of r^* is much less demanding.

For computation of r^* or of the Bartlett correction it is a great advantage to use symbolic computations for calculation of the derivatives and the cumulants, at least for validating the results. The computational issues are described in more detail in Bellio & Brazzale (2000) and in Bellio (1999, sect. 1.2), and computer algebra packages are reviewed in Kendall (1993) and in Stafford *et al.* (1994).

The r^* statistic does not involve the nuisance parameter, hence it is a similar test to the order considered. The Bartlett correction, however, involves the "true" parameter, but the asymptotic properties are kept if the estimate is plugged in. From an intuitive point of view the estimate under the hypothesis seems better, but the unrestricted estimate may also be used.

All of the approximations in section 3 involve moments or cumulants of log-likelihood derivatives, at least in the form of the information matrix. Analytic expressions for these may not exist or may be hard to calculate, and then it is of interest to estimate the quantities. The Bartlett correction only involves cumulants at the "true" parameter value, and these may be estimated from the data in case of independent replications; for $\tilde{\gamma}$ in (4) it is less clear whether such approximations are feasible. Recent work by Severini (1999) gives a method for this and moreover shows that the asymptotic properties of this empirical version are better than one would expect from the order of error of the estimates of $\hat{q}$ and $\hat{S}$ themselves.

3.5. Edgeworth expansions

Well established techniques of asymptotics result in correction terms to asymptotic approximations of first order. They are mainly based on the delta-method and Edgeworth expansions, and in terms of a density or distribution function of a standardized statistic they lead to additive expansions of the form

$$F_n = F_0 + n^{-1/2}F_1 + n^{-1}F_2 + \cdots,$$

where F_0 is the leading (normal) distribution, and the other F_is are densities or distribution functions for signed measures integrating to zero. A very informative review of such methods is found in Wallace (1958), and further development is due to many others, see, in particular, Bhattacharya & Ghosh (1978), Pfanzagl (1992, 1995), Amari (1985), Skovgaard (1986) and Ghosh (1994).

Although these methods are important as a technical basis and for investigation of theoretical properties of tests and estimators, they have never spread widely into statistical calculations, except for the Bartlett correction to the log-likelihood ratio test and perhaps for curvature considerations in non-linear regression (see Beale, 1960, Bates & Watts, 1988). One reason for this lack of success is that these expansions are essentially derived through Taylor series expansions about the true parameter value in the same way as the usual (first-order) asymptotic results, and hence they may break down in much the same cases. A related reason is that they may have a strange tail behaviour, for example giving negative probabilities, which is a symptom of the fact that they work best in the central region, and it is hard to know when they imply a numerical improvement. A more statistical reason is that much of the work has concentrated on distributions of estimators, and these are used only indirectly in statistical applications, namely to attach uncertainties or confidence intervals to estimates, or for testing. And then the question arises whether such procedures should start from the estimates when these are not sufficient. Fisher was concerned about this loss of information in the maximum likelihood estimate, and about its recovery by supplementation or conditioning, and the development of r^* follows this line of thinking. Fisher did not recommend formal testing, however, and was more concerned about relevance of the measure of precision attached to the estimate.

As a final comment to the use of these central expansions it may be noted that they do not lead to simpler expressions. This is well-known by researchers having tried to work with Edgeworth expansions or with the Bartlett corrections, for example.

3.6. *Variants and alternatives*

There are several versions of approximations to r^*, mostly for curved exponential families. The two methods of Jensen (1992, 1997) have very good asymptotic properties, keeping large deviation properties, and especially the second gives a technical shortcut to the distributional approximations, and is highly recommended for newcomers to the area. Both of the methods rely, however, on an embedding into a canonical exponential family together with a choice of sequence of tests of one degree of freedom, down to the model in question. The construction of this sequence of tests is the price paid for a presumably more accurate test.

An important parallel development is presented in a series of papers by Fraser and co-authors; see, in particular, Fraser & Reid (1988, 1993), Fraser (1990), Reid (1995, 1996) and Fraser *et al.* (1999). To define sample space derivatives they use certain pivots or approximate pivots, that is distribution constant functions of observations and parameters. For non-linear regression models these are standardized residuals, but in general cases there is no unique choice. Thus, the resulting inference depends not only on sufficient statistics but also on the structure of observations. This may be considered an advantage or a disadvantage depending on the point of view.

DiCiccio & Martin (1993) and Barndorff-Nielsen & Chamberlin (1991, 1994) develop other approximate versions of the r^*-type, using central expansions to some of the quantities involved, but giving very good numerical results.

Pierce & Peters (1992, 1994) present interesting detailed studies of r^* and relatives in canonical exponential families. They argue that the correction is composed of a "nuisance parameter adjustment" and an "information adjustment", thus providing more clear statistical interpretation of the formulae. They also investigate to what extent the inference depends on distributional properties rather than just on the observed likelihood function, a discussion highly relevant to the possible use of ancillaries, because of Birnbaum's theorem which effectively says that the conditionality principle implies the likelihood principle, see Birnbaum (1962).

Not much has been done along these lines for multiparameter testing. Fraser & Massam (1985) introduced a conditioning on the direction of a multivariate statistic and Skovgaard (1988) used the same technique on the score statistic to overcome certain technical difficulties, but this was not in connection with conditioning on approximate ancillaries. Closest to the present approach is probably Cheah *et al.* (1994) who also use such a directional conditioning for canonical exponential families.

4. Examples

Three examples follow: first a very simple and well-known example illustrating the usual extreme accuracy of saddlepoint and related approximations, then some linear normal models warning against the demanding task of multivariate approximations, and finally a more realistic use of the approximations in a non-linear bivariate normal model related to errors-in-variables models.

Example 1. Consider a pure location model with one observation and with an error distribution known to be the logistic; thus the density of Y is

$$f(y;\omega) = f_0(y-\omega), \quad f_0(x) = \exp(x)/(1+\exp(x))^2,$$

on the real line. For the hypothesis $\omega = 0$ the signed square root of the log likelihood ratio statistic is

$$r = \mathrm{sign}(y)\sqrt{2\log\frac{1+\exp(y)}{2} - y},$$

and Barndorff-Nielsen's statistic is

$$r^* = r - \frac{1}{r}\log\frac{\sqrt{f_0(y)/2}}{F_0(y) - 0.5},$$

where $F_0(y) = \exp(y)/(1+\exp(y))$ is the logistic distribution function.

The numerical results are shown in Table 1 where the observed p-values corresponding to various exact p-values for one-sided tests are given for r, r^* and the Bartlett corrected log-likelihood ratio statistic. The Bartlett correction factor is $\mathrm{EW} \approx 23/20$ when computed from the general asymptotic expression (16) which involves cumulants of order up to four in the uniform distribution. It is seen that all approximations do reasonably well, at least from a practical point of view, but the extreme accuracy of r^* is also apparent.

Example 2. We now consider test of the mean structure in a linear normal model. Here the usual F-test is exact but the chi-squared approximation to the log-likelihood ratio statistic is not. We use this class of models as a yardstick for what may, at best, be expected in non-linear normal models.

Let Y be an n-dimensional normally distributed random vector with mean vector μ and variance matrix $\sigma^2 I_n$ where I_n is the identity matrix. In the model we assume that $\mu \in L$, where L is a linear subspace of dimension $p-1$. Under the hypothesis $\mu \in H \subset L$, where H is a linear subspace of dimensions $p-q-1$. Thus the degrees of freedom are $\mathrm{df}_e = n-p+1$ for the error sum of squares and $\mathrm{df}_h = q$ for the hypothesis sum of squares.

Tables 2 and 3 show the results from two cases. In both cases w, w^* and w^{**} are shown together with the Bartlett corrected log-likelihood ratio test using either the exact mean or the general asymptotic approximation (16). In the first example the situation is not too demanding because a 4-dimensional parameter is being tested and there are 20 degrees of freedom for error.

Table 1. *One-sided test for location from one observation with logistic error distribution. Apparent p-values using approximations based on r, r^* and the Bartlett corrected log-likelihood ratio statistic, from an observation with given exact p-value*

	Apparent p using		
Exact p	r	r^*	Bartlett
0.4	0.387	0.401	0.395
0.2	0.172	0.203	0.189
$1.00 \cdot 10^{-1}$	$0.76 \cdot 10^{-1}$	$1.03 \cdot 10^{-1}$	$0.91 \cdot 10^{-1}$
$1.00 \cdot 10^{-2}$	$0.55 \cdot 10^{-2}$	$1.06 \cdot 10^{-2}$	$0.89 \cdot 10^{-2}$
$1.00 \cdot 10^{-3}$	$0.44 \cdot 10^{-3}$	$1.08 \cdot 10^{-3}$	$0.97 \cdot 10^{-3}$
$1.00 \cdot 10^{-4}$	$0.38 \cdot 10^{-4}$	$1.10 \cdot 10^{-4}$	$1.12 \cdot 10^{-4}$
$1.00 \cdot 10^{-5}$	$0.34 \cdot 10^{-5}$	$1.10 \cdot 10^{-5}$	$1.35 \cdot 10^{-5}$
$1.00 \cdot 10^{-6}$	$0.31 \cdot 10^{-6}$	$1.11 \cdot 10^{-6}$	$1.67 \cdot 10^{-6}$

Table 2. *Normal linear model with $n = 24$, $p = 5$, and $q = 4$. Apparent p-values using approximations based on w, w^*, w^{**}, the asymptotically Bartlett corrected log-likelihood ratio statistic (Bartlett), and the exactly Bartlett corrected log likelihood ratio statistic (w/EW), from observations with the given exact p-value*

	Apparent p using				
Exact p	w	w^*	w^{**}	Bartlett	w/EW
$1.00 \cdot 10^{-1}$	$0.63 \cdot 10^{-1}$	$0.98 \cdot 10^{-1}$	$1.00 \cdot 10^{-1}$	$1.29 \cdot 10^{-1}$	$1.00 \cdot 10^{-1}$
$1.00 \cdot 10^{-2}$	$0.42 \cdot 10^{-2}$	$0.97 \cdot 10^{-2}$	$0.99 \cdot 10^{-2}$	$1.60 \cdot 10^{-2}$	$0.99 \cdot 10^{-2}$
$1.00 \cdot 10^{-3}$	$0.29 \cdot 10^{-3}$	$0.95 \cdot 10^{-3}$	$0.99 \cdot 10^{-3}$	$1.97 \cdot 10^{-3}$	$0.98 \cdot 10^{-3}$
$1.00 \cdot 10^{-4}$	$0.20 \cdot 10^{-4}$	$0.94 \cdot 10^{-4}$	$0.99 \cdot 10^{-4}$	$2.39 \cdot 10^{-4}$	$0.97 \cdot 10^{-4}$

Table 3. *Normal linear model with $n = 40$, $p = 21$, and $q = 12$. Apparent p-values using approximations based on w, w^*, w^{**}, the asymptotically Bartlett corrected log-likelihood ratio statistic (Bartlett), and the exactly Bartlett corrected log likelihood ratio statistic (w/EW), from observations with the given exact p-value*

	Apparent p using				
Exact p	w	w^*	w^{**}	Bartlett	w/EW
$1.00 \cdot 10^{-1}$	$0.02 \cdot 10^{-1}$	$0.70 \cdot 10^{-1}$	$0.95 \cdot 10^{-1}$	$0.51 \cdot 10^{-1}$	$0.98 \cdot 10^{-1}$
$1.00 \cdot 10^{-2}$	$0.002 \cdot 10^{-2}$	$0.54 \cdot 10^{-2}$	$0.92 \cdot 10^{-2}$	$0.31 \cdot 10^{-2}$	$0.93 \cdot 10^{-2}$
$1.00 \cdot 10^{-3}$	$0.0002 \cdot 10^{-3}$	$0.42 \cdot 10^{-3}$	$0.89 \cdot 10^{-3}$	$0.19 \cdot 10^{-3}$	$0.86 \cdot 10^{-3}$
$1.00 \cdot 10^{-4}$	$0.00003 \cdot 10^{-4}$	$0.33 \cdot 10^{-4}$	$0.87 \cdot 10^{-4}$	$0.11 \cdot 10^{-4}$	$0.77 \cdot 10^{-4}$

The usual chi-squared approximation to w is not too bad, the asymptotic Bartlett correction works reasonably well, while both w^* and w^{**} are excellent.

The second example is much more demanding, testing a 12-dimensional hypothesis and furthermore with 9 nuisance parameters. The situation is not unrealistic, though, it may occur by testing interaction between two factors with 4 and 5 levels, respectively, for example. Here the usual chi-squared approximation to w is terrible, giving significant results far too often; in the extreme by a factor of more than 10,000. The asymptotic Bartlett correction is not too good but good enough for practical purposes, while w^* and w^{**} are far better, in particular the latter. Although it is unrealistic for non-linear models to use an exact Bartlett correction, it is interesting that even for this hard case there exists a Bartlett correction factor working very well, the asymptotic approximation (16) is just not good enough. But it should also be noticed that w^* and w^{**} are by far easier to calculate than the asymptotic Bartlett correction.

Example 3. As a more realistic example consider k groups indexed by $i = 1, \ldots, k$, and n_i independent bivariate measurements (X_{ij}, Y_{ij}) from each group. The model for (X_{ij}, Y_{ij}) is a bivariate normal distribution with arbitrary variance matrix Σ and with means

$$EX_{ij} = \mu_i, \quad EY_{ij} = \alpha + \beta\mu_i,$$

with all constants unknown. This example was inspired by several experiments in food science where foods were measured chemically (X) and sensorically by a panel of assessors (Y), and the relation between the two variables was in question. The model arises as a slight extension of the ultrastructural model by Dolby (1976). In our numerical example the model given above is the model under the hypothesis, while in the full model the bivariate means are fixed but arbitrary within groups. Since the full model is a linear exponential family the approximation to the sample space derivatives is not required, and the formula (13) may be used.

Table 4. *Test of linear relations between group means of two correlated normal random variables. Case 1 (see text) has three groups ($q = 1$) and case 2 has five groups ($q = 3$). Number of rejections in 10,000 simulations at various nominal levels of significance using approximations based on w, w^*, w^{**}. Expected number with an exact test is in the second column*

	Number of rejections						
		Case 1			Case 2		
Nominal p	Exact	w	w^*	w^{**}	w	w^*	w^{**}
0.99	9900	9912	9903	9899	9930	9901	9897
0.50	5000	5597	5110	5054	6184	5085	4999
0.10	1000	1521	1043	1006	1907	1083	1037
0.05	500	850	528	492	1159	582	542
0.01	100	240	122	113	373	131	114
0.001	10	47	13	10	67	11	9

Table 4 shows results from 10,000 simulations from each of two cases, both simulated from independent standard normal distributions of X and Y, with $\alpha = 1, \beta = 2$, and with $\mu_i = i$. Case 1 had three groups with five pairs in each so that $p = 9$ and $q = 1$. Case 2 had five groups with four pairs in each so that $p = 13$ and $q = 3$. The statistics investigated are w, w^* and w^{**} since it is quite cumbersome to work out the Bartlett correction factor for this test. Note that w^* is the square of r^* in case 1. The standard chi-squared approximation to the log-likelihood ratio statistic is not very good, but not as hopeless as for large q in the linear model. Both of the approximations w^* and w^{**} work very well.

5. Derivation of the test statistics

The present section reviews the derivation of the test statistics with emphasis on the technical principles employed. In particular, the method used in section 5.2 in combination with the Laplace integration and transformation techniques in sections 5.4 and 5.7 may make it possible to derive test statistics with similar properties based on estimating equations other than the derivative of the full log-likelihood.

The development of the methods, starting from Fisher's (1922, 1925) foundational likelihood papers and the likelihood ratio test, Neyman & Pearson (1928) and Wilks (1938), have separate inferential and mathematical foundations. The inferential line aims at recovering the loss of information in the maximum likelihood estimator, see Fisher (1925, 1934), Rao (1961) and Efron (1975). This is attempted through conditioning on ancillary or approximately ancillary statistics, the prototype example being the famous mirror likelihood by Fisher (1934). Cox (1958), Birnbaum (1962) and Basu (1964) pursue some foundational problems of conditioning and ancillarity, Pierce (1975) and Efron & Hinkley (1978) re-initiated Fisher's approach and made it more general and operational and further important steps in this direction were taken by Cox (1980), Barndorff-Nielsen (1980) and McCullagh (1984).

On the mathematical side the basis is Laplace integration techniques and, in particular, the saddlepoint methods introduced in statistics by Esscher (1932) and Daniels (1954).

Important mathematical generalizations of Laplace' integration method to problems with singularities or neighboring endpoints of integration were obtained by Bleistein (1966), possibly with essential contributions from others, see Daniels (1987). The relevance of these improvements for approximating distribution functions was noted by Lugannani & Rice (1980).

The mathematical and the inferential lines were merged in the important paper by Barndorff-Nielsen (1980) where the p^* formula was obtained, and even more strongly in his subsequent introduction of r^*; see Barndorff-Nielsen (1986). Approaches strongly related to the p^* formula were taken by Durbin (1980) and Field (1982).

5.1. Barndorff-Nielsen's formula

The remarkable p^*-formula was derived by Barndorff-Nielsen (1980). It is a formula for the conditional density of the maximum likelihood estimator, $\hat{\omega}$, given an (approximately) ancillary statistic A, and the expression is

$$p^*(\hat{\omega}|A=a) \approx c(a)|\hat{j}|^{1/2}\exp(l-\hat{l}), \tag{17}$$

where $c(a)$ is a normalizing constant, and the remaining notation needs some comments. First, for the formula to make sense it is required that the statistic $(\hat{\omega}, A)$ is an exactly sufficient statistic, such that we may write the log-likelihood function

$$l(\omega) = l(\omega;\, \hat{\omega},\, a)$$

with a similar notation for its derivatives, in particular for j. Second, l is short for $l(\omega)$ and $\hat{l}$ for $l(\hat{\omega})$. The formula (17) is exact for transformation models when A is the maximal invariant statistic, but the ancillary statistic may not be unique, even for transformation models, see McCullagh (1992). In the present connection it is, however, more important that as a general approximation in curved exponential families it has a relative error of order $O(n^{-1})$ uniformly in a large deviation region (see appendix). This holds true also with the approximation $c(a) = (2\pi)^{-p/2}$. In other words it is a very accurate approximation.

What is most remarkable about the formula is that it does not depend on any particular choice of the approximate ancillary statistic A. This is a somewhat deceptive property, however, because it means that without specifying the ancillary statistic we can calculate the conditional density of $\hat{\omega}$ at any particular data value, y, but not the complete density for any particular value a. For inferential purposes this is disappointing since it means that we can only calculate the density at the observed value of $\hat{\omega}$, and it appears that not much can be done on this basis. As we shall see even more surprisingly this is not true: only slightly more information is necessary to approximate tail probabilities, and the formula (17) does lead to accurate tail probabilities by use of advanced asymptotic techniques.

The formula is known as the "p^*-formula" and it has been a famous result since its appearance in 1980. It will be referred to as "Barndorff-Nielsen's formula" in the sequel. Proofs for its asymptotic validity may be found, for example, in Barndorff-Nielsen (1980), Barndorff-Nielsen & Cox (1994) and Barndorff-Nielsen & Wood (1998). Below we will give a brief heuristic argument based on a result which opens for generalizations to other estimators than the maximum likelihood estimator.

5.2. Density for estimators from estimating equations

Suppose that we want to estimate ω by a p-dimensional estimating equation

$$q(\omega;\, y) = 0,$$

which could be the derivative of the log-likelihood function. Let $Dq(\omega; y)$ denote the $p \times p$ matrix of partial derivatives of q, and $\widehat{Dq} = Dq(\hat{\omega}; y)$. Assume that for each v in the parameter space Ω we have a statistic $A(v, y)$, briefly denoted $A(v)$, and that we use $\hat{A} = A(\hat{\omega})$ as a supplementary statistic. We require that $\widehat{Dq}$ is a function of $(\hat{\omega}, \hat{A})$. To fix ideas think of the maximum likelihood estimator with $A(v) = j(v)$, such that $\hat{A}$ is the observed Fisher information, but the result below holds more generally.

Now under regularity conditions the simultaneous density of the estimate and $\hat{A}$ may be expressed as

$$f(\hat{\omega} = v, \hat{A} = a) = |Dq(\hat{\omega})| f(q(v) = 0, A(v) = a) \tag{18}$$

using f to denote the density of the statistic in question at the value indicated. The regularity conditions include the existence of certain densities, non-singularity of $\widehat{Dq}$, and unique existence of a solution to the estimating equation. In repeated sampling the latter condition usually holds except for a set with exponentially fast decreasing probability. Such details may convert the formula from being exact to being an approximation of high order of accuracy.

Formula (18) generalizes the similar result for the marginal density of the estimator from Skovgaard (1985), see also Pazman (1984), Skovgaard (1990) and Jensen & Wood (1998). In those papers $\widehat{Dq}$ might not be a function of $\hat{\omega}$ but the result is then easily obtained by taking $A = Dq(v)$ above and marginalizing. In relation to the maximum likelihood estimator the generalization is useful for handling distributional properties of the Fisher statistic $(\hat{\omega}, \hat{D}_2, \ldots, \hat{D}_k)$ for some fixed k, because it expresses the density in terms of the density of the log-likelihood derivatives at a fixed point. This latter statistic is a sum of i.i.d. random variables under repeated sampling and its distribution is usually much easier to handle exactly or approximately.

The proof of (18) will not be given here, we refer to the papers mentioned above, but the simple conceptual argument is as follows. Suppose there is a one-to-one transformation from $(\hat{\omega}, A(\hat{\omega}))$ to $(q(v), A(v))$, at least locally around $\hat{\omega} = v$. Then the matrix of partial derivatives of the transformation is a block triangular matrix of the form

$$\begin{pmatrix} \widehat{Dq} & ? \\ 0 & \text{identity} \end{pmatrix},$$

at the point $\hat{\omega} = v$. This matrix has determinant $|\widehat{Dq}|$ and the result (18) follows. Technical difficulties arise, however, because there may not be a strict functional relation between the two sets of statistics. This is the case with the Fisher statistic, for example. The "extra randomness" creeping into the transformation may be handled as in the referenced papers.

Obviously formula (18) has some perspectives of application outside maximum likelihood estimation, and as indicated in section 6 there may be possibilities of obtaining results analogous to Barndorff-Nielsen's using marginal, conditional or partial likelihoods, for example.

5.3. Derivation of Barndorff-Nielsen's formula

When $\hat{\omega}$ is the maximum likelihood estimator and $(\hat{\omega}, A)$ is sufficient formula (18) gives

$$\begin{aligned} f_\omega(\hat{\omega} = v, A = a) &= \exp(l(\omega) - l(\hat{\omega})) f_v(\hat{\omega} = v, A = a) \\ &= |\hat{j}| \exp(l(\omega) - l(v)) f_v(D_1(v) = 0, A = a) \\ &= |\hat{j}| \exp(l(\omega) - l(v)) f_v(D_1(v) = 0 | A = a) f_{\hat{\omega}}(A = a). \end{aligned}$$

If A is exactly ancillary we may divide through by the last factor to make the left hand side a conditional density as in Barndorff-Nielsen's formula. Further, the conditional likelihood given A is proportional to the unconditional likelihood, so, in particular, $D_1(v)$ and $D_2(v)$ equal the conditional score and observed information, respectively. Thus, in the f_v-distribution, the conditional mean of $D_1(v)$ is zero and the observed Fisher information is $-D_2(v)$ which is $\hat{j}$ at $D_1(v) = 0$ in the conditional distribution. Hence we may expect an excellent approximation to the conditional density of $D_1(v)$ at its mean by using the normal density with variance $\hat{j}$. This gives the factor $(2\pi)^{-p/2}|\hat{j}|^{-1/2}$ and we arrive at Barndorff-Nielsen's formula. When A is not exactly ancillary the arguments above hold only approximately.

The argument above essentially follows Skovgaard (1990). It may be noted that we did not explicitly use the construction of A allowed in (18) by defining a set of statistics, $A(\upsilon)$, and then using $A(\hat{\omega})$, but this is the way A is often constructed. This is true, for example, for the Efron–Hinkley ancillary which uses a standardized version of $D_2(\upsilon)$ evaluated at $\hat{\omega}$, see Efron & Hinkley (1978).

5.4. Laplace-type approximations

We now review the use of Laplace-type approximations—a useful class of approximations with large deviation properties, related to saddlepoint approximations. Their use and properties have been pointed out by Dinges (1986) who used the name "Wiener germs" for asymptotic expansions of this type, and useful results have been given also by Tierney & Kadane (1986), Tierney *et al.* (1989) and Jing & Robinson (1994), but also earlier related results have been presented, see, for example, Esscher (1932), Borovkov & Rogozin (1965), Woodroofe (1978), and Höglund (1979).

Mathematically the results rely on classical asymptotic approximations of integrals going back to Laplace, and more recent uniform expansions, see Bleistein (1966). In the present connection the integration result below is behind the conversion of Barndorff-Nielsen's formula to a tail probability as we shall see in the next section.

Definition 1

A sequence $(\tilde{f}_n)$ of real functions is called a *Laplace-type asymptotic approximation to the sequence* (f_n) *of densities on* $\mathbf{R}^p$, if there is a set K containing zero as an inner point, such that $\tilde{f}_n$ is defined on K and is of the form

$$\tilde{f}_n(x) = \left(\frac{n}{2\pi}\right)^{p/2} a(x)\exp(-nh(x)), \quad x \in K, \tag{19}$$

satisfying the following conditions:

(A) The asymptotic approximation

$$f_n(x) = \tilde{f}_n(x)(1 + O(n^{-1}))$$

holds uniformly for $x \in K$ as $n \to \infty$.

(B) For some $c > 0$ we have

$$P_n(X \notin K) = O(\exp(-cn))$$

as $n \to \infty$, where

$$P_n(X \in A) = \int_A f_n(x)\,dx.$$

(C) The functions a and h are respectively two and four times continuously differentiable in K, and h has a unique minimum at a point $x_0 \in \text{int}(K)$. At that point we have

$$h(x_0) = 0, \quad h''(x_0) \text{ is positive definite}, \quad a(x_0) = \sqrt{|h''(x_0)|},$$

where $h''(x_0)$ is the matrix of second derivatives of h at x_0, and $|h''(x_0)|$ is its determinant.

In the one-dimensional case condition (C) simply means that $h''(x_0)$ is positive. Because h is maximal at x_0, the condition $h'(x_0) = 0$ is automatically fulfilled. The saddlepoint approximation to a density of an average of independent replications of a random variables is of this kind, and so is Barndorff-Nielsen's formula although it is not immediately obvious from the expression (17). A series of statistics for which a Laplace-type approximation is available is asymptotically normal with asymptotic mean x_0 and asymptotic variance $(nh''(x_0))^{-1}$, but this approximation is usually not of sufficient quality since it only guarantees an additive error of order $O(n^{-1/2})$. The essential feature is that the exponential factor in (19) ensures that the probability mass concentrates around x_0 as $n \to \infty$, and consequently Taylor series expansions of functions around this point result in good approximations to integrals.

It is obvious that since an explicit expression is available for the (approximate) density, one-to-one transformations may be carried out by multiplication with the Jacobian and result in another Laplace-type approximation. Also, since the error is relative there is no problem in dividing such approximations, so conditioning does not alter the type of approximation either. Two types of computation are not trivial, however, namely marginalization and tail probability calculation, but it turns out that both may be handled by clever asymptotic methods, while still keeping a relative error of order $O(n^{-1})$ in a large deviation region. Both results are available in the literature, for reviews see Daniels (1987) and Jing & Robinson (1994), but we give the results briefly below, the tail integration result because we need it as background for its multivariate generalization, and the marginalization result for completeness to make it available in more explicit form than found in the references mentioned above.

When computing test probabilities we need to integrate the (one-dimensional) density of the test statistic from the observed value to infinity. Lugannani & Rice (1980) discovered that such probabilities could be approximated accurately using known mathematical techniques, and this had a vast impact on the use of saddlepoint approximations, although Esscher (1932) actually had results of similar quality, see the review paper by Daniels (1987).

Theorem 1

Let $(\tilde{f}_n)$ be a sequence of Laplace-type approximations to the densities (f_n) as in definition 1 with $p = 1$. Let $Q_n(x) = 1 - F_n(x)$ denote the right tail probability at x, where F_n is the distribution function. Consider the expression

$$\tilde{Q}_n(\hat{x}) = 1 - \Phi(\sqrt{n}\hat{z}) + \frac{1}{\sqrt{n}}\phi(\sqrt{n}\hat{z})\left(\frac{a(\hat{x})}{h'(\hat{x})} - \frac{1}{\hat{z}}\right), \tag{20}$$

where ϕ and Φ denote the standard normal density and distribution function, respectively, and

$$\hat{z} = \operatorname{sign}(\hat{x} - x_0)\sqrt{2h(\hat{x})}.$$

At $\hat{x} = x_0$ we should replace $\tilde{Q}_n$ by its limiting value

$$\tilde{Q}_n(0) = \frac{1}{2} + \frac{1}{\sqrt{2\pi n}}\left(\frac{a'(x_0)}{h''(x_0)} - \frac{1}{3}\frac{h^{(3)}(x_0)}{h''(x_0)^{3/2}}\right).$$

This expression is an approximation to the tail integral, with the asymptotic behaviour

$$\int_{\hat{x}}^{\infty} f_n(x)\,dx = \tilde{Q}_n(\hat{x})(1 + O(n^{-1}))$$

uniformly in $\hat{x}$ in some set containing x_0 as an inner point.

An alternative to $\tilde{Q}_n(\hat{x})$ with the same asymptotic properties is

$$\tilde{Q}_n^*(\hat{x}) = 1 - \Phi(z_n^*), \tag{21}$$

where

$$z_n^* = \sqrt{n}\hat{z} - \frac{1}{\sqrt{n}\hat{z}} \log\left(\frac{a(\hat{x})\hat{z}}{h'(\hat{x})}\right) \tag{22}$$

has the limiting value

$$\frac{-1}{\sqrt{n}}\left(\frac{a'(x_0)}{h''(x_0)} - \frac{1}{3}\frac{h^{(3)}(x_0)}{h''(x_0)^{3/2}}\right)$$

at $\hat{x} = x_0$.

Note two facts about this result. First, it is not necessary to write the n explicitly in the Laplace approximation to the density; if we simply absorb it into $h_n(x) = nh(x)$ and use the result as if $n = 1$ then we get the same result. Second, the result (21) states that the statistic z^*, derived from the observation $z = \hat{z}$, is standard normal with the given accuracy. The proof is omitted except that below we give the main idea of the proof of the version (21) which is the one giving Barndorff-Nielsen's statistic, r^*, from his p^*-formula, although he derived it differently. The line of proof given below follows Jing & Robinson (1994). In the special case of the saddlepoint approximation to an average, the result (20) is the one from Lugannani & Rice (1980) matching the technique from Bleistein (1966) or Temme (1982).

Proof. First we transform x to z by the (locally) increasing transformation satisfying $\frac{1}{2}z^2 = h(x)$. This gives the Jacobian $z(dz/dx) = h'(x)$ so the (approximate) density of z becomes

$$\left(\frac{n}{2\pi}\right)^{1/2} g(z)\exp(-(n/2)z^2) = \left(\frac{n}{2\pi}\right)^{1/2} \exp(-(n/2)(z - \log g(z)/nz)^2)(1 + O(n^{-1})),$$

where $g(z) = za(x)/h'(x)$. This is the normal density for the statistic z^* because the Jacobian of the transformation from z to z^* is $(1 + O(n^{-1}))$. This completes the proof since the density of z^* has been proved to be normal except for a relative error of order $(1 + O(n^{-1}))$ in a large deviation region.

For generalization to the multidimensional case we note that this remarkably simple proof consists of two variable transformations: the first to make the exponent an exact square, while the second is a small adjustment to remove the factor $g(z)$.

An important point for inferential purposes is that the information needed to approximate the tail probability is only $h(x)$, $a(x)$ and $h'(x)$ at the point $x = \hat{x}$ from which the tail probability is to be calculated. Thus, the approximation is locally defined, but does involve neighbouring points in the form of the derivative of h.

Although not used explicitly in the sequel we now review the useful result about marginalization of Laplace-type densities.

Theorem 2

Let $(\tilde{f}_n)$ *be a sequence of Laplace-type approximations to the densities* (f_n) *as in definition 1, and let* g: $\mathbf{R}^p \to \mathbf{R}^q$ *be a smooth mapping with differential* $g'(x)$ *of rank* $q < p$ *at* $x = x_0$. *Then a Laplace-type approximation holds for the density of* $Y = g(X)$, *and in a neighbourhood of* $y_0 = g(x_0)$ *the approximating density may be written*

$$\tilde{f}_n(y) = \left(\frac{n}{2\pi}\right)^{q/2} b(y)\exp(-nh(x_y)) \tag{23}$$

where $h(x_y) = \inf\{h(x): g(x) = y\}$,

$$b(y) = a(x_y)|G(x_y)|^{-1/2}|g'(x_y)G(x_y)^{-1}g'(x_y)^{\mathrm{T}}|^{-1/2}, \tag{24}$$

$$G(x_y) = h''(x_y) - \lambda(y)^{\mathrm{T}} g''(x_y), \tag{25}$$

and $\lambda(y)$ *is a q-vector arising as a Lagrangian multiplier given by* $h'(x_y) = \lambda(y)^{\mathrm{T}} g'(x_y)$, *and may be expressed as*

$$\lambda(y) = \{g'(x_y)h''(x_0)^{-1}g'(x_y)^{\mathrm{T}}\}^{-1}g'(x_y)h''(x_0)^{-1}h'(x_y),$$

for example. In the expression $\lambda(y)^{\mathrm{T}} g''(x_y)$ *above the q-vector* $\lambda(y)$ *is multiplied on the q-side of the* $q \times p \times p$ *array* $g''(x_y)$. *For the Laplace-type approximation to the density of Y to have relative error of order* $O(n^{-1})$ *it is sufficient that g is four times continuously differentiable in a neighbourhood of* x_0.

The result (23) may be found in Tierney *et al.* (1989) and in Jing & Robinson (1994), and related results in Borovkov & Rogozin (1965), Woodroofe (1978) and Daniels & Young (1991). The expression above for $b(y)$ may be useful because it only uses the function $g(x)$ without need of any supplementary parametrization of the level surfaces of $g(x)$. However, when such one is available it may be easier to use it. Suppose

$$t: \mathbf{R}^{p-q} \to \{x \in \mathbf{R}^p: g(x) = y\}$$

is a smooth parametrization of the level surface, and let u_0 be the crucial point with $t(u_0) = x_y$. Then $b(y)$ in (23) and (24) may alternatively be expressed as

$$b(y) = a(x_y)|g'(x_y)g'(x_y)^{\mathrm{T}}|^{-1/2}|t'(u_0)^{\mathrm{T}}t'(u_0)|^{1/2}\left|\frac{\partial^2 h}{\partial u^2}(u_0)\right|^{-1/2}. \tag{26}$$

5.5. *Test probabilities for one-dimensional interest parameter*

Consider Barndorff-Nielsen's formula (17). For independent replication this is a Laplace-type approximation although it is not immediately transparent. But the approximation $c(a) \sim (2\pi)^{-p/2}$ holds in general, and the *n*s are inherent in $|\hat{j}|^{1/2}$ as a factor $n^{p/2}$ and in $l - \hat{l}$ as a factor n when the log-likelihood is written as a function of the average of the sufficient statistics from each independent observation. Thus, $\hat{l} - l$ plays the role of $nh(x)$ and $|\hat{j}|^{1/2}$ replaces $a(x)$, and formally we take $n = 1$. Strictly speaking this is true in curved exponential families for which a minimal sufficient statistic of fixed dimension exists, but in general the important point is that a fixed deviation of $\hat{\psi} - \psi$ corresponds to an increase of the exponent in approximate proportion to n.

The technique is most transparent for the case $p = q = 1$ so that there are no nuisance parameters. Then, to calculate the right tail probability, say, from $\hat{\omega}$ we simply use theorem 1. Since $\frac{1}{2}r^2 = \hat{l} - l$ this immediately gives the statistic z^* from (22) as

$$z^* = r - \frac{1}{r}\log\left(\frac{|\hat{j}|^{1/2}r}{l'(\hat{\omega}) - l'(\omega)}\right), \tag{27}$$

which agrees with (1) and (2) for the case with no nuisance parameters. Theorem 1 now gives the result that the relative error is $O(n^{-1})$ in a large deviation region.

For the case with nuisance parameters, but still with $q = 1$, the idea is the same, but we have to reduce the dimension from p to q by a marginalization or conditioning. A convenient way is to use the following series of transformations,

$$\hat{\omega} \mapsto (\tilde{\nu}, \tilde{\lambda}) \mapsto (\tilde{\nu}, r) \mapsto (\tilde{\nu}, r^*) \mapsto r^*,$$

where $\tilde{\lambda}$ is the profile score vector. The first transformation could be bypassed but it will be useful when $q > 1$ and for the modified profile likelihood. The Jacobians for the first two transformations are

$$\frac{d\hat{\omega}}{d(\tilde{\nu}, \tilde{\lambda})} = |\tilde{D}'_1|^{-1} |\tilde{j}_{\nu\nu}|, \tag{28}$$

$$\frac{d(\tilde{\nu}, \tilde{\lambda})}{d(\tilde{\nu}, r)} = \frac{r}{[(\hat{l}' - \tilde{l}')^{\mathrm{T}} (\tilde{D}'_1)^{-1}]_{\psi}}, \tag{29}$$

as may be derived directly from the defining equations, see, for example, Barndorff-Nielsen & Cox (1994) or Skovgaard (1996). This leads to the density

$$\begin{aligned} f_{\omega}(\tilde{\nu}, r|a) &= (2\pi)^{-p/2} \exp(l - \tilde{l} - \tfrac{1}{2} r^2) |\tilde{j}_{\nu\nu}|^{1/2} \gamma \\ &= \{(2\pi)^{-(p-1)/2} \exp(l - \tilde{l}) |\tilde{j}_{\nu\nu}|^{1/2}\} (2\pi)^{-1/2} \exp(-\tfrac{1}{2} r^2 + \log \gamma), \end{aligned} \tag{30}$$

where γ is defined in (2). The transformation from r to r^* turns the factor outside the curly brackets into a standard normal density, as in the proof of theorem 1, and then a Laplace integration over $\tilde{\nu}$ removes the factor in curly brackets, because

$$\frac{\partial^2 (\tilde{l} - l)}{\partial \tilde{\nu}^2} = [j(\omega; \nu, r, a)]_{\nu\nu}$$

when evaluated at the data point (ν, r, a) where $\tilde{\nu} = \nu$ while r and a are kept fixed. This is the maximum point for the exponent $l - \tilde{l}$. In this way the approximate normal distribution of r^* is obtained.

We return to the case $q > 1$ in section 5.7.

5.6. *Sample space derivatives*

As is seen from expressions (27), (28) and (29) we need the sample space derivatives $\hat{l}' - \tilde{l}'$ and $\tilde{D}'_1$ to work out r^*. These are the derivatives obtained by moving the sufficient statistic such that $\hat{\omega} \mapsto \hat{\omega} + d\hat{\omega}$ while the ancillary, a, is kept fixed. In other words we move the observation along a level surface of the ancillary statistic. In general such sample space derivatives cannot be worked out without specifying the ancillary statistic but useful approximations can be obtained.

To be able to get a geometric picture consider a curved exponential family with log-likelihood function

$$l(\omega) = \theta(\omega)^{\mathrm{T}} t - \kappa(\theta(\omega))$$

apart from a data-dependent constant, where $t = t(y)$ is the canonical sufficient statistic and θ is the canonical parameter vector. The log-likelihood derivatives are

$$D_1(\omega) = \left(\frac{d\theta}{d\omega}\right)^{\mathrm{T}} (t - \tau(\omega)),$$

$$D_k(\omega) = \left(\frac{d^k\theta}{d\omega^k}\right)^{\mathrm{T}} (t - \tau(\omega)) + \text{non-random terms}, \quad k = 2, 3, \ldots,$$

where $\tau(\omega) = E_{\omega} t$. These are all linear functions of the vector $t - \tau$, projected on the different subspaces spanned by $d^k\theta / d\omega^k$. Their sample space derivatives are

$$D_k' = \left(\frac{d^k\theta}{d\omega^k}\right)^{\mathrm{T}} t',$$

which similarly are linear functions of t'. Differentiation of the log-likelihood equation $l(\hat{\omega};\, \hat{\omega},\, a) = 0$ immediately gives $\hat{D}_1' = \hat{j}$. This is the first of the observed balance relations, see Barndorff-Nielsen & Cox (1994, section 5.2), and the only sample space derivative which is known exactly. It means that we know the projection of t' on the space spanned by $d\theta/d\omega$ at $\hat{\omega}$, but not on the complementary subspace. The approximation to the sample space derivatives described below simply sets the projection of the sample space derivative on the complementary space to zero.

For a full exponential model, $d\theta/d\omega$ spans the entire space so $D_k(\omega)$ is a linear function of $D_1(\omega)$ for all $k > 1$, and the same applies to the sample space derivatives. This leads to the idea of regressing D_k on D_1 to obtain the change in D_k from that of D_1. The resulting approximation, which is not restricted to curved exponential families, is

$$\hat{D}_k' \approx \hat{\chi}_{k1}\hat{i}^{-1}\hat{D}_1' = \hat{\chi}_{k1}\hat{i}^{-1}\hat{j} \tag{31}$$

where χ_{k1} is the covariance between D_k and D_1. In this way we obtain an approximation to the sample space derivative of the (locally) sufficient statistic $(D_1, D_2, \ldots)$ at $\hat{\omega}$ and hence of any likelihood quantity. It is less obvious, but shown in Skovgaard (1996), that the required sample space derivatives can be calculated explicitly by summing the infinite Taylor series expansions, resulting in the approximations (3).

In some cases it is possible to work out the sample space derivatives exactly, first of all when $\hat{\omega}$ is sufficient so no ancillary statistic is needed. This is the case in a full exponential family for which the approximation above becomes exact. The result (5) is obtained by further simplification of (4), but they are identical in this case.

For the location model we use the fact that given the configuration statistic, $D_k(\hat{\omega};\, \hat{\omega},\, a)$ is constant for all k so by differentiation we get

$$\hat{D}_k' = -\hat{D}_{k+1}.$$

Again an infinite Taylor series summation gives the result leading to (7), for example,

$$\begin{aligned}\tilde{D}_1' &= \hat{D}_1' + (\tilde{\omega} - \hat{\omega})^{\mathrm{T}}\hat{D}_2' + \cdots \\ &= -(\hat{D}_2 + (\tilde{\omega} - \hat{\omega})^{\mathrm{T}}\hat{D}_3 + \cdots) \\ &= -\tilde{D}_2 = \tilde{j}\end{aligned}$$

The approximation (31) is not particularly related to any specific ancillary statistic. An alternative approach is to construct (possibly approximately) such a statistic and then try to work out the sample space derivatives. This is the approach taken by Jensen (1992, 1997) and by Fraser *et al.* (1999) and it generally results in approximations that depend on the structure of the model in other ways than through likelihood quantities and sufficient statistics.

5.7. *Multiparameter tests*

To derive the test statistic (8) or (9) for the case $q > 1$ we need a multivariate version of the tail probability approximation from theorem 1. It is not even obvious what corresponds to a tail probability in the multidimensional case, but the way it was done in the one-dimensional case suggests an analogue. Consider a statistic x with density of Laplace-type from definition 1. To obtain the tail probability of x we transformed x to the statistic z^* from (22) which was approximately normally distributed. We did this by first transforming to z with $z^2 = 2h(x)$ and then perturbed z to the more accurate z^*. We now consider the same approach for the multidimensional case with the intention of using the squared length of the resulting statistic as an approximately $\chi^2(q)$-distributed test statistic.

Assume first that in analogy with the proof of theorem 1 we have transformed x to the statistic z with density

$$f_n(z) = c_{n,p} g(z) \exp(-\tfrac{n}{2} z^{\mathrm{T}} z), \tag{32}$$

where $c_{n,p} = (n/2\pi)^{-p/2}$. We then perturb z to

$$z^* = z - \frac{1}{n}\delta(z) \tag{33}$$

with $\delta(z)$ chosen to satisfy $z^{\mathrm{T}}\delta(z) = \log g(z)$, such that

$$-\frac{n}{2}\left(z - \frac{1}{n}\delta\right)^{\mathrm{T}}\left(z - \frac{1}{n}\delta\right) = -\frac{n}{2} z^{\mathrm{T}} z + \log g(z) + O(n^{-1}).$$

A differentiable transformation of this kind may be written explicitly by choosing

$$\delta(z) = \log g(z) \frac{d \log g(z)/dz}{(d \log g(z)/dz)^{\mathrm{T}} z},$$

but actually we only need its existence to calculate the test statistic

$$w^* = \|z^*\|^2 = \|z\|^2 - \frac{2}{n} \log g(z) + O(n^{-2}),$$

in which we ignore the error term, and we have used the notation $\|z\|^2 = z^{\mathrm{T}} z$. To obtain a statistic which generalizes the one-dimensional version we may prefer the asymptotically equivalent approximation

$$w^* = \|z\|^2 \left(1 - \frac{\log g(z)}{n\|z\|^2}\right)^2.$$

In modified form this development will eventually lead to the statistics in (8) and (9). It is not clear which version will be best for applications; the first form is the one that arises with fewest approximations, but the latter form has the advantage of being non-negative.

There is still the problem, however, of reaching a statistic z with density (32). In principle it is simple because we just have to transform x smoothly to a vector of squared length $2h(x)$, and that can be done in many ways. The problem is to find a transformation for which we can also calculate the Jacobian, preferably in terms of locally defined quantities as in the one-dimensional case. The obvious choice

$$x \mapsto \frac{x}{\|x\|}\sqrt{2h(x)}$$

is not differentiable at $x = 0$ unless $d^2h(x)/dx^2$ equals the identity matrix at $x = 0$. The solution used here is to define length such that this condition is fulfilled. Hence we use

$$\Delta = \frac{d^2 h}{dx^2}(0)$$

as an inner product such that the corresponding squared length of x is

$$\|x\|_0^2 = x^{\mathrm{T}} \Delta x.$$

This results in the statistic

$$z = \frac{x}{\|x\|_0}\sqrt{2h(x)}$$

for which the exponent of the density has the factor $z^{\mathrm{T}}\Delta z$ instead of $z^{\mathrm{T}}z$, but this does not prevent the derivation of w^* as above, apart from trivial changes. With respect to the inner product Δ the Jacobian of the transformation from x to z is

$$\left|\frac{dx}{dz}\right| = \frac{\|x\|_0^p}{(2h(x))^{p/2-1}(dh/dx)^{\mathrm{T}}x}.$$

Using the new inner product in the derivation of w^* above, or equivalently using the normal distribution with variance $(n\Delta)^{-1}$, we finally get the approximate χ^2 statistic

$$w^* = \|z\|_0^2\left(1 - \frac{\log g(z)}{n\|z\|_0^2}\right)^2, \tag{34}$$

where $\|z\|_0^2 = 2h(x)$ and

$$g(z) = \frac{(x^{\mathrm{T}}\Delta x)^{p/2}|\Delta|^{-1/2}a(x)}{(2h(x))^{p/2-1}(dh/dx)^{\mathrm{T}}x}. \tag{35}$$

The transformation is not entirely locally defined because of the appearance of Δ. In our application $\Delta = \tilde{\bar{j}}^{-1}$ requires a further approximation to $\bar{j}$ when this is not directly computable. For this we use $\bar{j} \approx \bar{i}\bar{S}^{-1}\hat{j}\bar{i}^{-1}\hat{S}$, which is parametrization invariant, is exact when $\tilde{\psi} = \psi$, and gives a good approximation when the ancillary statistic is close to its mean; unconditionally the error is of order $O(1/n)$.

We now apply this technique to the derivation of a test statistic from Barndorff-Nielsen's p^*-formula in analogy with section 5.5. Throughout we ignore notationally the dependence on the ancillary statistic which is held fixed. The case without nuisance parameters follows exactly the method described, but for the more general case consider the density of $(\tilde{\nu}, \tilde{\lambda})$ obtained from Barndorff-Nielsen's formula and the Jacobian (28),

$$f_\omega(\tilde{\nu}, \tilde{\lambda}) \sim \{(2\pi)^{-(p-q)/2}|\tilde{j}_{\nu\nu}|^{1/2}\exp(l - \tilde{l})\}\{(2\pi)^{-q/2}a(\hat{\omega})\exp(-\tfrac{1}{2}w)\}, \tag{36}$$

where

$$a(\hat{\omega}) = |\tilde{j}_{\nu\nu}|^{1/2}|\hat{j}|^{1/2}|\tilde{D}_1'|^{-1}$$

corresponds to $a(x)$ in the Laplace-type density (19). The first factor in curly brackets in (36) is eventually going to disappear because it integrates to 1 by a Laplace integration over $\tilde{\nu}$. We cannot, however, do this integration right away because $a(\hat{\omega})$ depends on $\tilde{\lambda}$ through $\hat{\omega}$. The solution is to do the Laplace integration at the very end after having transformed the statistic $\tilde{\lambda}$ to (approximate) normality as described above. When the method described above is applied to the second factor in (36) we obtain the approximate chi-squared statistic

$$w^{**} = w - 2\log\gamma,$$

where γ is the expression (35) with $a(x)$ replaced by $a(\hat{\omega})$ from above,

$$\Delta = \frac{\partial^2}{\partial\tilde{\lambda}^2}(\hat{l} - \tilde{l}) = [\tilde{\bar{j}}^{-1}]_{\psi\psi},$$

and $(dh/dx)^{\mathrm{T}}x$ as

$$\left(\frac{\partial}{\partial\tilde{\lambda}}(\hat{l} - \tilde{l})\right)^{\mathrm{T}}\tilde{\lambda} = (\hat{l}' - \tilde{l}')^{\mathrm{T}}(\tilde{D}_1')^{-1}\tilde{D}_1.$$

Noting that

$$\tilde{\lambda}^{\mathrm{T}}\Delta\tilde{\lambda} = \tilde{D}_1^{\mathrm{T}}\tilde{\tilde{j}}^{-1}\tilde{D}_1$$

the statistic w^{**} above may be identified with the result given in (8). As previously mentioned the nuisance parameter estimate $\tilde{\nu}$ is integrated out by a Laplace integration after the conditional transformation of $\tilde{\lambda}$ to approximate normality. This step does not change the statistic w^{**}.

6. Discussion

Beside direct application of the results to various models, several possibilities of extensions are possible. One such extension aims at non-parametric analysis of a linear model. Consider the classical linear model

$$Y = X\beta + \sigma\epsilon,$$

where Y and ϵ are n-vectors, β is a p-vector of parameters, σ is a positive constant and X is an $n \times p$ matrix of fixed coefficients. If the elements of ϵ are independent with a known distribution the model is a transformation model. A maximal invariant statistic is the residual vector scaled to have length one. This statistic is exactly ancillary and the sample space derivatives may be calculated without need of approximations because derivatives with respect to estimates may be expressed in terms of derivatives with respect to the parameters. Thus, the application of the methods described in this paper is, in principle at least, straightforward.

When the distribution of the elements of ϵ is unknown we need to estimate it, and then the same method as above may be applied. The density may be estimated on the basis of the residuals, but some care is needed. For example, we cannot estimate the mean of the distribution if the model contains a constant term, because the residuals then sum to zero. Usually the mean, or some other location measure, of the error distribution is defined to be zero, however, precisely because we do have the constant term in the model, so we do not need it in the error distribution also. Similarly we use the scaled residual vector to estimate the error distribution up to a scaling constant. This use of residuals to estimate the error distribution is really the idea behind the bootstrap, but the empirical distribution of residuals often used in bootstrap applications, cannot be used here because it does not give a smooth density. An alternative is to use the empirical saddlepoint method, see Feuerverger (1989), which furthermore allows differentiation of the density. It does require considerable numerical work, however, to estimate the complete log-density, its first two derivatives and their moments, all of which are required to apply the present methods, but it would lead to an "estimated likelihood" or "saddlestrap" approach to non-parametric analysis of variance. One further challenging complication should be mentioned: the empirical saddlepoint method results in a distribution which is concentrated on the supporting interval of the observations, but we need an estimate which has support beyond these boundaries. In a neighbourhood of the mean the empirical saddlepoint approximation to the log-density will be concave, and towards the boundaries it will tend to infinity. If the assumption of a concave log-density is not unreasonable a natural modification is to extend the estimated log-density by straight lines continuing from the two turning points closest to the mean. This would give a density with exponentially decreasing tails, and hence a more robust approach than normal linear model analysis.

A potentially very powerful extension of the methods would be to start from an estimating equation other than the score function, obtain an analogue of Barndorff-Nielsen's formula, and continue as in the present paper. More specifically consider a particularly interesting class of examples, namely the method of restricted (or residual) maximum likelihood estimation in linear mixed models. Let Y be an n-dimensional normally distributed vector with mean vector $\mu(\beta)$ and variance matrix $\Sigma(\eta)$, say, where μ is a linear function of β. Let $t = t(y)$ denote the vector of residuals using ordinary least squares to obtain an initial estimate μ. The residual vector has a distribution which depends only on η. The REML estimating equations, see Patterson & Thompson (1971), for β and η may then be written

$$D_\beta \log f(y|t; \hat{\beta}, \hat{\eta}) = 0,$$

$$D_\eta \log f(t|\hat{\eta}) = 0,$$

where D_β and D_η denote the differentials with respect to the two parameter vectors. The REML estimates have for years, and for good reasons, been preferred to the maximum likelihood estimators, but good asymptotic approximations for their use for testing hypotheses about β have been lacking until recently, see Welham & Thompson (1997), Kenward & Roger (1997) and Elston (1998). Using the method from section 5.2 an approximation to the density may be derived, still with the log-likelihood difference appearing in the exponent when $\hat{\eta}$ is supplemented by an approximately ancillary function of $t(y)$. A problem is that the log likelihood difference $\hat{l} - \tilde{l}$ may not always be positive, because $\hat{l}$ is no longer the overall maximum of the full log likelihood function. Thus, part of this difference must be moved outside the exponent to ensure positivity and the use of Laplace integration methods. Obviously, all this is not straightforward, in particular not the derivation of resulting asymptotic properties, but it is equally obvious that the major part of the methods described here are not confined to strict maximum likelihood estimation.

In the derivation of the test, w^*, for a multidimensional hypothesis in section 5.7 there was a choice of transformation to make the exponent of the density approximation quadratic. Many such transformations exist and it is a natural question whether an approximation may be derived which avoids the quantity $\tilde{\tilde{j}}$, or which is better in other respects.

Finally, recall that the results concerning r^* and relatives deal with the null-distribution only. The initial motivation of conditional inference was recovery of information and relevance of sample space, but the results given here are only on the accuracy of the approximation of the null-distribution. Investigations of properties under the alternatives are missing.

Acknowledgements

I am grateful to Donald Pierce for fruitful discussions and for useful comments to a preliminary version of the paper, and to Ruggero Bellio for enlightening information on computational aspects and numerical results. Several helpful comments from the referees and from the editor are appreciated. The material in this paper was presented as invited lectures at the 17th Nordic Conference on Mathematical Statistics, Helsingør, Denmark, June 1998.

References

Amari, S.-I. (1985). *Differential geometric methods in statistics.* Lecture Notes in Statistics, **28**, Springer-Verlag, New York.

Barndorff-Nielsen, O. E. (1980). Conditionality resolutions. *Biometrika* **67**, 293–310.

Barndorff-Nielsen, O. E. (1986). Inference on full or partial parameters, based on the standardized signed log likelihood ratio. *Biometrika* **73**, 307–322.

Barndorff-Nielsen, O. E. (1991). Modified signed log likelihood ratio. *Biometrika* **78**, 557–563.

Barndorff-Nielsen, O. E. & Chamberlin, S. R. (1991). An ancillary invariant modification of the signed log likelihood ratio. *Scand. J. Statist.* **18**, 341–352.

Barndorff-Nielsen, O. E. & Chamberlin, S. R. (1994). Stable and invariant adjusted directed likelihoods. *Biometrika* **81**, 485–499.

Barndorff-Nielsen, O. E. & Cox, D. R. (1989). *Asymptotic techniques for use in statistics*. Chapman & Hall, London.

Barndorff-Nielsen, O. E. & Cox, D. R. (1994). *Inference and asymptotics*. Chapman & Hall, London.
Barndorff-Nielsen, O. E. & Sørensen, M. (1994). A review of some aspects of asymptotic likelihood theory for stochastic processes. *Intern. Statist. Rev.* **62**, 133–165.
Barndorff-Nielsen, O. E. & Wood, A. T. A. (1998). On large deviations and choice of ancillary for p^* and r^*. *Bernoulli* **4**, 35–63.
Bartlett, M. S. (1937). Properties of sufficiency and statistical tests. *Proc. Roy. Soc. London Ser. A* **160**, 268–282.
Basu, D. (1964). Recovery of ancillary information. *Sankhya Ser. A* **26**, 3–16.
Bates, D. M. & Watts, D. G. (1988). *Nonlinear regression analysis and its applications*. Wiley, New York.
Beale, E. M. L. (1960). Confidence regions in non-linear estimation, (with discussion). *J. Roy. Statist. Soc. Ser. B* **22**, 41–88.
Bellio, R. (1999). Likelihood asymptotics: applications in biostatistics. PhD thesis, University of Padova, Italy.
Bellio, R. & Brazzale, A. R. (1999). Higher-order likelihood-based inference in nonlinear regression. In *Proceedings of the 14th international workshop on statistical modelling* (eds H. Friedl, A. Berghold & G. Kauermann), 440–443. Technical University, Graz.
Bellio, R. & Brazzale, A. R. (2000). A computer algebra package for approximate conditional inference. *Statist. Comput.* to appear.
Bhattacharya, R. N. & Ghosh, J. K. (1978). On the validity of the formal Edgeworth expansion. *Ann. Statist.* **6**, 434–451.
Birnbaum, A. (1962). On the foundations of statistical inference (with discussion). *J. Amer. Statist. Assoc.* **57**, 269–306.
Bleistein, N. (1966). Uniform asymptotic expansions of integrals with stationary point near algebraic singularity. *Comm. Pure Appl. Math.* **19**, 353–370.
Borovkov, A. A. & Rogozin, B. A. (1965). On the multi-dimensional central limit theorem. *Theory Probab. Appl.* **10**, 55–62.
Cheah, P. K., Fraser, D. A. S. & Reid, N. (1994). Multiparameter testing in exponential models: third order approximations from likelihood. *Biometrika* **81**, 259–270.
Cox, D. R. (1958). Some problems connected with statistical inference. *Ann. Math. Statist.* **29**, 357–372.
Cox, D. R. (1980). Local ancillarity. *Biometrika* **67**, 279–286.
Daniels, H. E. (1954). Saddlepoint approximations in statistics. *Ann. Math. Statist.* **25**, 631–650.
Daniels, H. E. (1987). Tail probability approximations. *Intern. Statist. Rev.* **55**, 37–48.
Daniels, H. E. & Young, G. A. (1991). Saddlepoint approximation for the studentized mean, with an application to the bootstrap. *Biometrika* **78**, 169–179.
DiCiccio, T. J., Field, C. A. & Fraser, D. A. S. (1990). Approximations of marginal tail probabilities and inference for scalar parameters. *Biometrika* **77**, 77–95.
DiCiccio, T. J. & Martin, M. A. (1993). Simple modifications for signed roots of likelihood ratio statistics. *J. Roy. Statist. Soc. Ser. B* **55**, 305–316.
Dinges, H. (1986). Asymptotic normality and large deviations. *Proceedings of the 10th Prague conference on information theory, statistical decision functions and random processes*. Academia, Prague
Dolby, G. R. (1976). The ultrastructural relation: a synthesis of the functional and structural relations. *Biometrika* **63**, 39–50.
Durbin, J. (1980). Approximations for densities of sufficient estimators. *Biometrika* **67**, 311–333.
Efron, B. (1975). Defining the curvature of a statistical problem (with applications to second order efficiency) (with discussion). *Ann. Statist.* **3**, 1189–1242.
Efron, B. & Hinkley, D. V. (1978). Assessing the accuracy of the maximum likelihood estimator: Observed versus expected Fisher information. *Biometrika* **65**, 457–487.
Elston, D. A. (1998). Estimation of denominator degrees of freedom of F-distributions for assessing Wald statistics for fixed-effects factors in unbalanced mixed models. *Biometrics* **54**, 1085-1096.
Esscher, F. (1932). On the probability function in the collective theory of risk. *Scand. Actuar. J.* **15**, 175–195.
Feuerverger, A. (1989). On the empirical saddlepoint approximation. *Biometrika* **76**, 457-464.
Field, C. (1982). Small sample asymptotic expansions for multivariate M-estimates. *Ann. Statist.* **10**, 672–689.
Fisher, R. A. (1922). On the mathematical foundation of theoretical statistics. *Phil. Trans. Roy. Statist. Soc. Ser. A* **222**, 309–368.
Fisher, R. A. (1925). Theory of statistical estimation. *Proc. Cambridge Philos. Soc.* **22**, 700–725.
Fisher, R. A. (1934). Two new properties of mathematical likelihood. *Proc. Roy. Soc. Ser. A* **144**, 285–307.
Fraser, D. A. S. (1990). Tail probabilities from likelihoods. *Biometrika* **77**, 65–76.

Fraser, D. A. S. & Massam, H. (1985). Conical tests: observed levels of significance and confidence regions. *Statist. Hefte* **26**, 1–17.
Fraser, D. A. S. & Reid, N. (1988). On conditional inference for a real parameter: A differential approach on the sample space. *Biometrika* **75**, 251–264.
Fraser, D. A. S. & Reid, N. (1993). Third order asymptotic models: likelihood functions leading to accurate approximations to distribution functions. *Statist. Sinica* **3**, 67–82.
Fraser, D. A. S., Reid, N. & Wu, J. (1999). A simple general formula for tail probabilities for frequentist and Bayesian inference. *Biometrika* **86**, 249–264.
Ghosh, J. K. (1994). *Higher order asymptotics.* CBMS Regional Conference Series in Mathematics **4**.
Höglund, T. (1979). A unified formulation of the central limit theorem for small and large deviations from the mean. *Z. Wahrscheinlichkeitstheorie verw. Gebiete* **49**, 105–117.
Jensen, J. L. (1992). The modified signed likelihood statistic and saddlepoint approximations. *Biometrika* **79**, 693–703.
Jensen, J. L. (1993). A historical sketch and some new results on the improved log likelihood ratio statistic. *Scand. J. Statist.* **20**, 1–15.
Jensen, J. L. (1995). *Saddlepoint approximations.* Clarendon Press, Oxford.
Jensen, J. L. (1997). A simple derivation for r^* for curved exponential families. *Scand. J. Statist.* **24**, 33–46.
Jensen, J. L. & Wood, A. T. A. (1998). Large deviation and other results for minimum contrast estimators. *Ann. Inst. Statist. Math.* **50**, 673–695.
Jing, B. & Robinson, J. (1994). Saddlepoint approximations for marginal and conditional probabilities of transformed variables. *Ann. Statist.* **22**, 1115–1132.
Kendall, W. S. (1993). Computer algebra in probability and statistics. *Statist. Neerland.* **47**, 9–25.
Kenward, M. G. & Roger, J. H. (1997). Small sample inference for fixed effects from restricted maximum likelihood. *Biometrics* **53**, 983-997.
Lawley, D. N. (1956). A general method for approximating to the distribution of the likelihood ratio criteria. *Biometrika* **43**, 295–303.
Lugannani, R. & Rice, S. O. (1980). Saddlepoint approximation for the distribution of the sum of independent random variables. *Adv. Appl. Probab.* **12**, 475–490.
McCullagh, P. (1984). Local sufficiency. *Biometrika* **71**, 233–244.
McCullagh, P. (1987). *Tensor methods in statistics*. Chapman & Hall, London.
McCullagh, P. (1992). Conditional inference and Cauchy models. *Biometrika* **79**, 247–259.
McCullagh, P. & Cox, D. R. (1986). Invariants and likelihood ratio statistics. *Ann. Statist.* **14**, 1419–1430.
Møller, J. (1986). Bartlett adjustments for structured covariances. *Scand. J. Statist.* **13**, 1–15.
Neyman, J. & Pearson, E. S. (1928). On the use and interpretation of certain test criteria for purposes of statistical inference. *Biometrika* **20A**, 175–240; 263–294.
Pace, L. & Salvan, A. (1997). *Principles of statistical inference from a neo-Fisherian perspective*. World Scientific Publishing, Singapore.
Patterson, H. D. & Thompson, R. (1971). Recovery of inter-block information when block sizes are unequal. *Biometrika* **58**, 545–554.
Pazman, A. (1984). Probability distribution of the multivariate nonlinear least squares estimates. *Kybernetika* **20**, 209–230.
Pfanzagl, J. (1992). *Contributions to a general asymptotic statistical theory.* Lecture Notes in Statistics. **13**, Springer-Verlag, New York.
Pfanzagl, J. (1995). *Asymptotic expansions for general statistical models.* Lecture Notes in Statistics. **31**, Springer-Verlag, New York.
Pierce, D. A. (1975). Discussion to Efron (1975). *Ann. Statist.* **3**, 1219-1221.
Pierce, D. A. & Peters, D. (1992). Practical use of higher order asymptotics for multiparameter exponential families (with discussion). *J. Roy. Statist. Soc. Ser. B* **54**, 701–737.
Pierce, D. A. & Peters, D. (1994). Higher-order asymptotics and the the likelihood principle: one-parameter models. *Biometrika* **81**, 1–10.
Porteous, B. T. (1985). Improved likelihood ratio statistics for covariance selection models. *Biometrika* **72**, 97–101.
Rao, C. R. (1961). Asymptotic efficiency and limiting information. *Proc. 4th Berkeley Symp. Math. Statist. Probab.* **1**, 531–546. University of California Press, Berkeley.
Reid, N. (1995). The roles of conditioning in inference. *Statist. Sci.* **10**, 138–157.
Reid, N. (1996). Likelihood and higher-order approximations to tail areas: a review and annotated bibliography. *Canad. J. Statist.* **24**, 141–166.
Severini, T. (1999). An empirical adjustment to the likelihood ratio statistic. *Biometrika* **86**, 235–247.

Skovgaard, I. M. (1985). Large deviation approximations for maximum likelihood estimators. *Probab. Math. Statist.* **6**, 89–107.
Skovgaard, I. M. (1986). On multivariate Edgeworth expansions. *Intern. Statist. Rev.* **54**, 169–186.
Skovgaard, I. M. (1988). Saddlepoint expansions for directional test probabilities. *J. Roy. Statist. Soc. Ser. B* **50**, 269–280.
Skovgaard, I. M. (1989). A review of higher order likelihood inference. In: *Proceedings of the 47th Session of the International Statistical Institute. Vol. LIII* **3**, pp. 331–351. Paris, Aug 29–Sept 6. International Statistical Institute, Voorborg.
Skovgaard, I. M. (1990). On the density of minimum contrast estimators. *Ann. Statist.* **18**, 779–789.
Skovgaard, I. M. (1996). An explicit large-deviation approximation to one-parameter tests. *Bernoulli* **2**, 145–165.
Stafford, J. E., Andrews, D. F. & Wang, Y. (1994). Symbolic computation: a unified approach to studying likelihood. *Statist. Comput.* **4**, 235–245.
Temme, N. M. (1982). The uniform asymptotic expansion of integrals related to the cumulative distribution functions. *Siam J. Math. Anal.* **13**, 239–253.
Tierney, L. & Kadane, J. B. (1986). Accurate approximations for posterior moments and marginal densities. *J. Amer. Statist. Assoc.* **81**, 82–86.
Tierney, L., Kass, R. E. & Kadane, J. B. (1989). Approximate marginal densities of nonlinear functions (Corrigenda 78, 233–234). *Biometrika* **76**, 425–433.
Wald, A. (1941). Asymptotically most powerful tests of statistical hypotheses. *Ann. Math. Statist.* **12**, 1–19.
Wald, A. (1943). Tests of statistical hypotheses concerning several parameters when the number of observations is large. *Trans. Amer. Math. Soc.* **54**, 426–482.
Wallace, D. L. (1958). Asymptotic approximations to distributions. *Ann. Math. Statist.* **29**, 635–654.
Welham, S. J. & Thompson, R. (1997). Likelihood ratio tests for fixed model terms using residual maximum likelihood. *J. Roy. Statist. Soc. Ser. B* **59**, 701–714.
Wilks, S. S. (1938). The large-sample distribution of the likelihood ratio for testing composite hypotheses. *Ann. Math. Statist.* **9**, 60–62.
Woodroofe, M. (1978). Large deviations of likelihood ratio statistics with applications to sequential testing. *Ann. Statist.* **6**, 72–84.

Received August 1999, in final form June 2000

Ib Skovgaard, Department of Mathematics and Physics, The Royal Veterinary and Agricultural University, Thorvaldsensvej 40, DK-1871 Frederiksberg C, Denmark.

Appendix: further notation and concepts

This appendix contains some exponential family notation, followed by some remarks on the profile score vector and on large deviation asymptotic approximations.

For exponential families we use the notation θ for the canonical parameter and $t = t(y)$ for the corresponding canonical sufficient statistic. The density may then be written

$$f(y;\, \theta) = b(y)\exp\{\theta^{\mathrm{T}} t(y) - \kappa(\theta)\},$$

where $\kappa(\theta)$ is the logarithm of the normalizing constant with derivatives $\tau(\theta) = d\kappa/d\theta = E_\theta t(y)$ and $\Sigma(\theta) = d^2\kappa/d\theta^2 = \mathrm{var}_\theta\, t(y)$.

The profile score vector is the derivative of the profile log-likelihood,

$$\tilde{\lambda} = (\partial/\partial\psi_0) l(\tilde{\omega})$$

which for theoretical computations is conveniently expressed as the Lagrangian (vector) multiplier from maximization of the log-likelihood under the restriction $\psi = \psi_0$. This holds more generally when $\psi(\omega)$ is a smooth, but possibly nonlinear function of ω; then the estimating equation becomes

$$D_1(\omega) - \lambda^{\mathrm{T}} \frac{d\psi}{d\omega} = 0,$$

which under the restriction $\psi(\omega) = \psi_0$ has the solution $(\tilde{\omega}, \tilde{\lambda})$ with $\tilde{\lambda}$ being the profile score. This is why the score test is sometimes referred to as the Lagrange multiplier test.

For asymptotic approximations in the present connection an important distinction is between additive errors and relative errors in a large deviation region. This distinction is best understood in the case of an average, $\bar{y}$, of n independent replications, $y_1, \ldots, y_n$, with mean zero, as a prototype of a statistic tending to zero at the rate $1/\sqrt{n}$. If K is a fixed set containing zero as an inner point then $K/\sqrt{n}$ is called a region of normal deviations, since it is of constant size in terms of the limiting probability measure, whereas K itself is a set of large deviations because it increases at the rate of $\sqrt{n}$ in terms of the standardized statistic. Taking a tail probability $Q_n(y) = P(\bar{y} \geqslant y)$ as example, standard asymptotic expansions such as Edgeworth expansions give approximations that may be shown to be

$$Q_n(y) + O(n^{-s/2})$$

uniformly, for some integer s (or slightly better). This is an additive error and even if it holds for any y it is, of course, not of interest when the true value is smaller than the error. Since the tail probability usually tends to zero at exponential rate in y this happens "just outside" a normal deviation region. In contrast, asymptotic approximations with relative error in a large deviation region equal

$$Q_n(y)\{1 + O(n^{-s/2})\}$$

uniformly in some large deviation region. In practice this means that the error decreases in the tails of the distribution in proportion to the tail probability, and numerical examples clearly show the superior tail behaviour of such approximations. Note, however, that for fixed n it is not possible to distinguish normal and large deviations, the concepts really deal with sequences of sets or deviations. But the techniques leading to large deviation properties are of a different nature than the classical expansions which usually employ Taylor series expansions around the mean of the distribution. This shows up in the numerical behaviour, even for fixed n.

Chapter 12

On the use of a parameter dependent group structure for inference

Introduction by D A S Fraser
University of Toronto

12.1 Introduction

Statistical inference has two somewhat different routes for assembling information concerning the value of a parameter of interest: one route leads to likelihood and the other to p-values. These inference summaries, p-value and likelihood, are intrinsically different, and neither in general can be calculated from the other. The p-value route, however, is often more immediately meaningful and the likelihood route more accessible. The long-term development for these however has typically involved many steps, some steps forward, some difficulties, some correcting steps, then steps forward again. It is with great pleasure that I have this opportunity to introduce the significant paper, Skovgaard *et al.* (2001), for the volume of selected papers by Ib Skovgaard and coauthors. The paper addresses a key step in the development of likelihood.

Group structure is found in many familiar statistical models such as the location model $f(y-\theta)$, the regression model $\sigma^{-n}f\{(y-X\beta)/\sigma\}$, and more general models in which data value arises as a transformation of error value. These transformation models are easily enriched by adding a parameter ψ attached to the error distribution f, and even further generalized by having the parameter ψ also affect how the transformations are applied to the error variable.

We present an overview of this evolution of marginal likelihood likelihood.

12.2 Evolution with marginal likelihood

Likelihood for the full parameter θ of a model is widely available when the model is accessible in probability density form $f(y;\theta)$, and given as $L(\theta) = cf(\theta; y^0)$ where an arbitrary constant c is needed to avoid including information known to be irrelevant to the parameter θ. The likelihood is typically used in logarithmic form $\ell(\theta; y^0) = a + \log L(\theta; y^0)$ in most theoretical analyses.

Serious challenges for likelihood emerge when likelihood is needed for a component parameter say $\psi = \psi(\theta)$, where θ is say (ψ, λ) and λ is a nuisance parameter. A first step towards such likelihood for ψ is given by the long-standing profile likelihood function $L_{\mathrm{P}}(\psi) = L(\psi, \hat{\lambda}_\psi)$ where $\hat{\lambda}_\psi$ is the maximum likelihood value for λ given the value ψ for the interest parameter. Many difficulties have emerged concerning the accuracy and usefulness of this primitive likelihood, starting with whether it is even a likelihood function.

Modifications of this profile likelihood have been proposed to address these difficulties and the term marginal likelihood was used in Fraser (1967) for the transformation type models mentioned above, and then examined more fully in Fraser (1968), initially for models with interest parameter affecting only the error distribution (§4.2, §4.6) and then affecting also the application of the group to the error (§4.7); both of these use quotient differentials from calculus rather than the less accessible quotient measures from measure theory; quotient differentials in any case are widely appropriate for computation and heuristics.

For an interest parameter that affects just the error distribution, this marginal likelihood has been found to perform effectively and fully, and typically has second order accuracy. However, for models where the interest parameter also affects the application of the group, difficulties have emerged as noted in Fraser (1968), §11-2-4 starting with (11-33), where the marginal likelihood can be seen to "depend on the particular choice of variable to represent the response". Clearly this makes the marginal likelihood for such cases somewhat exploratory and at times indeterminant. This has also been addressed in Fraser and Reid (1989) where a rotationally symmetric information metric is proposed for use at the observed data point: this resolves most difficulties with the use of the particular marginal likelihood. In the paper here the authors also note the possible dependence on the choice of response presentation and shows in a particular context that the dependence can be catastrophic in a sense of providing no information concerning the interest parameter; this offers serious concerns to use of marginal likelihood unless particular choices are made for response presentation.

Approximations for marginal and conditional likelihoods have been developed by Fraser and Reid (1989) and Cox and Reid (1987), and then extended in Fraser (2003). As part of this we find that problems connected with the choice of response presentation have been recognized and resolved, with or without group structure; and also that difficulties have been recognized and resolved in the case of rotation in the contours of the interest parameter; see Davison *et al.* (2014), Fraser (2016) and Fraser *et al.* (2016).

12.3 Evolution with statistical theory

12.3.1 *Accuracy for the full parameter: exponential model*

Consider a statistical model $f(y;\theta)$ with p-dimensional parameter θ, n-dimensional variable y, and d-dimensional interest parameter ψ, initially all assumed continuous; and let y^0 be the observed data. Models don't always come prepackaged in a convenient form, but if a full model likelihood is wanted then the density version $f(y;\theta)$ provides it immediately as

$$L^0(\theta) = cf(y^0;\theta). \tag{12.1}$$

If a distribution is wanted beyond say likelihood information, then typically we need a variable say $u(y)$ of the same dimension as the parameter and want then the appropriate derived model for that variable. If the initial model $f(y;\theta)$ is an exponential model $\exp\{k(\theta)+\varphi'(\theta)u(y)\}H(y)$ then a full third order approximate model is immediately available as

$$g(u;\varphi) = \frac{\exp(k/n)}{(2\pi)^{p/2}}\exp(-r^2/2)|\jmath_{\varphi\varphi}(\hat{\varphi})|^{-1/2}, \tag{12.2}$$

where $-r^2/2 = \ell(\varphi;u) - \ell(\hat{\varphi};u)$ is the familiar log-likelihood ratio quantity and $\jmath_{\varphi\varphi}(\hat{\varphi})$ is the observed information matrix with subscripts denoting differentiation; both are immediately computable from the log-likelihood function. The preceding density is called the saddlepoint approximation (Daniels, 1954) and has extraordinary accuracy for computation. An alternative version is available by differentiating the maximum likelihood equation $\ell_\varphi(\hat{\varphi};u) = 0$ with respect to u obtaining $du = |\jmath_{\varphi\varphi}(\hat{\varphi})|d\hat{\varphi}$ and thus

$$g^*(\hat{\varphi};\varphi) = \frac{\exp(k/n)}{(2\pi)^{p/2}}\exp(-r^2/2)|\jmath_{\varphi\varphi}(\hat{\varphi})|^{1/2}; \tag{12.3}$$

in this latter form the augmented probability element $|\jmath_{\varphi\varphi}(\hat{\varphi})|^{1/2}d\hat{\varphi}$ is parameterization invariant. And also then $g^*(\hat{\varphi};\varphi)$ holds in much wider generality as Barndorff-Nielsen's p^* formula for regular models whenever likelihood is available as a function of the maximum likelihood variable (Barndorff-Nielsen, 1983).

12.3.2 *Accuracy for the full parameter: regular model*

Now for the full model $f(y;\theta)$ with n greater than the parameter dimension p suppose there is a quantile version $y = y(\theta; z)$ of the model available, as might be used for simulations, with z having some fixed distribution $h(z)$. Let $V = \partial y/\partial\theta|_{y^0,\hat{\theta}^0}$ be the $n \times p$ matrix recording the gradient of the response variable with respect to parameter change, evaluated at the data point y^0 and its associated maximum likelihood value $\hat{\theta}^0$. If we then define a new parameterization

$$\varphi(\theta) = \ell_{;y}(\theta; y^0)V, \tag{12.4}$$

we can fully analyze the given model with data as if it were an exponential model with canonical parameter φ and observed data $u^0 = 0$; this retains full third order accuracy for all the usual third order calculations (Fraser and Reid, 1995; Fraser *et al.*, 2010).

12.3.3 *Accuracy for a component parameter*

Now consider a d-dimensional interest parameter $\psi = \psi(\theta)$ with complementing nuisance parameter say λ. For a full regular model $f(y;\theta)$ we can then work with the tangent exponential model form described in §3.2 with canonical parameter $\varphi(\theta)$ and observed data $u^0 = 0$, let $\psi = \psi_0$ be a value for testing. Then from Fraser and Reid (1995), and Fraser *et al.* (2010) we have the existence of ancillary contours cross sectional to the observed nuisance parameter profile contour $\mathcal{L}^0 = \{u : \hat{\lambda}_{\psi_0} = \hat{\lambda}^0_{\psi_0}\}$; we abbreviate $\hat{\lambda}_{\psi_0} = \tilde{\lambda}$ and $\hat{\varphi}_{\psi_0} = \tilde{\varphi}$. Integration along the ancillary contours can be achieved using the integration normalization implicit in the Barndorff-Nielsen formula (12.3); this gives the following marginal distribution for the canonical variable u as restricted to the d-dimensional plane $\mathcal{L}^0$ and labelled s:

$$\frac{\exp(k/n)}{(2\pi)^{d/2}}\exp\{\ell(\tilde{\varphi};\hat{\varphi}) - \ell(\hat{\varphi};\hat{\varphi})\}|\jmath_{\varphi\varphi}(\hat{\varphi})|^{-1/2}|\jmath_{(\lambda\lambda)}(\tilde{\varphi})|^{1/2}ds; \tag{12.5}$$

for this the nuisance parameter uses a rescaling to agree fully with the constrained full exponential parameterization φ,

$$|\jmath_{(\lambda\lambda)}(\psi_0, \tilde{\lambda}^0; s)| = |\jmath_{\lambda\lambda}(\psi_0, \tilde{\lambda}^0; s)||X(\psi_0, \tilde{\lambda}^0)|^{-2}, \tag{12.6}$$

and $X(\varphi)$ is the $p \times (p-d)$ Jacobian $\partial\varphi/\partial\lambda$ of φ with respect to λ.

12.3.4 *Likelihood for a component parameter*

Suppose now that we want likelihood for the component parameter ψ. The observed data point y^0 has a corresponding parameter value $\hat{\varphi}^0$ which lies on the profile contour $C = \{\hat{\varphi}_\psi\}$ for the interest parameter ψ. If we change ψ_0 the constrained maximum likelihood value $\tilde{\varphi}$ moves on the contour C; and when $\tilde{\varphi}$ moves on the contour C the tangent plane may or may not change. If it changes we will refer to the parameter ψ as rotating at $\tilde{\varphi}$ and otherwise refer to the parameter as being linear at the point $\tilde{\varphi}$. An important aspect of the sample space nuisance parameter contour $\mathcal{L}^0 = \{u : \hat{\lambda}_{\psi_0} = \hat{\lambda}^0_{\psi_0}\}$ is that it is perpendicular to the tangent plane to C at the tested value ψ_0.

Now consider likelihood for the d-dimensional parameter ψ. This initially seems routine: just take the likelihood evaluation of (12.5) at the observed data and note that the dependence on ψ_0 is contained in the ingredients $\ell(\tilde{\varphi}; \hat{\varphi})$ and $\jmath_{(\lambda\lambda)}(\tilde{\varphi})$. But also there can be possible rotation of the parameter ψ and this can provide an implicit dependence in the scaling of the variable u on the plane $\mathcal{L}^0$. From this we have that the apparent likelihood expression from (12.5) may not be parameterization invariant.

This lack of invariance can be avoided by using a parameterization $\bar{\varphi}' = \varphi' T$ that has a rotationally symmetric information matrix $\hat{\jmath}^0_{\bar{\varphi}\bar{\varphi}} = I$, say the identity matrix I at the observed data point (Fraser, 2003). Accordingly we would use $\hat{\jmath}^0_{\varphi\varphi} = T'T$ where T is say a positive upper triangular square-root of the observed information matrix. We then have $L^0(\psi) = cg(s^0)$ where (12.6) is calculated in terms of the parameter $\bar{\varphi}$, and ψ replaces ψ_0. Examples of this extended likelihood are recorded in Fraser (2003).

12.4 Discussion

Marginal likelihood has evolved over an extended period: Fraser (1967, 1968), Cox and Reid (1987), Barndorff-Nielsen (1987), Fraser and Reid (1989), Fraser (2003), and many more. Likelihood extensions are continuing and we should be reminded of the risks and difficulties as part of this process, as in this paper by Skovgaard *et al.*.

Acknowledgements

This research has received support from the National Science and Engineering Research Council of Canada and the Senior Scholars Funding of York University.

ON THE USE OF A PARAMETER DEPENDENT GROUP STRUCTURE FOR INFERENCE

IB M. SKOVGAARD[1], TORBEN MARTINUSSEN[1] & STINA W. ANDERSEN[2]
[1]The Royal Veterinary and Agricultural University, Denmark
[2]University of Copenhagen, Denmark

Published in
Data Analysis From Statistical Foundations A.K.M.E. Saleh, ed., Nova Scientific Publishers, Huntington, NY, 2001 53 – 63.

SUMMARY

In a simple transformation model a fixed probability measure, P_0, is transformed by a group, G, of transformations on the sample space thus generating a model indexed by the parameter $g \in G$. If we start with a model P_ψ indexed by another parameter instead of starting from P_0, the group parameter may be eliminated by marginalizing to the maximal invariant statistic under the group. This statistic has a distribution depending only on ψ and may therefore be a good basis for inference on ψ. Complications arise if the group is allowed to depend on ψ so that the maximal invariant statistic then also depends on ψ. Fraser has advocated the use of the group structure and the maximal invariant statistic for inference about ψ even in this more complicated setting, but although the invariant statistic as well as its distribution are free of the nuisance parameter it is not obvious how to construct estimators or tests concerning ψ. In the present paper it is suggested how to use the group structure to construct an estimating equation for ψ in the situation with independent observations from distributions with fixed ψ and varying group parameters. Consistency follows under regularity conditions when an estimating equation of the proposed type exists.

Key words and phrases: Estimating equation, group action, invariance, likelihood, nuisance parameter, orthogonal likelihood.

1. INTRODUCTION

Let Y denote a random variable on $E \subseteq \mathbf{R}^N$ with distribution $P_{\psi,g}$ where ψ and g are unknown parameters. The model is assumed to be a transformation model in the parameter $g \in G$, where G is a group; more precisely the group acts on the sample space via the bijective smooth transformations $\gamma_g : E \to E$ satisfying $\gamma_{gg'} = \gamma_g \circ \gamma_{g'}$ for all $g, g' \in G$. Now $P_{\psi,g} = \gamma_g(P_\psi)$ so that the model is induced by the transformations starting from the distributions $P_\psi = P_{\psi,e}$ where e is the unit element in G. The interest parameter is ψ and g is a nuisance parameter.

The maximal invariant statistic $t(y)$ under the group of transformations has a distribution which depends only on ψ, and this statistic is ideal for inference about ψ, while inference about g is typically based on the conditional distribution of Y given $t(Y)$. This is the case, for example, in mixed linear models where the estimate of variance parameters are obtained from the likelihood based on the maximal invariant statistic. As another simple example, let $Y = (Y_1, \dots, Y_N)$ with the components being independent Gamma distributed with shape parameter ψ and scale factor $g > 0$, corresponding to the transformations $y \mapsto gy$. A version of the maximal invariant statistic is

$$T = (Y_1/\bar{Y}, \dots, Y_N/\bar{Y})$$

where $\bar{Y} = N^{-1} \sum_i Y_i$. In this example the group action does not depend on ψ so the statistic T as well as its distribution depend only on ψ. Standard procedures for inference about ψ now start from the marginal distribution of T and the effect of the nuisance parameter g is completely eliminated.

However, in more general cases the group action may depend on the interest parameter, so we write $\gamma_{\psi,g}$ for the transformation. Then also the statistic $T = t(y; \psi)$ depends on ψ, but still it has a distribution depending only on ψ so it may be worth looking for methods of inference about ψ based on T. But now it is not clear how to proceed although any test for the hypothesis $\psi = \psi_0$ based on a critical region of the form $T \in C(\psi_0)$ is a similar test. Fraser (1967, 1968, 1972, 1979) attempts to take advantage of the structure by introducing possible definitions of a "likelihood" function for the interest parameter; we return to one of these in Section 3 after some background on group theoretic decomposition of measures in Section 2. The properties of this likelihood are not clarified and in Section 4 we argue that independent replications with fixed interest parameter and varying nuisance parameter, as in Neyman & Scott (1948) may be appropriate for exploring consistency and efficiency in such situations. In Section 5 we try to set up an estimating equation leading to consistent estimation, but efficiency remains an open problem. Finally in Section 6 we compare the methods in a few simple examples.

2. DISTRIBUTION OF THE MAXIMAL INVARIANT

We now consider the group action in more detail with the particular aim of deriving a density for the maximal invariant statistic. For this purpose we need the decomposition of the underlying measure into measures on the conditional subspaces of E and a measure on the space of the invariant statistic. Since everything here is done for fixed ψ we suppress ψ in the notation. For a precise and general account we refer to Andersson (1982).

An *orbit* in the sample space under the action of G is a set of the form $\{\gamma_g(y) : g \in G\}$. Orbits are equivalence classes on E so the set of orbits is a quotient space denoted E/G. The orbit projection $\pi : E \to E/G$ maps y to its orbit which is a maximal invariant statistic under the action of G in the sense that any invariant statistic is a function of $\pi(y)$.

We shall now express the distribution of the maximal invariant as a density with respect to a measure on E/G. Such a density may then in principle be used for inferential purposes. The measure on E/G that we shall use is the quotient measure, see Andersson (1982) for details. A key assumption is that the action is proper which means that the mapping $(g, y) \mapsto (\gamma_g(y), y)$ is proper, meaning that the inverse image of any compact set is compact. Let β be a right invariant measure on G (such a measure always exists on a locally compact group) and suppose λ is a measure on E being relatively invariant with multiplier Δ_G^{-1}, that is,

$$\gamma_g(\lambda) = \Delta(g)\lambda, \tag{2.1}$$

where $\Delta : G \to \mathbf{R}_+$ is the modular function of the group, that is the density of a left invariant measure on G with respect to a right invariant measure on G. By definition the modular function satisfies $\Delta(g_1 g_2) = \Delta(g_1)\Delta(g_2)$; in particular $\Delta(e) = 1$. Then there exists a measure λ/β, called the quotient measure, on E/G and a set of measures $(\beta_\omega)_{\omega \in E/G}$ on E with $\operatorname{supp}(\beta_\omega) = \pi^{-1}(\omega)$, such that $\{(\beta_\omega)_{\omega \in E/G}, \lambda/\beta\}$ is a decomposition of λ with respect to π, that is,

$$\int_E h(y)\, d\lambda(y) = \int_{E/G} \left(\int_{\pi^{-1}(\omega)} h(y)\, d\beta_\omega(y) \right) d(\lambda/\beta)(\omega)$$

for any integrable function h on E. The measures $(\beta_\omega)_{\omega \in E/G}$ are given by

$$\int_{\pi^{-1}(\pi(z))} h(y)\, d\beta_{\pi(z)}(y) = \int_G h(\gamma_g(z))\, d\beta(g), \quad z \in E.$$

Suppose now that Y has density f with respect to a measure μ which is relatively invariant with multiplier χ_0. There exists a positive function η on E such that

$$\eta(\gamma_g(y)) = \chi_0(g)\Delta(g)\eta(y),$$

and, with this function, the measure $\lambda = (1/\eta)\mu$ satisfies (2.1). It then follows (see Andersson, 1982) that $\pi(Y)$ has density

$$k(\omega) = \eta(y) \int_G f(\gamma_g(y))\chi_0(g)\, \Delta(g)\, d\beta(g), \qquad \text{for any } y \in \pi^{-1}(\omega), \tag{2.2}$$

at $\omega \in E/G$ with respect to the quotient measure λ/β. Note that the measure $\Delta(g)d\beta(g)$ is a left invariant measure on G.

3. FRASER'S ORTHOGONAL LIKELIHOOD

For the situation with a group action depending on the interest parameter, ψ, Fraser has suggested the "orthogonal likelihood"; see Fraser (1967; 1968, p. 205; 1972 p. 903; 1979, p. 258). For reference here we use the 1972-paper [F72] in which this likelihood function of ψ is denoted L_1. Other likelihood functions are suggested in the same paper, but apart from the usual profile likelihood they require further structure and will not be considered here.

The expression for the likelihood L_1 is given in [F72] but its conceptual definition as a marginal or conditional density with respect to some measure on some space is elusive since it is derived as a marginal density of the orbit projection, $\pi(Y)$, with respect to a measure depending on the position of Y on the orbit.

Fraser considers a further structural element in the model, namely a ψ-depending transformation from the random variable considered above to the observation. In the present context such a transformation is redundant however, since it preserves the structure described above. On the other hand such a transformation may change the likelihood L_1, because of its dependence on Jacobians with respect to the Lebesgue measure on the sample space, see expression (3.2) further below.

Fraser implicitly assumes the action to be free, that is $\gamma_g(y) = \gamma_{g'}(y)$ for any y, g, g' implies that $g = g'$, where we still suppress the dependence on ψ from the notation. With some choice of reference point, $\rho_\omega \in E$ say, on any orbit $\omega \in E/G$, a free action allows a parametrization of the orbit by the group through the mapping

$$\phi_\omega : G \to E, \qquad \phi_\omega(g) = \gamma_g(\rho_\omega).$$

The inverse of ϕ_ω maps the orbit to the group; in [F72] the inverse image, $\phi^{-1}_{\pi(y)}(y)$ is denoted $[y]$, a notation which will be used in the sequel.

Consider the Jacobian of the group action mapping the reference point to $y \in E$,

$$J_N(y) = \left|D\gamma_g(\rho_{\pi(y)})\right|, \text{ at } g = [y],$$

where $D\gamma_g$ denotes the differential of γ_g and $|\cdot|$ denotes the determinant. Correspondingly we define the Jacobian on $G \subseteq \mathbf{R}^Q$,

$$J_Q(g) = \left|\frac{\partial gh}{\partial h}\right|, \text{ at } h = e.$$

Fraser shows that an invariant measure under the group action on the sample space is $J_N(y)^{-1}dy$, and that the measure $J_Q(g)^{-1}dg$ is a left-invariant measure on G. The parametrization ϕ_ω of the orbit ω transforms the left-invariant measure on G to a G-invariant measure on the orbit in E. Now consider the density f of Y with respect to dy; or equivalently the density $f(y)J_N(y)$ with respect to the invariant measure.

When we use the invariant measure on E we may take

$$\eta(y) = \Delta(\phi_{\pi(y)}^{-1}(y))$$

as the function η from the previous section, since $\chi_0(g) = 1$ and

$$\eta(\gamma_g(y)) = \Delta(g\phi_{\pi(y)}^{-1}(y)) = \Delta(g)\eta(y)$$

as required. Also $\eta(\rho_\omega) = 1$ for any orbit $\omega \in E/G$ so the density of $\pi(Y)$ at $\omega \in E/G$ from (2.2) becomes

$$k(\omega) = \int_G f(\phi_\omega(g))J_N(\phi_\omega(g))J_Q(g)^{-1}\, dg, \tag{3.1}$$

which is identical to the function k defined on page 900 in [F72]. Thus, with this choice of the function η, k is the density of $\pi(Y)$ with respect to the quotient measure, λ/β, between the invariant measures $J_N(y)^{-1}dy$ and the right invariant measure $(J_Q(g)^{-1}/\Delta(g))dg$.

Another way to represent the marginal distribution of the orbit projection, $\pi(Y)$, is with respect to the measure that decomposes the geometric (Lebesgue) measure dy on E with respect to the geometric measures on the orbits inherited from the Euclidean geometry. The geometric measure on the orbit ω may be written in terms of the parametrization ϕ_ω as $|D\phi_\omega(g)|_\circ dg$ where the generalized determinant $|D\phi_\omega(g)|_\circ$ is defined as

$$|D\phi_\omega(g)|_\circ = \left(|(D\phi_\omega(g))^T D\phi_\omega(g)|\right)^{1/2},$$

see Tjur (1980) and page 902 in [F72]. Now suppose that the orbits may be represented as elements in a Euclidean space, that is $\pi(y) \in \Omega \subseteq \mathbf{R}^{N-Q}$, where π is assumed to be a smooth mapping. Then we may compare the quotient space decomposition

$$\int_A f(y)\, dy = \int_\Omega \int_{A_\omega} f(\gamma_g(\rho_\omega)) \frac{J_N(\gamma_g(\rho_\omega))}{J_Q(g)}\, dg\, d(\lambda/\beta)(\omega),$$

where $A \in E$ and $A_\omega = A \cap \pi^{-1}(\omega)$, with the geometric decomposition (see Tjur, 1980)

$$\int_A f(y)\, dy = \int_\Omega \int_{A_\omega} f(\gamma_g(\rho_\omega)) \frac{|D\phi_\omega(g)|_\circ}{|D\pi(\gamma_g(\rho_\omega))|^\circ}\, dg\, d\omega,$$

where $|D\pi(y)|^\circ = |D\pi(y)D\pi(y)^T|^{1/2}$. This shows that the density of the quotient measure with respect to the Euclidean measure on Ω is

$$\frac{d(\lambda/\beta)(\omega)}{d\omega} = \frac{|D\phi_\omega(g)|_\circ}{|D\pi(\gamma_g(\rho_\omega))|^\circ} \frac{J_Q(g)}{J_N(\gamma_g(\rho_\omega))}$$

and hence that the right hand side is independent of g.

Re-introducing ψ into the notation we may now write Fraser's orthogonal likelihood as

$$L_1(\psi; y) = \frac{k(\pi(y); \psi)J_Q([y]; \psi)\, |D\phi_{\pi(y)}([y]; \psi)|_\circ}{J_N(y; \psi)}. \tag{3.2}$$

which is almost the density of the orbit projection with respect to the Euclidean measure, the only difference being that the divisor $|D\pi(y)|^{\circ}$ is missing. By omitting this divisor the orthogonal likelihood becomes independent of the particular representation of the orbit $\pi(y)$, but it is hard to interpret the result as a density. The word 'orthogonal' refers to the likelihood as a 'density' with respect to a local measure which is the local quotient (at y) between the geometric measure on E and the geometric measure on the orbit. A good property of the orthogonal likelihood is that it is independent of the choice of reference point, besides being independent of representation of the orbit projection. However, since it uses the geometric measure on the sample space it does become sensitive to one-to-one transformations of the observation space.

4. VARYING NUISANCE PARAMETERS

Rather than pursuing a philosophical discussion of possible merits of L_1 and alternatives we want to set up a framework for judging the performance of this and other procedures. One possible approach is to see if the information adds up in a logical way when independent experiments are performed. For this purpose we keep the interest parameter, ψ, fixed but allow the nuisance parameter to vary from experiment to experiment. This seems to be the most natural way to replicate the information contained in a single experiment with unknown nuisance parameter.

Thus, we now consider the setup from the famous paper by Neyman & Scott (1948): a sequence of independent observations, $Y_1, \ldots, Y_n$; each $Y_i \in E$ with a distribution P_{ψ,g_i} from a the same family $P_{\psi,g}$ as in the previous sections with a group action that may depend on ψ. Thus, ψ is the constant interest parameter while the nuisance parameters, g_i vary with i.

5. ESTIMATING EQUATION FOR THE INTEREST PARAMETER

Consider an estimating equation for ψ of the form

$$\sum_{i=1}^{n} h(\psi, y_i) = 0.$$

Under regularity conditions this leads to a consistent estimate of ψ if the equation is unbiased, that is if

$$\mathrm{E}_{\psi,g_i} h(\psi, Y_i) = 0$$

for all ψ and g_i. There are further conditions, in particular that the mean is not zero when evaluated under $P_{\psi',g}$ with $\psi' \neq \psi$, but for now the unbiasedness is our primary concern.

Let ℓ denote the log-likelihood from a single component, that is

$$\ell(\psi, g; y) = \log f(y; \psi, g),$$

where g and y are now used in place of g_i and y_i for arbitrary i. Consider the corresponding components of the score function,

$$\dot{\ell}_\psi(\psi, g; y) = \frac{\partial}{\partial \psi}\ell(\psi, g; y)$$

and

$$\dot{\ell}_g(\psi, g; y) = \frac{\partial}{\partial g}\ell(\psi, g; y)$$

Now define the function

$$m(\psi, g; \pi_0) = \mathrm{E}_{\psi,g}\{\dot{\ell}_\psi(\psi, g; Y)|\pi(Y; \psi) = \pi_0\}. \tag{5.1}$$

If possible, it would be desirable to use $m(\psi, g_i; \pi_i)$, where $\pi_i = \pi(y_i; \psi)$, as the ith component of the estimating equation, but it depends on the unknown g_i. A good property of m is that it has mean zero, because

$$\begin{aligned} \mathrm{E}_{\psi,g} m(\psi, g; \pi(\psi, Y)) &= \mathrm{E}_{\psi,g}\{\mathrm{E}_{\psi,g}\{\dot{\ell}_\psi(\psi, g; Y)|\pi(\psi, Y)\}\} \\ &= \mathrm{E}_{\psi,g}\dot{\ell}_\psi(\psi, g; Y) = 0. \end{aligned}$$

Another good property is that $m(\psi, g; \pi_i)$ depends on y_i only through π_i so it is orthogonal to $\dot{\ell}_g$ at any value of g_i. This is true for any function of π_i, $\xi(\pi_i)$ say, with mean zero, since

$$\mathrm{E}_{\psi,g_i}\{\xi(\pi_i)\dot{\ell}_g(\psi, g_i; Y_i)\} = \frac{\partial}{\partial g_i}\mathrm{E}_{\psi,g_i}\{\xi(\pi_i)\} = 0,$$

because the last mean is independent of g_i.

To get rid of g in equation (5.1) we basically try to estimate it, but it has to be done with care to preserve unbiasedness. Note first that there is no harm in treating ψ as known when we estimate g for this purpose. Second, we may as well use the conditional distribution given π_i to avoid introducing more randomness (and hence variance) than necessary. Thus, *assume* that for each fixed ψ we can find an unbiased estimate of $m(\psi, g; \pi_i)$ in the conditional distribution given π_i, that is we assume the existence of a function $h(\psi; y_i)$ satisfying

$$\mathrm{E}_{\psi,g}\{h(\psi; Y_i)|\pi(\psi, Y_i) = \pi_0\} = m(\psi, g; \pi_0) \tag{5.2}$$

for all ψ, g and π_0. Our proposal for the estimating equation is then

$$\sum_{i=1}^{n} h(\psi; y_i).$$

That the equation is unbiased follows from the computation

$$\begin{aligned} \mathrm{E}_{\psi,g_i} h(\psi; Y_i) &= \mathrm{E}_{\psi,g_i}\{\mathrm{E}_{\psi,g_i}\{h(\psi; Y_i)|\pi(\psi, Y_i)\}\} \\ &= \mathrm{E}_{\psi,g_i} m(\psi, g_i; \pi(\psi, Y_i)) = 0. \end{aligned}$$

A similar computation for the variance yields

$$\mathrm{var}_{\psi,g_i} h(\psi; Y_i) = \mathrm{var}_{\psi,g_i}\{m(\psi, g_i; \pi(\psi, Y_i))\} + \mathrm{E}_{\psi,g_i}\{\mathrm{var}_{\psi,g_i}\{h(\psi; Y_i)|\pi(\psi, Y_i)\}\}$$

where the second term represents the price paid for having to estimate $m(\psi, g_i; \pi_i)$.

The method discussed in the present section may be compared to the proposal by Neyman & Scott (1948) to subtract the estimated mean from the profile score statistic. This process only eliminates the nuisance parameters in special cases, however. A drawback of the present method is, on the other hand, that it may be difficult, or in practice impossible, to calculate the conditional mean of the score statistic.

6. SOME SIMPLE EXAMPLES

Example 1. Consider the following well known example from Neyman and Scott (1948). Let Y_{i1} and Y_{i2} be independent normal random variables with mean μ_i and standard deviation σ, the latter being the parameter of interest. Let $D_i = Y_{i2} - Y_{i1}$ and let $\hat{\mu}_i = (Y_{i1} + Y_{i2})/2$. The group generating the parameter μ_i is the set of translations of the two observations by the same amount, and the maximal invariant statistic is D_i. The score function for σ^2 from n pairs of observations is

$$\sum_i \dot{\ell}_{\sigma^2}(\sigma^2, \mu_i; Y_i) = -\frac{n}{\sigma^2} + \frac{1}{\sigma^4}\left(\frac{\sum D_i^2}{4} + \sum(\hat{\mu}_i - \mu_i)^2\right)$$

so the profile score, obtained by insertion of $\hat{\mu}_i$ for μ_i, is

$$-\frac{n}{\sigma^2} + \frac{1}{4\sigma^4}\sum D_i^2,$$

which has mean $-n/(2\sigma^2)$ and leads to an estimate converging in probability to $\sigma^2/2$.

In this case the group action, and hence the maximal invariant statistic D_i does not depend on the parameter of interest so Fraser's likelihood reduces to the marginal likelihood from $D = (D_1, \ldots, D_n)$ which solves the problem. Since the maximal invariant statistic does not depend functionally on σ^2 its density gives the same likelihood no matter which underlying measure is chosen.

The conditional mean of $\dot{\ell}_{\sigma^2}$ given the maximal invariant statistic is

$$m(\sigma^2, \mu_i, D_i) = -\frac{1}{2\sigma^2} + \frac{1}{4\sigma^4}D_i^2,$$

which leads to the estimating equation

$$-\frac{n}{2\sigma^2} + \frac{1}{4\sigma^4}\sum_i D_i^2 = 0,$$

which is independent of $\mu_1, \ldots, \mu_n$ and agrees with the score equation from the marginal likelihood based on the differences.

Example 2. Let $Y_1, \ldots, Y_n$ be independent, Y_i normally distributed on $\mathbf{R}^k$ with mean $\mu_i = a(\psi) + X_\psi \beta_i$ and variance $\sigma_i^2 I_k$, where a is a known function, I_k is the $k \times k$ identity matrix, and X_ψ is a $k \times p$ design matrix. The group element is $g_i = (\beta_i, \sigma_i) \in \mathbf{R}^p \times \mathbf{R}_+$. For fixed i the group action on the sample space is the mapping

$$\gamma_{g,\psi}(y) = a(\psi) + X_\psi \beta + \sigma(y - a(\psi)),$$

where $g = (\beta, \sigma)$, and the group multiplication is

$$gh = (\beta_g, \sigma_g)(\beta_h, \sigma_h) = (\beta_g + \sigma_g \beta_h, \sigma_g \sigma_h).$$

For fixed ψ let $\hat{e}_\psi(y_i) = y_i - a(\psi) - X_\psi \hat{\beta}_\psi$ denote the vector of residuals from the ordinary least squares estimation for the ith component. The ith component of the maximal invariant statistic is the normalized residual vector $\hat{e}_\psi(y_i)/||\hat{e}_\psi(y_i)||$. Some crucial quantities are $J_Q(g) = \sigma^{p+1}$ for $g = (\beta, \sigma)$ and $J_N(y_i) = ||\hat{e}_\psi(y_i)||^n$ if the reference point is chosen as the normalized residual vector. Further computations lead to Fraser's orthogonal likelihood

$$L_1(\psi) = \prod_i ||\hat{e}_\psi(y_i)||^{-k+p+1}$$

which in some sense adjusts the degrees of freedom compared to the profile likelihood

$$L_P(\psi) = \prod_i ||\hat{e}_\psi(y_i)||^{-k}.$$

The marginal densities of the maximal invariant with respect to either the quotient measure or the Euclidean measure on the space of orbits, Ω, should not be considered as possible "likelihood functions". The quotient measure is not unique, but if we use the version with density $k(\omega)$ from expression (3.1) with the present choice of reference point we get, in agreement with [F72, p. 906],

$$k(\pi(y); \psi) = \text{const} \cdot |X_\psi^T X_\psi|^{-n/2},$$

where the constant may depend on the observations but not on ψ. This function of ψ does not at all reflect how well the data fit the model given by ψ. The density of the standardized residual vector with respect to the Euclidean measure on Ω, which is a sphere in a linear subspace, is constant, so this density reflects the fit of the ψ-model equally badly.

Let us turn now to the method of estimating equations from Section 5. The score function from one component, y_i, with respect to a coordinate, ψ_j say, of the interest parameter is

$$\dot{\ell}_{\psi_j} = \frac{1}{\sigma_i^2}(y - a(\psi) - X_\psi \beta)^T (\dot{a}(\psi) + \dot{X}_j \beta),$$

where $\dot{a}(\psi)$ is the vector of derivatives of $a(\psi)$ and $\dot{X}_j = (\partial/\partial\psi_j) X_\psi$. The conditional mean of $\dot{\ell}_{\psi_j}$ given the maximal invariant is

$$m_j(\psi, \beta, \sigma; y_i) = \frac{1}{\sigma_i^2 ||\hat{e}_\psi(y_i)||} \hat{e}_\psi(y_i)^T (\dot{a}(\psi) + \dot{X}_j \beta) \sqrt{2} \frac{\Gamma((k-p+1)/2)}{\Gamma((k-p)/2)}.$$

If $k-p>1$ an unbiased estimate of m in the conditional distribution given the maximal invariant is

$$h(\psi;y_i)=\|\hat{\epsilon}_\psi(y_i)\|^{-2}\hat{\epsilon}_\psi(y_i)^T(\dot{a}(\psi)+\dot{X}_j\beta)\sqrt{2}\frac{(k-p-2)\Gamma((k-p+1)/2)}{\Gamma((k-p)/2)}.$$

To compare the methods consider the generalization by which k and p may vary with i. More specifically we may think of combining the estimates, $\bar{y}_i$, of a mean from samples of different size and with different variances; this case corresponds to $p=0$ and $a(\psi)=(\psi,\ldots,\psi)^T$ and was considered by Neyman & Scott (1948) and by Kalbfleisch & Sprott (1970), among others.

By differentiation of the profile log-likelihood we get the maximum likelihood estimate as the solution to the equation

$$\sum_i k_i(\bar{y}_i-\psi)s_i^{-2}=0$$

where $s_i^2=\|\hat{\epsilon}_\psi(y_i)\|^2/k_i$ is the variance estimate for the group when ψ is held fixed.

Differentiation of L_1 yields

$$\sum_i(k_i-1)(\bar{y}_i-\psi)s_i^{-2}=0,$$

and the estimating equation based on $h(\psi;y)$ yields the equivalent result

$$\sum_i(k_i-1)(\bar{y}_i-\psi)s_i^{-2}=0.$$

Neyman and Scott consider estimating equations of the form $\sum_i c_i(\bar{y}_i-\psi)s_i^{-2}=0$ and find optimal weights, c_i, in terms of asymptotic variance and thereby obtain the equation

$$\sum_i(k_i-2)(\bar{y}_i-\psi)s_i^{-2}=0.$$

They only prove the optimality for $k_i\geq 8$, however, and for sufficiently small k_i the result does not hold.

Note that all the estimating equations derived from the methods considered above use weights proportional to the inverse variance estimates. A representation of the maximal invariant statistic is

$$\left(\frac{\bar{y}_1-\psi}{s_1},\ldots,\frac{\bar{y}_n-\psi}{s_n}\right),$$

so a linear estimating equation based on this statistic uses weights proportional to the estimated standard errors. A linear function of the maximal invariant statistic would, however, lead to inference by which the nuisance parameters were eliminated. Although other than linear functions may be considered, it seems a reasonable conjecture that efficient estimation of ψ cannot be based entirely on the maximal invariant statistic.

7. DISCUSSION

None of the approaches discussed here lead to inference for ψ based on the maximal invariant statistic alone and it is still an open question whether a group structure for the nuisance parameter generally provides possibilities for inference which are beneficial in terms of concrete properties like efficiency, for example. It is certainly true that tests based on the maximal invariant statistic are similar, but examples suggest that they do not in general make efficient use of the data, and even when they do there is still no general method available for their construction. Fraser has been the chief advocator of the use of the group structure for inference even when the group structure depends on the parameter. It is our impression that there are still several open problems and possibilities in this direction that deserve further attention.

REFERENCES

Andersson, S. (1982). Distributions of maximal invariants using quotient measures. *Annals of Statistics* **10**, 955–961.

Fraser, D.A.S. (1967). Data transformations and the linear model. *Annals of Mathematical Statistics* **38**, 1456–1465.

Fraser, D.A.S. (1968). *The structure of inference.* Wiley, New York.

Fraser, D.A.S. (1972). The determination of likelihood and the transformed regression model. *Annals of Mathematical Statistics* **43**, 896–916.

Fraser, D.A.S. (1979). *Inference and linear models.* McGraw-Hill, New York.

Kalbfleisch, J. D. & Sprott, D. A. (1970). Application of likelihood methods to models involving large numbers of parameters. *Journal of the Royal Statistical Society, Series B* **32**, 175–208.

Neyman, J. & Scott, E. L. (1948). Consistent estimates based on partially consistent observations. *Econometrica* **16**, 1–32.

Tjur, T. (1980). *Probability based on Radon measures.* Wiley, New York.

Chapter 13

Efficient estimation of fixed and time-varying covariate effects in multiplicative intensity models

Introduction by Yanqing Sun and Seunggen Hyun

University of North Carolina, University of South Carolina Upstate

13.1 Introduction

Let T be the failure time of an event of interest and X a p-dimensional vector of the relevant covariates. The proportional hazards regression model or the Cox model postulates that the conditional hazard function of T given the covariate X is of the form $\lambda(t \mid X) = \lambda_0(t)\exp(X^\top\beta)$, where $\lambda_0(t)$ is an unspecified baseline hazard function and β is a p-dimensional vector of the regression parameters. The Cox model has been extensively studied and widely used in many fields. The key assumption in the Cox model is that the ratio of the hazard functions for two different covariate values is constant in t. This assumption can be violated, for example, when treatment effects change over time. Various useful alternatives and extensions have been made to the Cox model to enrich the collection of data modeling methods. A natural extension of the Cox model is to allow the regression parameters to depend on time, that is, $\lambda(t \mid X) = \lambda_0(t)\exp\{X^\top\beta(t)\}$, where $\beta(t)$ is a vector of time-varying regression functions. This extension benefits from the easy interpretation of $\beta(t)$ similar to the Cox model. The Cox model with time-varying regression coefficients has been studied by Zucker and Karr (1990), Murphy and Sen (1991), Martinussen *et al.* (2002), Cai and Sun (2003), and Yu and Lin (2010), among others. Martinussen *et al.* (2002), referred to as MSS hereafter, developed a set of theory for the Cox model with time-varying coefficients. They investigated both nonparametric and semiparametric time-varying Cox models, and the estimation and inference procedures developed by MSS are enormously useful in dynamic modeling of event history data. It is important to note that MSS developed the methods under the general framework of the multiplicative intensity model and hence they have applications beyond the hazard model for failure times. Further, MSS developed the statistical methods for semiparametric models where the coefficients of some covariates are time-varying while others are constants. This extension adds great flexibility and can be used to build models that improve the estimation efficiency. MSS is one of the early papers studying time-varying effects in event history analysis and has influenced the development in this area.

We hope to complement MSS by elaborating on its presentation and practice. We discuss MSS methods for multivariate counting processes where the intensities of different counting processes may follow different time-varying Cox models. An immediate application of this extension is to competing risks data, allowing estimation and comparison of the cause-specific hazard functions. The MSS algorithm is different from the local linear partial likelihood method of Cai and Sun (2003) and Yu and Lin (2010) which utilizes local polynomial techniques Fan and Gijbels (1996). We conducted a small simulation study comparing MSS algorithm with the local linear partial likelihood estimation of Cai and Sun (2003). The simulation also examines a bandwidth selection method for the two approaches.

13.2 MSS approach for multivariate intensity models

We describe an extension of MSS methods in context of multivariate multiplicative intensity models. Let $\{N_{ki}(t), i = 1, \ldots, n, k = 1, \ldots, K\}$ be multivariate counting processes with the intensities $\lambda_{ki}(t) = Y_i(t)\lambda_k\{t \mid X_i(t)\}$, where $Y_i(t)$ is a predictable process taking values 1 or 0 indicating whether individual i is at risk at time t, and $X_i(t) \in R^p$ are predictable covariate processes. Under the nonparametric time-varying Cox model, the intensities take the form

$$\lambda_{ki}(t) = Y_i(t) \exp\{X_i(t)^\mathsf{T} \beta_k(t)\}, \tag{13.1}$$

where $\beta_k(t)$ is a p-dimensional vector of time-varying regression functions. MSS methods are developed for $K = 1$.

Let $\Lambda_{ki}(t) = \int_0^t \lambda_{ki}(s)ds$ and $M_{ki}(t) = N_{ki}(t) - \Lambda_{ki}(t)$. It follows from Aalen and Johansen (1978) that $\{M_{ki}(t), i = 1, \ldots, n, k = 1, \ldots, K\}$ are orthogonal square integrable martingales. Denote $N_k(t) = \{N_{k1}(t), \ldots, N_{kn}(t)\}^\mathsf{T}$, $\lambda_k(t) = \{\lambda_{k1}(t), \ldots, \lambda_{kn}(t)\}^\mathsf{T}$, $\Lambda_k(t) =$

$\{\Lambda_{k1}(t), \ldots, \Lambda_{kn}(t)\}^\mathsf{T}$, and $M_k(t) = \{M_{k1}(t), \ldots, M_{kn}(t)\}^\mathsf{T}$. We organize the covariates into design matrix $X(t) = \{Y_1(t)X_1(t), \ldots, Y_n(t)X_n(t)\}^\mathsf{T}$.

The MSS estimating procedure for each $B_k(t) = \int_0^t \beta_k(s)\,ds$ under model (13.1) is obtained by updating the iteration step $\tilde{B}_k^{(m+1)} = g(\tilde{B}_k^{(m)})$. The function $g(\cdot)$ is defined by

$$g(\tilde{B}_k)(t) = \int_0^t \tilde{\beta}_k(s)ds + \int_0^t \tilde{A}(s)^{-1}X(s)^\mathsf{T}dN_k(s) - \int_0^t \tilde{A}(s)^{-1}X(s)^\mathsf{T}\tilde{\lambda}_k(s)ds, \tag{13.2}$$

where $\tilde{A} = A_{\tilde{\beta}_k}$ with $A_{\beta_k}(t) = X(t)^\mathsf{T}W(t)X(t)$ and $W(t) = \text{diag}\{\lambda_{ki}(t)\}$ is an $n \times n$ diagonal matrix. Here $\tilde{\beta}_k(t) = \int h^{-1}K\{(s-t)/h\}\,d\tilde{B}_k(s)$ is a simple kernel estimator of $\beta_k(t)$, where $K(\cdot)$ is a continuously differentiable kernel with support $[-1, 1]$ satisfying $\int K(u)du = 1$ and $\int uK(u)du = 0$, and h is the bandwidth. The MSS estimator $\widehat{B}_k(t)$ of $B_k(t)$ is from the last iteration and $\widehat{\beta}_k(t) = \int h^{-1}K\{(s-t)/h\}\,d\widehat{B}_k(s)$ is the estimator of $\beta_k(t)$, obtained by the kernel smoothing of $\widehat{B}_k(t)$.

The asymptotic properties established in Theorem 1 of MMS apply to each estimator $\widehat{B}_k(t)$ for $k = 1, \ldots, K$. Following arguments similar to those of MSS, Sun *et al.* (2008) showed the following weak convergence:

$$\left[\sqrt{n}\{\widehat{B}_1(t) - B_1(t)\}, \ldots, \sqrt{n}\{\widehat{B}_K(t) - B_K(t)\}\right] \xrightarrow{\mathcal{D}} \{U_1(t), \ldots, U_K(t)\},$$

in $D[0,\tau]^{Kp}$, where $U_1(t), \ldots, U_K(t)$, $0 \le t \le \tau$, are independent mean zero Gaussian martingales with covariance matrix $\Sigma_k(t) = \int_0^t a_k^{-1}(s)\,ds$. Here $a_k(t) = E\left[Y_i(t)X_i(t)X_i(t)^\mathsf{T}\exp\{X_i(t)^\mathsf{T}\beta_k(t)\}\right]$, $D[0,\tau]^{Kp}$ is the product space of Kp copies of $D[0,\tau]$, and $D[0,\tau]$ is the Skorohod space of functions on $[0,\tau]$ that are right continuous with a left limit. The MSS results can also be established for the semiparametric time-varying Cox model where the intensities take the forms $\lambda_{ki}(t) = Y_i(t)\exp\{X_i(t)^\mathsf{T}\beta_k(t) + Z_i(t)^\mathsf{T}\gamma_k\}$, for $i = 1, \ldots, n$ and $k = 1, \ldots, K$, where $Z_i(t)$ is a q-dimensional vector of covariates and γ_k is a q-dimensional vector of parameters.

This extension to multivariate multiplicative intensity models is useful in analysis of competing risks data. Let T_i be the failure time, C_i the censoring time and V_i the cause of failure. Define $\tilde{T}_i = \min(T_i, C_i)$ and $\delta_i = I(T_i \le C_i)$, where $I(\cdot)$ is the indicator function. Let $N_{ki}(t) = \delta_i I(V_i = k)I(\tilde{T}_i \le t)$ be the counting process to indicate whether an event of type k has been observed by time t for subject i. Then $\{N_{ki}(t), i = 1, \ldots, n, k = 1, \ldots, K\}$ are multivariate counting processes with intensities $\lambda_{ki}(t) = Y_i(t)\lambda_k\{t \mid X_i(t)\}$ for $i = 1, \ldots, n$ and $k = 1, \ldots, K$, where $\lambda_k\{t \mid X_i(t)\}$ is the conditional cause-specific hazard function due to cause k. The extension to $K > 1$ allows comparison and inference about the cause-specific hazard functions for different causes. For instance, Sun *et al.* (2008) developed estimation and hypothesis testing procedures for stain-specific vaccine efficacy to prevent clinically significant infection due to different types of the infecting strains.

13.3 Bandwidth choice and numerical comparison

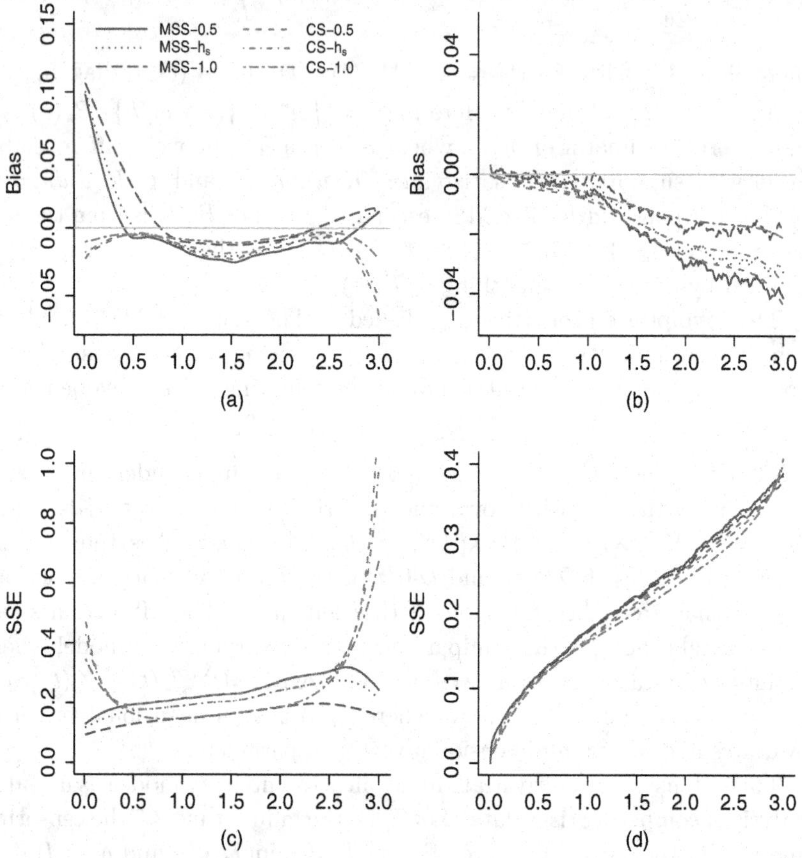

Fig. 13.1 (a) The biases of MSS and CS estimators for $\beta(t)$ using three different bandwidths; (b) the biases of the estimators for the cumulative regression function $B(t) = \int_0^t \beta(s)\, ds$; and their sample standard errors in (c) and (d), respectively.

We conducted a small simulation study to compare the finite sample performance of MSS estimator and the estimator of Cai and Sun (2003), referred as CS hereafter. Both procedures use kernel smoothing. Cross-validation is commonly used for bandwidth selection in the nonparametric function estimation literature. Here we examine an alternative method of bandwidth selection based on the formula $h = C\hat{\sigma}_T {n_o}^{-1/3}$, where n_o and $\hat{\sigma}_T$ are respectively the number of the observed failure times and their estimated standard error, and C is a constant ranging from 2 to 5. A larger C can be used if the distribution of the sampling times is skewed or sparse. Similar bandwidth selection methods have been investigated for nonparametric density estimation and for semiparametric failure time regression in Jones *et al.* (1991) and Zhou and Wang (2000), among others. It is worthwhile to note that a similar formula was adopted for recurrent event data by Qi *et al.* (2016).

As an illustration, we consider the following time-varying Cox model

$$\lambda(t \mid X) = 0.4 \exp\{(-0.2 + 0.05t)X\}, \tag{13.3}$$

where X is a Bernoulli random variable with success probability 0.4 and $\beta(t) = -0.2 + 0.05t$. We consider the simple random right censoring $C \wedge \tau$, where $\tau = 3$ and C is exponentially distributed with parameter specified so that approximately 40% of the observations are censored. We use the Epanechnikov kernel $K(x) = 0.75(1 - x^2)I\{|x| \leq 1\}$. Figure 13.1 displays the bias (Bias) and sample standard error (SSE) of the MSS and CS estimators for $n = 500$, and $h = 0.5, 1.0$ and using the data-driven bandwidth $h_s = 5\hat{\sigma}_T {n_o}^{-1/3}$ based on 1000 independent samples. The MSS estimator of $\beta(t)$ tends to have larger bias near zero and smaller variance near the boundaries. The larger bias of MSS estimator is possibly because it is a locally constant smooth of $\widehat{B}(t)$ and the smaller variance near boundaries could be attributed to using the recursive formula (13.2), where the third term can stabilize the variance. The two procedures, however, tend to show similar performance in estimation of $B(t)$. The data-driven bandwidth $h_s = 5\hat{\sigma}_T {n_o}^{-1/3}$ seem to work well.

Acknowledgment

The work of Yanqing Sun was partially supported by NIAID NIH award R37AI054165 and National Science Foundation grant DMS-1513072.

Efficient Estimation of Fixed and Time-varying Covariate Effects in Multiplicative Intensity Models

TORBEN MARTINUSSEN
The Royal Veterinary and Agricultural University

THOMAS H. SCHEIKE
University of Copenhagen

IB M. SKOVGAARD
The Royal Veterinary and Agricultural University

ABSTRACT. The proportional hazards assumption of the Cox model does sometimes not hold in practise. An example is a treatment effect that decreases with time. We study a general multiplicative intensity model allowing the influence of each covariate to vary non-parametrically with time. An efficient estimation procedure for the cumulative parameter functions is developed. Its properties are studied using the martingale structure of the problem. Furthermore, we introduce a partly parametric version of the general non-parametric model in which the influence of some of the covariates varies with time while the effects of the remaining covariates are constant. This semiparametric model has not been studied in detail before. An efficient procedure for estimating the parametric as well as the non-parametric components of this model is developed. Again the martingale structure of the model allows us to describe the asymptotic properties of the suggested estimators. The approach is applied to two different data sets, and a Monte Carlo simulation is presented.

Key words: cumulative regression functions, Cox-model, martingales, multiplicative intensity, non-parametrics, semi-parametrics, time-varying coefficients

1. Introduction

The Cox regression model is a standard tool for analysing survival data, or more generally event history data; see Andersen *et al.* (1993, ch. 7). Letting $N_i(t)$ be the counting process associated with the ith individual the intensity of the Cox model is assumed to be

$$\lambda_i(t) = Y_i(t)\lambda_0(t)\exp(x_i(t)^{\mathrm{T}}\beta),$$

where $Y_i(t)$ is a predictable process and $x_i(t) \in \mathbf{R}^p$ is a predictable covariate vector. For right-censored life-time data $Y_i(t)$ is 0 or 1 and indicates whether the individual is at risk or not at time t. The baseline $\lambda_0(t)$ gives the intensity for an individual with covariates equal to 0. If the covariates are time independent the model stipulates that the hazard rates for different values of the covariates are proportional. This is often violated when a treatment effect is investigated. It therefore is of interest to investigate the more general intensity model

$$\lambda_i(t) = Y_i(t)\exp(x_i(t)^{\mathrm{T}}\beta(t)), \tag{1}$$

where the effects of covariates are allowed to vary with time. When the first covariate is constant and equal to one ($x_{i1}(t) = 1$) the model contains a baseline $\lambda_0(t)$ that is parametrized as $\exp\{\beta_1(t)\}$. This model has been studied by a number of authors, e.g. Zucker & Karr (1990), Murphy & Sen (1991), Hastie & Tibshirani (1993), Grambsch & Thearneau (1994), and Fahrmeir & Klinger (1998). Murphy & Sen (1991) study a histogram sieve estimator of the cumulative time-varying effects. Practically, the histogram sieve estimator may be difficult to use since one needs to choose a suitable number of time segments and endpoints; see Murphy (1993) for an example. Grambsch and Therneau point out that smoothing of the Schoenfeld residuals may be used to study the functional form of covariate effects. It is of interest to note that our approach may be seen as an improvement of their approach; we return to this in the discussion.

We consider a new likelihood based estimation procedure for the cumulative parameter functions $\int_0^t \beta(s)\,ds$ based on finding a solution to the score equation. Using the fixed point theorem, it is shown that the equation has $n^{1/2}$-consistent solution that is also efficient in the sense of Bickel *et al.* (1993). The properties of the resulting estimator are derived using the martingale structure of the problem. This new estimation procedure ties in nicely with the efficient one-step estimation procedure for Euclidean parameters, see the improvement procedure for semiparametric models in Pfanzagl (1982) and Bickel *et al.* (1993) for a general treatment. More specific applications to martingales and counting processes are given in Sasieni (1992) and in Greenwood & Wefelmeyer (1990, 1991). Notice, however, that the one-step improvement procedure is not immediately applicable to our problem because there is not sufficient information to obtain a $\sqrt{n}$-consistent estimate of β; our iteration step must be combined with smoothing, and $\sqrt{n}$-consistency is obtained for the cumulated parameter processes only.

The general time-varying Cox regression type model (1) is extremely flexible, but may be too large a model for small and medium sized samples when several covariates are considered. Further, some effects may not depend on time and should therefore not be reported or fitted as general non-parametric regression functions. We therefore also study the semiparametric version of (1),

$$\lambda_i(t) = Y_i(t)\exp(x_i(t)^{\mathrm{T}}\beta(t) + z_i(t)^{\mathrm{T}}\gamma) \tag{2}$$

where $x_i(t) \in \mathbf{R}^p$ and $z_i(t) \in \mathbf{R}^q$ are predictable covariate vectors. For this semiparametric model we can apply the same ideas giving us a likelihood based iterative estimation procedure shown to be efficient. The model has not been studied in detail before. Our estimation procedure is novel and makes an otherwise difficult estimation problem rather straightforward.

In sections 2 and 3, we present the estimators for the models (1) and (2) respectively, and derive the asymptotic distribution of the estimators as well as show their efficiency. Section 4 considers some tests for time-dependence of the regression coefficients. An application to the malignant melanoma data (Andersen *et al.*, 1993), and to some myocardial infarction data are given in section 5. We also present a Monte Carlo simulation study illustrating the performance of the estimators. Finally, section 6 contains some closing remarks.

2. Non-parametric models

Consider n independent and identically distributed counting processes, $N_i(t)$, with intensity given by (1). As a particular example we may think of Cox's model for right-censored survival data with time-dependent covariates and parameters, where $Y_i(t)$ is zero if the individual is dead or censored before time t, and the exponential factor is the hazard at time t.

The log-likelihood function is

$$\ell(\beta) = \sum_i \left\{ \int_0^\infty x_i(t)^{\mathrm{T}}\beta(t)\,dN_i(t) - \int_0^\infty Y_i(t)\exp\left(x_i(t)^{\mathrm{T}}\beta(t)\right)dt \right\},$$

apart from a constant non-informative factor due to the processes Y_i and x_i.

To estimate the vector function β we differentiate the log-likelihood with respect to this parameter, for the moment assuming differentiability of ℓ. If γ denotes another vector function, the differential is

$$D\ell(\beta)[\eta] = \sum_i \left\{ \int_0^\infty x_i(t)^{\mathrm{T}} \eta(t)\, dN_i(t) - \int_0^\infty Y_i(t) \exp\left(x_i(t)^{\mathrm{T}} \beta(t)\right) x_i(t)^{\mathrm{T}} \eta(t)\, dt \right\},$$

which is a linear approximation to $\ell(\beta + \eta) - \ell(\beta)$ with error of order $O(\|\eta\|^2)$ as $\|\eta\| \to 0$ where we use $\|\cdot\|$ to denote the norm with respect to coordinates as well as time. Thus, the score equation, setting the above expression to zero for any function η, may formally be written

$$D\ell_t(\beta) = \sum_i \left\{ x_i(t)\, dN_i(t) - Y_i(t) \exp\left(x_i(t)^{\mathrm{T}} \beta(t)\right) x_i(t)\, dt \right\} = 0, \quad t > 0. \tag{3}$$

The same equation may be derived from a quasi-likelihood point of view, using only considerations of means and variances. Given the past, the counting process $dN_i(t)$ has conditional mean and variance $\mathrm{E}_t\, dN_i(t) = \mathrm{var}_t\, dN_i(t) = \lambda_i(t)\, dt$, so the quasi-likelihood equation

$$\sum_i \frac{\partial \mathrm{E}_t\, dN_i(t)}{\partial \beta} \{\mathrm{var}_t\, dN_i(t)\}^{-1} (dN_i(t) - \mathrm{E}_t\, dN_i(t)) = 0$$

simplifies to (3). The equation may also be written

$$X(t)^{\mathrm{T}} \Lambda(t) (\Lambda(t))^{-1} (dN(t) - \lambda(t)\, dt) = 0$$

where $\lambda(t)$ is the n-vector with ith element $\lambda_i(t) = Y_i(t) \exp(x_i(t)^{\mathrm{T}} \beta(t))$, $N(t) = (N_1(t), \ldots, N_n(t))^{\mathrm{T}}$, $X(t) = (x_1(t), \ldots, x_n(t))^{\mathrm{T}}$ and $\Lambda(t) = \mathrm{diag}(\lambda_i(t))$ with $\mathrm{diag}(\cdot)$ denoting the $n \times n$ diagonal matrix with the given elements.

Equation (3) will be the starting point for our estimation although it has no solution as it is written here, because the first term represents a pure jump process while the second is absolutely continuous. Starting from an initial estimate, $\tilde{\beta}$ say, we compute a single Newton–Raphson step towards the (imaginary) solution of the equation. For this we need the second derivative

$$D^2 \ell_t(\tilde{\beta}) = -\sum_i Y_i(t) \exp\left(x_i(t)^{\mathrm{T}} \tilde{\beta}(t)\right) x_i(t) x_i(t)^{\mathrm{T}}\, dt = -\tilde{A}(t)\, dt,$$

where

$$A_\beta(t) = \sum_i Y_i(t) \exp\left(x_i(t)^{\mathrm{T}} \beta(t)\right) x_i(t) x_i(t)^{\mathrm{T}}$$

is a $p \times p$ matrix and $\tilde{A} = A_{\tilde{\beta}}$. Hence the iteration step is

$$\begin{aligned} \tilde{\beta}_{\mathrm{new}}(t) &= \tilde{\beta}(t) + \tilde{A}(t)^{-1} D\ell_t(\tilde{\beta}) \\ &= \tilde{\beta}(t) + \tilde{A}(t)^{-1} X(t)^{\mathrm{T}}\, dN(t) - \tilde{A}(t)^{-1} X(t)^{\mathrm{T}} \tilde{\lambda}(t)\, dt, \end{aligned} \tag{4}$$

where $\tilde{\lambda}$ is λ evaluated with $\beta = \tilde{\beta}$.

It is not a good idea to try to iterate towards a solution for each time point t, because the information about any particular time point is limited so not even consistency can be obtained. To stabilize the solution smoothing is needed. In this work we focus on the cumulative regression coefficients

$$B(t) = \int_0^t \beta(s)\,ds.$$

We integrate the linearized equation (4) to estimate the cumulative regression coefficients and get the iteration step $\tilde{B}^{(k+1)} = g(\tilde{B}^{(k)})$ where

$$g(\tilde{B})(t) = \int_0^t \tilde{\beta}(s)\,ds + \int_0^t \tilde{A}(s)^{-1}X(s)^{\mathrm{T}}\,dN(s) - \int_0^t \tilde{A}(s)^{-1}X(s)^{\mathrm{T}}\tilde{\lambda}(s)\,ds, \tag{5}$$

and introduce smoothness of the underlying regression coefficients through the estimation of $\beta(t)$. For simplicity, $\tilde{\beta}(t)$ is taken to be a simple kernel estimator of $\beta(t)$, that is,

$$\tilde{\beta}(t) = \int b^{-1}K\Big(\frac{u-t}{b}\Big)d\tilde{B}(u),$$

with K a continuous differentiable kernel with support $[-1,1]$ satisfying

$$\int K(u)\,du = 1, \qquad \int uK(u)\,du = 0.$$

Let $\bar{\beta}(t)$ denote the similarly smoothed derivative of $B(t)$, that is,

$$\bar{\beta}(t) = \int b^{-1}K\Big(\frac{u-t}{b}\Big)\,dB(u)$$

and denote $f(\bar{\beta})$ by $\bar{f}$ for any function f of β. Further, let A be A_β evaluated at true parameter.

Theorem 1

Assume that (i) subjects are independent and identically distributed with bounded covariates and

$$\sup_{t\in[0,\tau]} \|n^{-1}A(t) - a(t)\| \xrightarrow{P} 0$$

where a is non-singular with continuous components, (ii) $\beta(t)$ is three times continuously differentiable and that (iii) $b \sim n^{-\alpha}$ where $1/8 < \alpha < 1/4$. Then with probability tending to 1 as $n \to \infty$, (5) has a solution $g(\hat{B}) = \hat{B}$ such that $\|\hat{B} - B\| = O_p(n^{-1/2})$. Further,

$$n^{1/2}(\hat{B} - B) \xrightarrow{\mathcal{D}} U \quad \text{as } n \to \infty$$

in $D[0,\tau]^p$, where U is a mean-zero Gaussian martingale with covariance function

$$\Sigma(t) = \int_0^t a^{-1}(u)\,du. \tag{6}$$

Proof. We first establish a basic result needed in the rest of the proof. We can decompose the counting process as

$$dN(t) = \lambda(t)\,dt + dM(t), \tag{7}$$

where M is a (vector) martingale. By use of the martingale central limit theorem and condition (i), it may be seen that

$$n^{1/2}\int_0^{\cdot} A(s)^{-1}X(s)^{\mathrm{T}}\,dM(s) \xrightarrow{\mathcal{D}} U \quad \text{as } n \to \infty \tag{8}$$

where U is a mean-zero Gaussian martingale with covariance function given by (6). The latter point follows since the predictable variation process of the martingale in (8) is

$$n\left\langle \int_0^t A(s)^{-1}X(s)^{\mathrm{T}}\,dM(s)\right\rangle = n\int_0^t A(s)^{-1}X(s)^{\mathrm{T}}\mathrm{diag}(\lambda_i(s))X(s)A(s)^{-1}\,ds$$
$$= \int_0^t (n^{-1}A(s))^{-1}\,ds$$

which converges in probability to the expression (6).

2.1. *Consistency*

We will now establish condition (A1) and (A2) of lemma 1 in the appendix with $R_n = cn^{-\delta}$, $2\alpha < \delta < 1/2$. By a Taylor series expansion it is seen that

$$\bar{\beta}(t) - \beta(t) = \int b^{-1}K\left(\frac{u-t}{b}\right)dB(u) - \beta(t)$$
$$= O(b^2),$$

where the bound is uniform in t if the third derivative of β is uniformly bounded. Using the decomposition (7) and a Taylor series expansion, we get

$$g(B)(t) = \int_0^t \bar{\beta}(s)\,ds + \int_0^t \bar{A}(s)^{-1}X(s)^{\mathrm{T}}\,\mathrm{diag}\Big(\exp\left(x_i(s)^{\mathrm{T}}(\beta(s) - \bar{\beta}(s))\right) - 1\Big)\bar{\lambda}(s)\,ds$$
$$+ \int_0^t \bar{A}(s)^{-1}X(s)^{\mathrm{T}}\,dM(s)$$
$$= \int_0^t \bar{\beta}(s)\,ds + \int_0^t \bar{A}(s)^{-1}X(s)^{\mathrm{T}}\mathrm{diag}(\bar{\lambda}_i(s))X(s)\{\beta(s) - \bar{\beta}(s)\}\,ds$$
$$+ \int_0^t \bar{A}(s)^{-1}X(s)^{\mathrm{T}}\,dM(s) + O(\|\beta - \bar{\beta}\|^2)$$
$$= B(t) + \int_0^t \bar{A}(s)^{-1}X(s)^{\mathrm{T}}\,dM(s) + O(\|\beta - \bar{\beta}\|^2).$$

Hence,

$$\frac{\|g(B) - B\|}{n^{-\delta}} \le \|n^{\delta-1/2}n^{1/2}\int_0^t \bar{A}(s)^{-1}X(s)^{\mathrm{T}}\,dM(s)\| + O(n^{\delta}\|\beta - \bar{\beta}\|^2)$$

and the right-hand side is seen to converge to zero, because $\delta - 1/2 < 0$ and $\delta - 4\alpha < 0$.

We now turn to condition (A1). By Taylor series expansions, we have

$$g(B_1) - g(B_2) = \int h_s(\bar{\beta}_1(s)) - h_s(\bar{\beta}_2(s))\,ds + \int (\bar{A}_1(s)^{-1} - \bar{A}_2(s)^{-1})X(s)^{\mathrm{T}}\,dM(s) \tag{9}$$

where

$$h_s(\bar{\beta}_j) = \bar{A}_j(s)^{-1}X(s)^{\mathrm{T}}\,\mathrm{diag}(\exp(\eta_{ij}(s)) - \eta_{ij}(s) - 1)\bar{\lambda}_j(s)$$

with $\eta_{ij}(s) = x_i(s)^{\mathrm{T}}(\beta(s) - \bar{\beta}_j(s))$ for $j = 1, 2$. By the mean value theorem, and since h_s is a vector-valued function which vanishes at the true point and which has first differential $h_s'(\beta(s)) = 0$ at the same point,

$$\|h_s(\bar{\beta}_1) - h_s(\bar{\beta}_2)\| \le \sup_{\beta^*}\|h_s'(\beta^*)\|\|\bar{\beta}_1 - \bar{\beta}_2\|,$$

where β^* is between $\bar{\beta}_1$ and $\bar{\beta}_2$. A further application of the mean value theorem shows that

$$\|h_s(\bar{\beta}_1) - h_s(\bar{\beta}_2)\| \le \sup_{\beta^\dagger}\|h_s''(\beta^\dagger)\|\|\bar{\beta}_1 - \bar{\beta}_2\|\|\beta^\dagger - \beta\|,$$

where $\|\beta^\dagger - \bar{\beta}\| \le \sup\{\|\bar{\beta} - \bar{\beta}_1\|, \|\bar{\beta} - \bar{\beta}_2\|\}$. From the assumptions it follows that the second derivative h''_s is bounded. Hence

$$\begin{aligned}\|h_s(\bar{\beta}_1) - h_s(\bar{\beta}_2)\| &\le \| \int b^{-2} K_d\Big(\frac{u-t}{b}\Big)(B_1(u) - B_2(u))\,du\| \; O(\|\beta^\dagger - \bar{\beta}\|)\\ &\le b^{-1}\|B_1 - B_2\| \; O(\|\beta^\dagger - \beta\|),\end{aligned}$$

where K_d is the derivative of K. Since

$$b^{-1}\|\beta^\dagger - \bar{\beta}\| \le b^{-2}\max\|B_j - B\|,$$

which tends to zero in probability because $2\alpha - \delta < 0$, this part of $\|g(B_1) - g(B_2)\|$ will eventually be smaller than any fraction of $\|B_1 - B_2\|$. The second term of (9) is handled in a similar way. Considering only the one dimensional case, and with

$$C(t) = X(t)^{\mathrm{T}}\operatorname{diag}(x_{i1}(t))A(t)X(t),$$

we have (MacRae, 1974)

$$\begin{aligned}(\bar{A}_1(t)^{-1} - \bar{A}_2(t)^{-1}) &= -A(t)^{-1}C(t)(\bar{\beta}_1(t) - \bar{\beta}_2(t))A(t)^{-1}\\ &\quad + (\bar{\beta}_1(t) - \bar{\beta}_2(t))O(\|\beta^\dagger - \bar{\beta}\|)\end{aligned}$$

since the second derivative of $A(t)^{-1}$ (as a function of β) is bounded. For the leading term we get

$$\begin{aligned}&\int A(t)^{-1}C(t)b^{-1}\int K\Big(\frac{u-t}{b}\Big)d(B_1 - B_2)(u)A(t)^{-1}X(t)^{\mathrm{T}}\,dM(t),\\ &= \int A(t)^{-1}C(t)b^{-2}\int K_d\Big(\frac{u-t}{b}\Big)(B_1 - B_2)(u)\,du\; A(t)^{-1}X(t)^{\mathrm{T}}\,dM(t),\\ &= b^{-2}\int\int A(t)^{-1}C(t)K_d\Big(\frac{u-t}{b}\Big)A(t)^{-1}X(t)^{\mathrm{T}}\,dM(t)(B_1 - B_2)(u)\,du,\\ &\le \|B_1 - B_2\| \int\Big|\int b^{-2}A(t)^{-1}C(t)A(t)^{-1}K_d\Big(\frac{u-t}{b}\Big)X(t)^{\mathrm{T}}\,dM(t)\Big|\,du.\end{aligned}$$

The last integral is a martingale integral and we can show that it converges uniformly to 0 by use of Lenglart's inequality (Andersen *et al.*, 1993). Its predictable variation is

$$\begin{aligned}&= b^{-4}\int A(t)^{-1}C(t)A(t)^{-1}X(t)^{\mathrm{T}}A(t)X(t)A(t)^{-1}C(t)A(t)^{-1}K_d^2\Big(\frac{u-t}{b}\Big)\,dt\\ &\le b^{-4}\int A(t)^{-1}C(t)A(t)^{-1}C(t)A(t)^{-1}K_d^2\Big(\frac{u-t}{b}\Big)\,dt\end{aligned}$$

being of order $b^{-3}n^{-1}$ in probability. Since this term just has to tend to zero in probability the rate bounds for b are sufficient.

Thus, the mapping is a contraction and the existence of a solution $\hat{B}$ follows from lemma 1 in the appendix. By use of the martingale central limit theorem and lemma 2 in the appendix, it further follows that this solution is $n^{1/2}$-consistent.

2.2. *Asymptotic normality*

The asymptotic distribution of $\hat{B}$ is obtained from (5) starting from $\hat{B}$. Arguing as in the beginning of the proof, we have

$$\hat{B}(t) = B(t) + \int_0^t \hat{A}(s)^{-1}X(s)^{\mathrm{T}}\,dM(s) + O(\|\beta - \hat{\beta}\|^2).$$

If
(a) $n^{1/2}\int_0^t(\hat{A}(s)^{-1} - A(s)^{-1})X(s)^{\mathrm{T}}\,dM(s) = o_p(1)$

such that $\hat{A}$ can be replaced by the predictable A, and

(b) $n^{1/2}O(\|\beta - \hat{\beta}\|^2) = o_p(1)$,

the proof follows from (8). To show (b), it suffices to choose b such that

$$\|\hat{\beta} - \beta\| = o_p(n^{-1/4}). \tag{10}$$

To this end we split the error of $\hat{\beta} - \beta$ into a bias part and a random part,

$$\hat{\beta}(t) - \beta(t) = \int b^{-1}K\Big(\frac{u-t}{b}\Big)d(\hat{B}(u) - B(u)) + \bar{\beta}(t) - \beta(t)$$

Hence

$$\|\hat{\beta} - \beta\| \le O(b^{-1}\|\hat{B} - B\|) + O(b^2)$$

and (10) is seen to be met with

$$b = n^{-\alpha}, \quad 1/8 < \alpha < 1/4.$$

To show that (a) is valid, we Taylor expand the matrix function (considering only the one dimensional case) and obtain for the leading term of (a):

$$\begin{aligned}
&\int A(t)^{-1}C(t)b^{-2}\int K_d\Big(\frac{u-t}{b}\Big)(\hat{B} - B)(u)\,du\,A(t)^{-1}X(t)^{\mathrm{T}}\,dM(t),\\
&= b^{-2}\int\int A(t)^{-1}C(t)K_d\Big(\frac{u-t}{b}\Big)A(t)^{-1}X(t)^{\mathrm{T}}\,dM(t)(\hat{B} - B)(u)\,du,\\
&\le \|\hat{B} - B\|b^{-2}\bigg\|\int\bigg|\int A(t)^{-1}C(t)A(t)^{-1}K_d\Big(\frac{u-t}{b}\Big)X(t)^{\mathrm{T}}\,dM(t)\bigg|\,du\bigg\|,
\end{aligned}$$

Since $\hat{\beta}$ is a smoothed version of $\hat{B}$ the above change of integrals effectively smoothes the martingale rather than $\hat{B}$ and the martingale central limit theorem applies. By use of Lenglart's inequality it is seen that (a) holds, and the proof is complete.

Remark 1. The estimator $\hat{B}$ is efficient in that the variance (6) attains the variance bound calculated from the efficient influence operator given by Sasieni (1992). The variance of the histogram sieve estimator considered by Murphy & Sen (1991) agrees with (6) and is thus also efficient. Note, however, that our result appears different due to the different parametrization of the baseline.

Remark 2. Considering Aalen's additive intensity model (Aalen, 1980):

$$\lambda_i(t) = Y_i(t)(x_i(t)^{\mathrm{T}}\beta(t)),$$

the score equation similar to the one derived in this paper is

$$X^{\mathrm{T}}(\Lambda(t))^{-1}(dN(t) - X\beta(t)\,dt) \tag{11}$$

where $\Lambda(t) = \mathrm{diag}(Y_i(t)x_i(t)^{\mathrm{T}}\beta(t))$. Solving this equation gives the efficient estimator of Huffer & McKeague (1991). It is possible to solve (11) directly whereas iteration and smoothing is needed for the multiplicative model we study. Efficiency of the estimator based on (11) was discussed in Sasieni (1992) and in Greenwood & Wefelmeyer (1990, 1991).

Remark 3. Consistent estimates of the variance function $\Sigma(t)$ are provided either by

$$n\int_0^t \hat{A}(s)^{-1}\,ds$$

or by the optional variation process

$$n\int_0^t \hat{A}(s)^{-1}X(s)^{\mathrm{T}}\operatorname{diag}(dN(s))X(s)\hat{A}(s)^{-1}$$

Simultaneous Hall–Wellner confidence bands over the period from $[0,\tau]$ are given by

$$\hat{B}(t)\pm C_\gamma\hat{\Sigma}(\tau)^{1/2}\left(1+\frac{\hat{\Sigma}(t)}{\hat{\Sigma}(\tau)}\right)$$

where C_γ is the upper γ-quantile of $\sup_{s\in[0,1/2]}|W^O(s)|$ with $W^O(s)$ the standard Brownian bridge (Hall & Wellner, 1980).

Remark 4. Estimates of the regression functions may be found by smoothing of the estimated cumulative regression functions:

$$\hat{\beta}_j(t)=\int b_j^{-1}K\left(\frac{t-u}{b_j}\right)d\hat{B}_j(u),$$

where b_j is an appropriate bandwidth for smoothing $\hat{B}_j$, $j=1,\ldots,p$. Derivation of asymptotic properties may then be pursued along the lines of Ramlau-Hansen (1983).

Remark 5. From the proof, it is seen that with a probability tending to one the solution is unique within a ball of radius $O(n^{-\delta})$ from B where $2\alpha<\delta<1/2$, $1/8<\alpha<1/4$.

3. Semiparametric models

Consider the semiparametric model

$$\lambda_i(t)=Y_i(t)\exp(x_i(t)^{\mathrm{T}}\beta(t)+z_i(t)^{\mathrm{T}}\gamma)$$

where $x_i(t)\in\mathbf{R}^p$ and $z_i(t)\in\mathbf{R}^q$ are predictable covariate vectors. Define matrices $X(t)=(x_1(t),\ldots,x_n(t))$ and $Z(t)=(z_1(t),\ldots,z_n(t))$. To ease notation we show explicit dependence of time in the following only when we wish to emphasize it. For fixed γ the score equation for $\beta(t)$ is

$$X^{\mathrm{T}}\{dN-\lambda\,dt\}=0.$$

Now, a Taylor expansion around an initial set of estimates $(\tilde{\beta},\tilde{\gamma})$ gives

$$(\beta-\tilde{\beta})\,dt=(X^{\mathrm{T}}DX)^{-1}X^{\mathrm{T}}\{dN-\tilde{\lambda}\,dt-DZ(\gamma-\tilde{\gamma})\,dt\} \tag{12}$$

where $D=\tilde{\Lambda}(t)=\operatorname{diag}(\tilde{\lambda}_i)$. The score equation for γ after a linear Taylor expansion is

$$Z^{\mathrm{T}}\{dN-\tilde{\lambda}\,dt-DX(\beta-\tilde{\beta})\,dt-DZ(\gamma-\tilde{\gamma})\,dt\}=0. \tag{13}$$

Inserting 12 into 13 and solving for γ gives the updating step for γ

$$g_\gamma(\tilde{\gamma})=\tilde{\gamma}+\left(\int_0^\tau Z^{\mathrm{T}}GDZ\,dt\right)^{-1}\int_0^\tau Z^{\mathrm{T}}G(dN-\tilde{\lambda}\,dt). \tag{14}$$

where $G(t) = (I - DX(X^{\mathrm{T}}DX)^{-1}X^{\mathrm{T}})$. Inserting (14) into (12) gives the the updating step for B:

$$g_B(\tilde{B})(t) = \int_0^t \tilde{\beta}(s)\,ds + \int_0^t (X^{\mathrm{T}}DX)^{-1}X^{\mathrm{T}}\{dN - \tilde{\lambda}\,ds - DZ(g_\gamma(\tilde{\gamma}) - \tilde{\gamma})\,ds\}. \tag{15}$$

Before giving the results for the semiparametric model we need some definitions. Let

$$C_1(t) = (n^{-1}\int_0^t Z^{\mathrm{T}}GDZ\,ds)^{-1}, \quad C_2(t) = \int_0^t (X^{\mathrm{T}}DX)^{-1}X^{\mathrm{T}}DZ\,ds,$$
$$C_3(t) = \int_0^t (X^{\mathrm{T}}DX)^{-1}\,ds$$

with limits in probability $c_1(t)$, $c_2(t)$ and $c_3(t)$ respectively. Finally define

$$\Sigma(t) = \begin{pmatrix} c_1(\tau) & -c_1(\tau)c_2(t)^{\mathrm{T}} \\ -c_2(t)c_1(\tau)^{\mathrm{T}} & c_3(t) + c_2(t)c_1(\tau)c_2(t)^{\mathrm{T}} \end{pmatrix}.$$

Theorem 2

Under the assumptions of theorem 1 and with $c_1(t), c_2(t)$ and $c_3(t)$ existing and yielding a non-singular matrix $\Sigma(t)$, (14) and (15) have $n^{1/2}$-consistent solutions $(g_\gamma(\hat{\gamma}) = \hat{\gamma}, g_B(\hat{B}) = \hat{B})$ with a probability tending to 1 as $n \to \infty$. Further,

$$n^{1/2}(\hat{\gamma} - \gamma) \xrightarrow{\mathscr{D}} V \quad \text{as } n \to \infty,$$

where V is a mean-zero normal with variance $\Sigma_{11}(\tau)$, and

$$n^{1/2}(\hat{B} - B) \xrightarrow{\mathscr{D}} W \quad \text{as } n \to \infty$$

in $D[0,\tau]^p$, where W is a mean-zero Gaussian process with variance $\Sigma_{22}(\cdot)$. The covariance between V and $W(t)$ is $\Sigma_{12}(t)$.

Proof. We focus only on the distributional properties of the estimators $(\hat{\gamma}, \hat{B})$; other aspects of the proof follow the proof of theorem 1. Rewriting the expressions for the estimators and using a Taylor-series expansion, we get

$$n^{1/2}(\hat{\gamma} - \gamma) = \left(n^{-1}\int_0^\tau Z^{\mathrm{T}}GDZ\,dt\right)^{-1} n^{-1/2}\int_0^\tau Z^{\mathrm{T}}G\,dM = C_1(\tau)M_1(\tau)$$

where

$$M_1(t) = n^{-1/2}\int_0^t Z^{\mathrm{T}}G\,dM,$$

and

$$\begin{aligned} n^{1/2}(\hat{B}(t) - B(t)) &= n^{1/2}\int_0^t (X^{\mathrm{T}}DX)^{-1}X^{\mathrm{T}}\,dM - n^{1/2}\int_0^t (X^{\mathrm{T}}DX)^{-1}X^{\mathrm{T}}DZ\,ds(\hat{\gamma} - \gamma) \\ &= M_2(t) - C_2(t)C_1(\tau)M_1(\tau) \end{aligned}$$

where

$$M_2(t) = n^{1/2}\int_0^t (X^{\mathrm{T}}DX)^{-1}X^{\mathrm{T}}\,dM.$$

Now, proceeding as in theorem 1 it follows that the non-predictable integrands can be replaced by predictable integrands, and that the additional error terms will converge to 0 when multiplied by $n^{1/2}$. Therefore, the martingale central limit theorem implies that

$$(M_1, M_2)^{\mathrm{T}} \xrightarrow{\mathscr{D}} U = (U_1, U_2) \quad \text{as } n \to \infty$$

in $D[0,\tau]^{(p+q)}$, where U is a mean-zero Gaussian martingale. Hence $n^{1/2}(\hat{\gamma} - \gamma)$ converges in distribution towards a mean-zero normal V with a variance given as the limit in probability of

$$C_1(\tau)\langle M_1\rangle(\tau)C_1(\tau)^{\mathrm{T}} = \left(n^{-1}\int_0^\tau Z^{\mathrm{T}}GDZ\,dt\right)^{-1} = C_1(\tau),$$

and

$$n^{1/2}(\hat{B}(t) - B(t)) \xrightarrow{\mathscr{D}} U_2(t) - c_2(t)c_1(\tau)U_1(\tau) \tag{16}$$

where the covariance function of the right-hand side of (16) is given as the limit in probability of

$$\langle M_2(t) - C_2(t)C_1(\tau)M_1(\tau)\rangle = \int_0^t (X^{\mathrm{T}}DX)^{-1}\,ds + C_2(t)C_1(\tau)C_2(t)^{\mathrm{T}} + o_p(1)$$

since $\langle M_2, M_1\rangle(t)$ converges in probability to zero. Finally, the covariance function is given as the limit of $\langle C_1(\tau)M_1(\tau), -C_2(t)C_1(\tau)M_1(\tau)\rangle = -C_1(\tau)C_2(t)^{\mathrm{T}}$.

Remark 6. The suggested estimator $\hat{\gamma}$ for γ is efficient, as its variance attains the variance bound calculated from the efficient influence operator given in Sasieni (1992).

Remark 7. McKeague & Sasieni (1994) analyse the semiparametric additive risk model
$\lambda_i(t) = Y_i(t)(x_i(t)^{\mathrm{T}}\beta(t) + z_i(t)^{\mathrm{T}}\gamma)$
where $x_i(t) \in \mathbf{R}^p$ and $z_i(t) \in \mathbf{R}^q$ are predictable covariate vectors. Applying the estimation principle presented here for the semiparametric model (2) to the semiparametric additive risk model yields the estimation procedure of McKeague and Sasieni.

Remark 8. In the following section we propose a test for whether or not an effect of a covariate is time-varying. The test is based on the asymptotic description of the non-parametric components of the semiparametric model. Therefore, note that a consistent estimator of the variance function of the process $n^{1/2}(\hat{B}(t) - B(t))$ is

$$\int_0^t (X^{\mathrm{T}}DX)^{-1}\,ds + C_2(t)C_1(\tau)C_2(t)^{\mathrm{T}}.$$

4. Test for time-varying effects

In this section we briefly outline a test for whether or not an effect of a covariate is time-varying. We consider the semiparametric regression model

$$\lambda_i(t) = Y_i(t)\exp(x_i(t)^{\mathrm{T}}\beta(t) + z_i(t)^{\mathrm{T}}\gamma),$$

where $x_i(t) = (x_{i1}(t), \ldots, x_{ip}(t)) \in \mathbf{R}^p$ and $z_i(t) \in \mathbf{R}^q$ are predictable covariate vectors, and wish to test the hypothesis H_0: $\beta_p(t) \equiv \beta_p$ vs the alternative that $\beta_p(t)$ is varying with time. If this hypothesis can be investigated with a test we can carry out successive tests to investigate which components have time dependent effects. Many authors have considered tests for time dependence of covariate effects, see Murphy (1993) and Grambsch & Therneau (1994) and references therein. These tests are primarily concerned with tests aimed at validating the Cox-regression model, which is a special case of the general semiparametric model considered in this paper. Murphy (1993) proposed a test for the hypothesis considered in this work based on sieve estimators, and derived the asymptotic distribution of the test-statistic. Grambsch & Therneau (1994) suggested a test for the proportional hazard assumption in the Cox-regression model vs a specific alternative. This limits the usefulness of their procedure since it is rarely known in practice which deviations to look for.

A test for H_0 may be based on the following process

$$n^{1/2}\left(\hat{B}_p^w(t) - \hat{\beta}_p \int_0^t w(s)\,ds\right) \tag{17}$$

where $\hat{B}_p^w(t) = \int_0^t w(s)d\hat{B}_p(s)$ with $w(t)$ denoting a stochastic weight function and with $\hat{B}_p(t)$ the estimator of $B_p(t)$ obtained before simplifying the model, while $\hat{\beta}_p$ is computed under the null hypothesis. It may be shown that (17), under the null hypothesis, converges towards a mean-zero Gaussian process, see Martinussen & Scheike (1999) for a similar construction. One may then perform a maximal deviation test based on (17). The limiting distribution of (17) is, however, complicated and the distribution of the maximal deviation test statistic needs to be simulated. Alternatively, one may use the suggestion by Khmaladze (1980), see also Andersen *et al.* (1993, pp. 464–469).

5. Examples

5.1. Application to malignant melanoma data

We apply the methods to a data set given in Andersen *et al.* (1993). It concerns survival with malignant melanoma (cancer of the skin). In the period 1962–77, 205 patients had their tumor removed and were followed until the end of 1977. The time variable is time since operation and the number of deaths in the considered period was 57. We focus on the covariates: sex of the patient (coded 1 for male and 0 for female), tumor thickness and presence/absence of ulceration in the tumor (coded 1 for presence and 0 for absence). We initially treated all covariates and the baseline non-parametrically as in model (1). We computed the suggested estimator using the Epanechnikov kernel with bandwidth $b = 16$ months for smoothing. Figure 1 shows the estimator of the cumulative regression functions together with pointwise 95% confidence bands based on the optional variation process, see the remarks to theorem 1. The effect of ulceration changes at about two years which is also seen in the baseline showing the non-parametric function for a female patient with no ulceration and zero log (thickness). The reason for the abrupt change is that the first 18 deaths are all patients with ulceration, and then the nineteenth death is a female patient with no ulceration. The effect of sex and log (thickness) seems to be constant with time. We hence treated the baseline and ulceration non-parametrically while sex and log (thickness) were treated parametrically with constant effect.

The estimates of the time-independent parameters, with estimated standard errors in brackets, are 0.32 (0.23) and 0.55 (0.15), for sex and log (thickness) respectively. These estimates are also added to Fig. 1, showing that the straight line estimates are largely contained within the confidence limits around the non-parametric estimates. Figure 1 also contains the straight line estimate for ulceration (obtained from the model where ulceration, sex and log (thickness) were treated parametrically with constant effect) which in the first four to five years falls outside the confidence limits supporting that the influence of ulceration varies

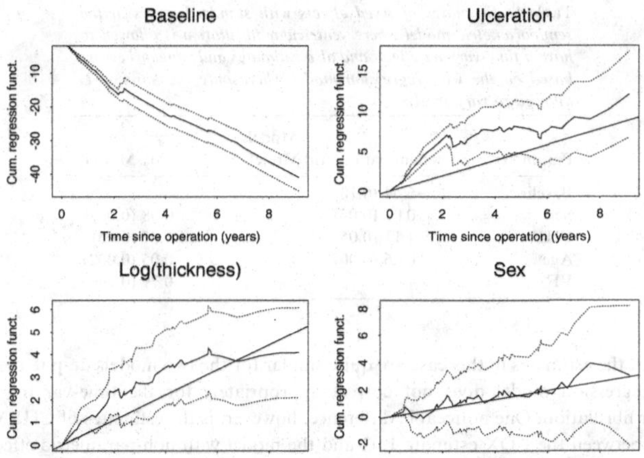

Fig. 1. Estimates of the cumulative regression functions with 95% pointwise confidence limits. Straight line estimates are obtained from model (2).

with time. Ulceration and log (thickness) are significant risk factors while sex is non-significant. The non-parametrics estimates in the semiparametric model gave estimates that are almost identical with the purely non-parametric effects shown on Fig. 1. Uniform Hall–Wellner bands were constructed but are not shown and were very wide, thus indicating that we are at the limit of the data when trying to estimate as many as four non-parametric effects with only 57 deaths.

5.2. Application to myocardial infarction data

The TRACE study group (see e.g. Jensen *et al.*, 1997) has studied the prognostic importance of various risk factors on mortality for approximately 6000 patients with myocardial infarction. The patients had various risk factors recorded such as age, sex (male = 1), congestive heart failure (CHF) (present = 1) and ventricular fibrillation (VF) (present = 1). We use time since myocardial infarction as our time-scale. The total number of deaths in a 3-year period after the infarction was 2020, and of these, 649 took place within the first month. Some risk factors were expected to have strongly time-varying effects, in particular ventricular fibrillation. We fitted a model with time-varying effects and with a bandwidth that varied with time. Allowing for a small bandwidth ($b = 0.4$) just after the infarction where lots of information is available, and increasing approximately linearly to a larger bandwidth ($b = 0.9$) 3 years into the study.

The estimates of the cumulative effects clearly indicate that most effects do not have strongly time-varying effects, only ventricular fibrillation appears to have a strongly time-varying effect, with a great prognostic impact the first month where after its effect vanishes. Uniform Hall–Wellner confidence limits were added to the estimates indicating that only ventricular fibrillation had a significantly time-varying effect. We therefore fitted a semi-parametric version of the model where only ventricular fibrillation had a time-varying effect. Additionally, we fitted a regular Cox regression model, and estimates are given in Table 1.

Table 1. *Estimates of fixed effects with standard errors in the semiparametric model where ventricular fibrillation is allowed to have a time-varying effect, and also estimates and standard errors based on the Cox regression model where only the baseline is allowed to vary freely*

	Models	
Parameter	Semiparametric Model	Cox Model
Baseline	−6.64 (0.18)	—
Sex	0.09 (0.05)	0.08 (0.05)
CHF	1.11 (0.05)	1.02 (0.05)
Age	0.06 (0.002)	0.05 (0.002)
VF	—	0.79 (0.07)

Note that the estimates in this case are quite similar for the two models despite the fact that the Cox regression model does not correct appropriately for the time-varying effect of ventricular fibrillation. One noticeable difference, however, is the estimates of CHF where the difference between the COX estimate 1.02 and the model with non-parametric effect for VF 1.11 is almost two standard deviations. The effect of ventricular fibrillation based on the semiparametric model is not shown but is very similar to the one given in the model where all effects are non-parametric. The semiparametric model provides simple interpretable estimates of non-varying effects and an accurate description of time-varying effects, as well as assurance that the non-varying effects are not seriously biased.

5.3. Simulations

We conducted a small simulation study to investigate if the finite sample properties of the estimator are well described by the derived asymptotic distribution. The main interest focus on the performance of the standard errors of the cumulative effects.

We simulated survival times based on the intensity

$$\lambda_i(t) = \exp(-0.5 + 0.5 * \sin((t + 1.5))X_{i1} + 0.5 * \sin(t)X_{i2})$$

where X_{i1} and X_{i2} are independent Bernoulli variables with probability equal to 0.5. The bandwidth was chosen to be 1.5 and we used the Tukey kernel $(\cos(x\pi) + 1)/2$. We simulated 1000 repetitions of 200 independent identically distributed realizations that were censored at time 3. We also made 1000 repetitions for sample size 100, and got a pretty well-behaved estimator, except for some difficulties for larger time-points. We therefore report the results of sample size 200. Approximately 10% of the survival times were censored.

The initial baseline estimate was started at the log occurrence-exposure rate. Final estimates were reached at convergence.

The integrated squared bias were 1.6×10^{-3} and 2.5×10^{-3}, and the integrated mean squared error were 0.39 and 0.40 for the two non-parametric covariate effects, respectively. The estimator performed quite well and essentially provides an unbiased estimate of the cumulative regression functions.

To investigate the performance of the asymptotic standard errors we have computed pointwise coverage probabilities that are depicted in Fig. 3c and d for the two non-parametric covariate effects. Figures 3a and b gives a pointwise description of the bias.

It appears that the asymptotic standard errors provide an excellent description of the variability in the present study. We therefore conclude that the standard errors will have some merit for small sample data.

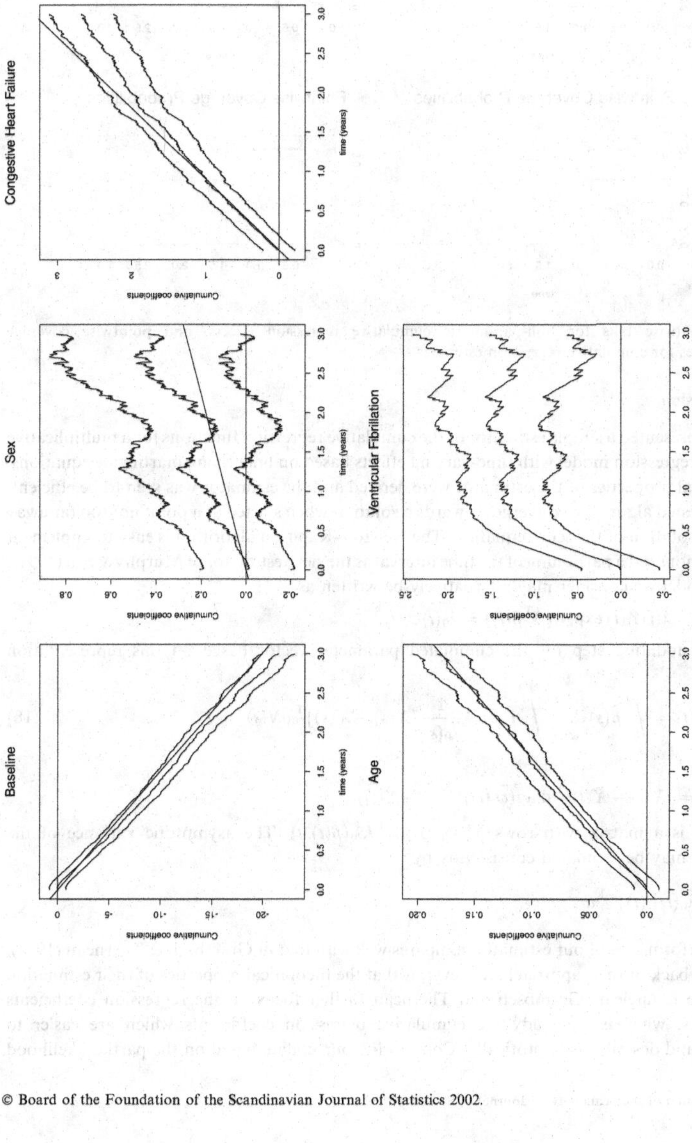

Fig. 2. Estimates of the cumulative regression functions with 95% uniform Hall–Wellner confidence bands.

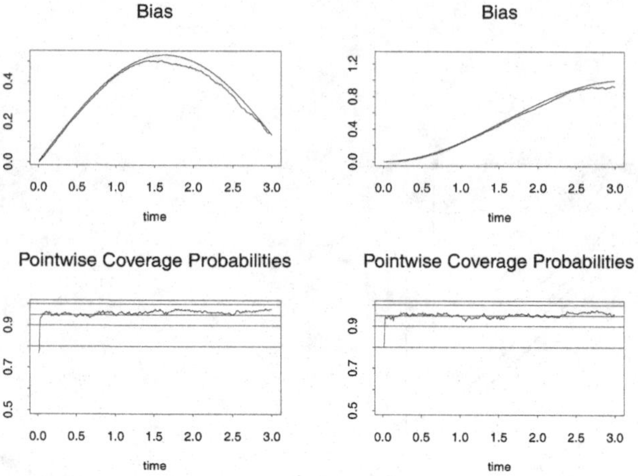

Fig. 3. Pointwise bias for non-parametric cumulative regression effects, and pointwise coverage probabilities for cumulative regression estimates.

6. Discussion

We have presented a simple estimator of the cumulative regression functions for a multiplicative intensity regression model with time-varying effects based on underlying martingale equations. Theoretical properties of the estimator were derived and the estimator was seen to be efficient. The suggested algorithm converges towards a solution when started at a point not too far away from the solution of the score equation. The Newton–Raphson algorithm is easy to implement and does not need a partioning of the time interval as the sieve estimator of Murphy & Sen (1991).

The model we consider may alternatively be written as

$$\lambda_i(t) = \lambda(t)Y_i(t)\exp(x_i(t)^{\mathrm{T}}\beta(t)) = \lambda_0(t)\phi_i(t),$$

say. The updating step for the cumulated parameter vector based on this representation becomes

$$g(\tilde{B})(t) = \int_0^t \tilde{\beta}(s)\,ds + \int_0^t \tilde{\Gamma}(s)^{-1}\frac{1}{\tilde{\lambda}_0(s)}\{X(s) - \bar{X}(s)\}^{\mathrm{T}}\,dN(s) \tag{18}$$

where

$$\tilde{\Gamma}(t) = (X(t) - \bar{X}(t))^{\mathrm{T}}\operatorname{diag}(\tilde{\phi}_i(t))(X(t) - \bar{X}(t)),$$

and $\bar{X}(t)$ is a matrix with rows $\sum_i \tilde{\phi}_i(t)x_i(t)^{\mathrm{T}}/S_0(\tilde{\beta}(t),t)$. The asymptotic variance of the estimator may be estimated consistently by

$$\int_0^{\cdot}\{\hat{\lambda}_0(t)\hat{\Gamma}(t)\}^{-1}\,dt.$$

The performance of our estimator compares well with that of Grambsch & Therneau (1994). One drawback of their approach, however, is that the theoretical properties of their estimation procedure is unclear. Grambsch and Therneau further focus on the regression coefficients themselves, whereas we study the cumulative regression coefficients which are easier to estimate and describe asymptotically. Considering our update based on the partial likelihood

(18) and the scaled Schoenfeld residual of Grambsch and Therneau it appears that these are equivalent. Note, however, that the Schoenfeld residuals are based on fitting a Cox regression model with fixed effects whereas our residuals are based on a time-varying model. The smoothing of the scaled Schoenfeld residuals that are used to provide a rough guess of the functional form can therefore be understood, in this framework, as a one-step improvement of an initial time-constant estimator.

In addition to the analysis of the Cox model with time-varying effects we studied a semiparametric submodel where some regression effects are constant over time. This model is often more sensible to implement for small to medium sample sizes and further, can summarize effects more clearly. We proposed estimators for the non-parametric as well as parametric components of the model and derived their asymptotic distribution, which showed that the estimation procedure was efficient.

As usual the choice of the bandwidth will influence the results. Cross-validation and plug-in estimates of the bandwidth are both a possibility. For the TRACE data it seems obvious to have a time-varying bandwidth to reflect that more information is available early in the time period. In the semi-parametric version of the model, the parametric components appears to be rather stable to different choices of the bandwidth. We tried three different choices 0.5, 1 and 1.5 for the TRACE study and got almost identical results. Further research is needed to provide theoretical guidance for choice of bandwidth.

Acknowledgments

We are grateful to a referee and an associate editor for their helpful comments. We thank the TRACE study group for letting us use their data. The second author was supported by NIH grant 5 R01 CA54706-09.

References

Aalen, O. (1980). A model for non-parametric regression analysis of counting processes. In *Mathematical statistics and probability theory* (eds W. Klonecki, A. Kozek & J. Rosinski). Lecture Notes in Statistics **2**, 1- 25. Springer-Verlag, New York.

Andersen, P. K., Borgan, Ø., Gill, R. & Keiding, N. (1993). *Statistical models based on counting processes.* Springer-Verlag, New York.

Bickel, P. J., Klaassen, C. A. J., Ritov, Y. & Wellner, J. A. (1993). *Efficient and adaptive estimation for semiparametric models.* Springer-Verlag, New York.

Fahrmeir, L. & Klinger, A. (1998). A nonparametric multiplicative hazard model for event history analysis. *Biometrika* **85**, 581–592.

Friedman, A. (1970). *Foundations of modern analysis.* Dover Publications, New York.

Grambsch, P. M. & Therneau, T. M. (1994). Proportional hazards tests and diagnostics based on weighted residuals. *Biometrika* **81**, 515–526.

Greenwood, P. E. & Wefelmeyer, W. (1990). Efficiency of estimators for partially specified filtered models. *Stochastic Process. Applic.* **36**, 353–370.

Greenwood, P. E. & Wefelmeyer, W. (1991). Efficient estimating equations for nonparametric filtered models. In *Statistical inference in stochastic processes*, 107–141. Marcal Dekker, New York.

Hall, W. & Wellner, J. (1980). Confidence bands for a survival curve from censored data. *Biometrika* **67**, 133–143.

Hastie, T. & Tibshirani, R. (1993). Varying-coefficient models (with discussion). *J. Roy. Statist. Soc. Ser. B* **55**, 757–796.

Huffer, F. W. & McKeague, I. W. (1991). Weighted least squares estimation for Aalen's additive risk model. *J. Amer. Statist. Assoc.* **86**, 114–129.

Jensen, G. V., Torp-Pedersen C., Hildebrandt P., Kober, L., Nielsen, F. E., Melchior, T., Joen, T. & Andersen P. K. (1997). Does in-hospital ventricular fibrillation affect prognosis after myocardial infarction? *Euro. Heart J.* **18**, 919–924.

Khmaladze, E. V. (1981). Martingale approach to the goodness of fit tests. *Theory Probab. Appl.* **26**, 246–265.
MacRae, E. C. (1974). Matrix derivatives with an application to an adaptive linear decision problem. *Ann. Statist.* **2**, 337–346.
Martinussen, T. & Scheike, T. H. (1999). A semi-parametric additive regression model for longitudinal data. *Biometrika* **86**, 691–702.
McKeague, I. W. & Sasieni, P. D. (1994). A partly parametric additive risk model. *Biometrika* **81**, 501–514.
Murphy, S. A. (1993). Testing for time dependent coefficient in Cox's regression model. *Scand. J. Statist.* **20**, 35–50.
Murphy, S. A. & Sen, P. K. (1991). Time-dependent coefficients in a Cox-type regression model. *Stochastic Process. Applic.* **39**, 153–180.
Pfanzagl, J. (1982). *Contributions to a general asymptotic statistical theory*. Lecture Notes in Statistics **13**, Springer-Verlag, New York.
Ramlau-Hansen, H. (1983). Smoothing counting process intensities by means of kernel functions. *Ann. Statist.* **11**, 453–466.
Sasieni, P. D. (1992). Information bounds for the additive and multiplicative intensity models (disc: P263–265). In *Survival analysis: state of the art* (eds J.P. Klein and P. K. Goel) 249–263. Kluwer, Dordrecht, Boston, London.
Zucker, D. M. & Karr, A. F. (1990). Nonparametric survival analysis with time-dependent effects: a penalized partial likelihood approach. *Ann. Statist.* **18**, 329–353.

Received June 2000, in final form April 2001

Torben Martinussen, Department of Mathematics and Physics, The Royal Veterinary and Agricultural University, Thorvaldsensvej 40, DK-1871 Frederiksberg C, Denmark.
E-mail: torbenm@dina.kvl.dk

Appendix

Consider a sequence of estimating equations

$$g_n(\omega) = \omega \tag{19}$$

for a parameter $\omega \in \Omega$ where Ω is a normed vector space. We want to specify conditions for the asymptotic existence of a solution converging to a particular (true) point ω_0. When we state that a result holds asymptotically we mean that for any $\varepsilon > 0$ there exists an integer n_0 such that

$$P_n(\text{result holds}) > 1 - \varepsilon$$

for all $n > n_0$.

The following lemmas specify conditions under which the fixed point theorem may be used for our purpose. The fixed point theorem states that if a mapping, g say, is a contraction of the metric space, E say, into itself, it has a unique fixed point: $g(\omega) = \omega$; see for example Friedman (1970, sect. 3.8). That g is a contraction means that there is a constant $\theta < 1$ such that

$$\text{dist}(g(\omega'), g(\omega)) \le \theta \text{ dist } (\omega', \omega)$$

for all ω', ω in E.

As a final preparation define the closed ball

$$B(\omega_0, R) = \{\omega \in \Omega : \|\omega - \omega_0\| \le R\}.$$

The first lemma gives the basic argument.

Lemma 1

Assume that for some bounded sequence R_n and some $\theta < 1$,
(A1) asymptotically, the mapping g_n is a θ-contraction on $B_n = B(\omega_0, R_n)$, that is

$$\|g_n(\omega') - g_n(\omega)\| \le \theta\|\omega' - \omega\|, \quad \omega', \omega \in B_n,$$

(A2) we have

$$\frac{\|g_n(\omega_0) - \omega_0\|}{R_n} \to 0$$

in probability as $n \to \infty$.

Then asymptotically, there exists a unique point, $\hat{\omega}_n$ *in* B_n *such that*

$$g_n(\hat{\omega}_n) = \hat{\omega}_n.$$

Proof. Assumption (A1) directly states that g_n is a contraction on B_n, so for the application of the fixed point theorem it only remains to show that asymptotically g_n maps B_n into itself. When the two assumptions hold, any $\omega \in B_n$ satisfies

$$\begin{aligned}\|g_n(\omega) - \omega_0\| &\leq \|g_n(\omega) - g_n(\omega_0)\| + \|g_n(\omega_0) - \omega_0\| \\ &\leq \theta\|\omega - \omega_0\| + \frac{\|g_n(\omega_0) - \omega_0\|}{R_n} R_n,\end{aligned}$$

which is eventually smaller than $\theta R_n + ((1 - \theta)/2)R_n$, say, for all $\omega \in B_n$.

Since lemma 1 establishes uniqueness of the solution it is an advantage to apply the result with the sequence R_n decreasing as slowly as possible, usually at best being constant. It is the verification of condition (A1) that limits our possibility of letting R_n decrease slowly, whereas condition (A2), becomes easier. From another point of view, namely that of rate of convergence, it is desirable to verify the conditions with R_n decreasing as rapidly as possible, usually at best approaching the rate of $n^{-1/2}$. In this direction the verification of assumption (A2) is limiting. In the following lemma we introduce a rate, r_n which is more rapid than R_n. The typical case is $r_n = 1/\sqrt{n}$ while R_n is constant or slowly decreasing.

Lemma 2

Assume that the conditions of lemma 1 *hold for the sequence* R_n. *Let* r_n *be a sequence with* $r_n/R_n \to 0$ *for which the following assumption holds:*

(C) The sequence $(g_n(\omega_0) - \omega_0)/r_n$ *is asymptotically tight, that is, for any* $\varepsilon > 0$ *there exists a* C *and an* n_0 *such that*

$$P\left(\frac{\|g_n(\omega_0) - \omega_0\|}{r_n} \leq C\right) > 1 - \varepsilon$$

for all $n > n_0$.

Then the unique solution $\hat{\omega}_n$ *to* (19) *in* B_n *is* r_n*-consistent, that is, the sequence* $(\hat{\omega}_n - \omega_0)/r_n$ *is asymptotically tight.*

Proof. Let $\varepsilon > 0$ be given and let C be as stated in the lemma. As in the proof of lemma 1 we get

$$\|g_n(\omega) - \omega_0\| \leq \theta\|\omega - \omega_0\| + \frac{\|g_n(\omega_0) - \omega_0\|}{cr_n} cr_n,$$

for any $c > 0$ and $\omega \in B_n$. Now choose c such that $C/c \leq (1 - \theta)/2$ implying that the right hand side is smaller than cr_n for all $\omega \in B(\omega_0, cr_n)$ with probability at least $1 - \varepsilon$ for all $n > n_0$. Thus the ball with radius cr_n is eventually mapped into itself, and since $cr_n/R_n \to 0$ the mapping is also eventually a contraction on this set. Hence the fixed point theorem shows the existence of a solution within $B(\omega_0, cr_n)$.

Chapter 14

Diffusion-type models with given marginal distribution and autocorrelation function

Introduction by Helle Sørensen
University of Copenhagen

14.1 Contributions

I am very pleased to have the opportunity to acknowledge Ib Skovgaard for his contributions to statistics. I am indebted to Ib for what he taught me, in particular about applied statistics, so although the current paper is part of Ib's theoretical work, I will treat it from an applied point of view. The paper will be referred to as BSS in the following.

Given a stationary time series, a good model should capture the invariant distribution as well as the auto-correlation structure. The BSS paper is exactly about selection of appropriate models in these respects.

The scene is that of stochastic differential equations (SDEs). SDEs are used as building blocks in many stochastic dynamical systems, as they incorporate random disturbances into deterministic ODE systems in a natural way. An SDE is specified by its drift function and diffusion function. In some applications these functions are proposed by theory or experience, but more often they must be decided upon empirically.

The simplest set-up in BSS is the following: Assume that the drift is linear, implying that the auto-correlation function is an exponential function, and consider a pre-specified marginal distribution. Then the paper provides a diffusion function such that the corresponding diffusion has the specified marginal distribution as its invariant distribution. By considering sums of diffusions it is possible to obtain auto-correlation functions that are weighted sums of exponentials, corresponding to several time scales, still maintaining a pre-specified invariant distribution. Mean-reverting, but not linear, drift functions and multivariate models are covered more briefly.

The BSS paper includes a long list of marginal distributions with explicit expressions for the corresponding diffusion function. Moreover, the class of attainable distributions is extended with the use of saddlepoint approximations to also cover distributions like the variance-gamma and the the normal-inverse Gaussian distributions. This draws on Ib's experience and published work in saddlepoint approximation.

As described, the BSS paper is an article on statistical modelling: the empirical auto-correlation is used to identify the number of components in the model, and the empirical marginal distribution is used to define an appropriate class of marginal distributions. The two parts of the model can be estimated separately, at least in principle (see below), whereas estimation using the complete structure of the model is much more involved (Sørensen, 2004, 2012).

A natural question therefore arises: Why do we need a model at all? One keyword, in my opinion, is *simulation.* With a completely specified model it is possible to simulate artificial data and study properties of data, estimators, test statistics, *etc.* As a particularly important application of simulations we can quantify the precision of estimators by parametric bootstrap methods. This point will be illustrated below.

14.2 Application and simulation

In this section I use a data series from neuroscience to illustrate the model framework and the usefulness of a complete probability model for the data generating mechanism.

The data comes from an experiment with turtles. Briefly, the turtles were killed, the carapace was removed from the animals, and the membrane potential was recorded from isolated moto-neurons, which are still active even though the animal is dead. See Berg and Ditlevsen (2003) for more details. I use the membrane potential from a single neuron. A data window has been selected such that it is reasonable to assume stationarity. The number of observations is $n = 4701$ corresponding to a time interval of 470 ms. The time series is shown in the top panel of Fig. 14.1.

I will be particularly interested in the mean, standard deviation and skewness of the stationary distribution. The sample estimates are

$$\bar{x} = -64.4, \quad \mathrm{SD}_x = 3.39, \quad \mathrm{Skewness}_x = 0.340. \tag{14.1}$$

These point estimates should be supplied with measures of uncertainty such as standard errors or confidence intervals. I will use parametric bootstrap simulation for that purpose and thus need an appropriate model to sample from.

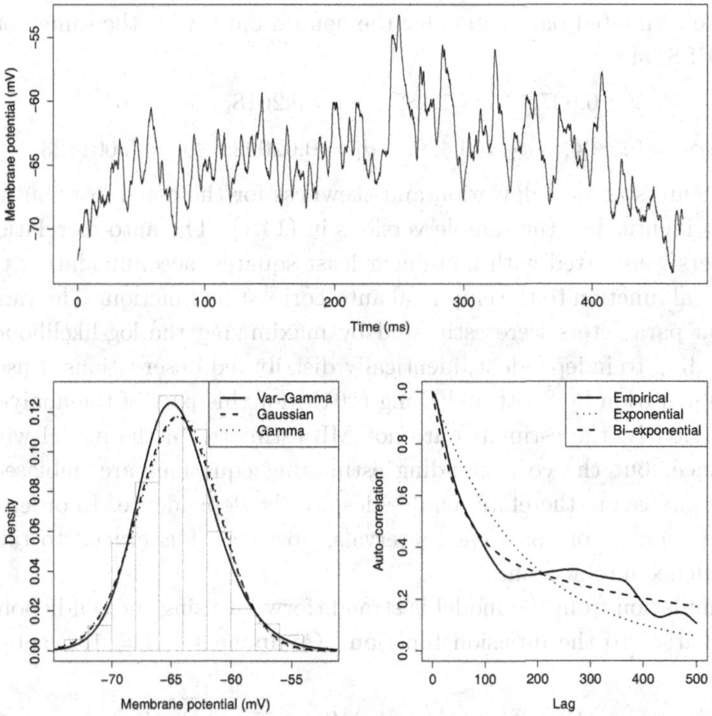

Fig. 14.1 The neuron data: The time series of membrane potential (top); histogram with three fitted densities (bottom left); and the empirical auto-correlation function with exponential and bi-exponential fit (bottom right).

In line with the BSS paper I start by examining the marginal distribution and the auto-correlations. The bottom left panel of Fig. 14.1 shows a histogram of the data together with three densities: a Gaussian density (dashed), a gamma density reflected at zero (dotted), and a variance-gamma density (solid). The bottom right panel shows the empirical auto-correlations (solid) as well as two fitted auto-correlation functions: an exponential (dotted) and a weighted sum of two exponentials, in the following referred to as bi-exponential (dashed). The variance-gamma density approximates the histogram slightly better than the Gaussian and gamma densities, and the bi-exponential function fits the empirical version reasonably well, so I choose to model the data as a sum of two variance-gamma diffusions, see Examples 3.1 and 4.2 in BSS. Notice that the model allows for skewness, and that the same model was used for the wind data in BSS (§6).

The estimated parameters for the neuron data, with the same notation as in BSS, are

$$\hat{\lambda} = 6.927, \quad \hat{\alpha} = 1.182, \quad \hat{\beta} = 0.2618, \quad \hat{\delta} = -67.13,$$
$$\hat{\phi}_1 = 0.5804, \quad \hat{\phi}_2 = 0.420, \quad \hat{\theta}_1 = 0.02092, \quad \hat{\theta}_2 = 0.001823.$$

The mean, standard deviation and skewness for the fitted distribution are almost identical to the sample versions in (14.1). The auto-correlation parameters were fitted with non-linear least squares, accommodating the bi-expontial function to the empirical auto-correlation function. The variance-gamma parameters were estimated by maximizing the log-likelihood corresponding to independent, identically distributed observations; I used the implementation by Scott and Dong (2015) for this part of the analysis.

Obviously, the estimates are not ML estimates in the model with dependence, but the corresponding estimating equations are unbiased and the estimates are therefore reliable despite the dependence. In order to get standard errors or confidence intervals, however, it is crucial to take the dependence into account.

Simulation from the model is straightforward using the saddlepoint approximation to the diffusion function. Component i, $i = 1, 2$, solves the SDE

$$dX_{i,t} = -\theta_i(X_{i,t} - \mu_i)\, dt + \sqrt{\tilde{v}_i(X_{i,t})}\, dW_{i,t},$$

where $\tilde{v}_i$ depends on parameters $(\phi_i\lambda, \alpha, \beta, \phi_i\delta)$ as explained in Example 3.1 in BSS, and

$$\mu_i = \phi_i\delta + \frac{\beta\phi_i\lambda}{\sqrt{\alpha^2 - \beta^2}}.$$

I used the Euler scheme with time resolution equal 0.01 ms and kept only every tenth simulated value in order to get the same observation frequency as for the observed data. Finally, I computed $X_{1,t} + X_{2,t}$ for each $t = 1, \ldots, n$. I repeated the procedure 2000 times and thus obtained 2000 simulated data series.

Fig. 14.2 shows kernel estimates for the density (left) and empirical auto-correlations (right) for a random sample of 20 simulated data series. The data generating functions ("true" functions) are shown in black. There is a considerable variation between samples, in particular for the auto-correlation function which for many data series looks exponential rather than bi-exponential.

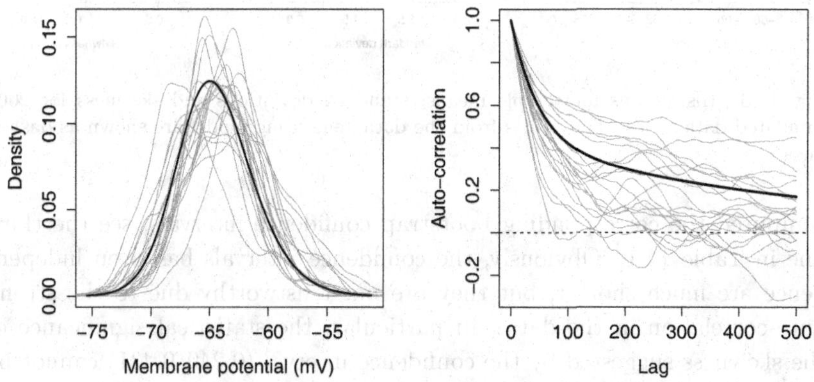

Fig. 14.2 Data summaries for 20 simulated data series: kernel estimates of the density (left) and empirical auto-correlations (right). The invariant density and auto-correlation function for the data generating model are shown in black.

Recall the interest in the sample moments in (14.1). In order to get confidence intervals I computed the sample moments for each of the 2000 simulated data series. Histograms for the simulated sample moments are shown in Fig. 14.3 with vertical red lines indicating the moments from the data generating model. The distribution of the standard deviation is slightly asymmetric, and the distribution of skewness is slightly biased.

I used the so-called basic bootstrap method (Davison and Hinkley, 1997, p. 194) for computation of 95% confidence intervals, *i.e.* $(2\xi - \xi^\star_{0.975}, 2\xi - \xi^\star_{0.025})$ where ξ is the point estimate of interest (sample mean/standard deviation/skewness) and $\xi^\star_p$ is the pth quantile in the bootstrap distribution for ξ. The confidence intervals are listed in the second line in Table 14.1 (bold). In particular, and in accordance with the histogram in Fig. 14.1, there is no evidence of skewness in the invariant distribution.

As argued, computation of reliable confidence intervals requires a complete model, not only the marginal distribution. In order to illustrate this, I also simulated 2000 datasets each consisting of 4701 independent, identically distributed observations from the variance-gamma distribution and

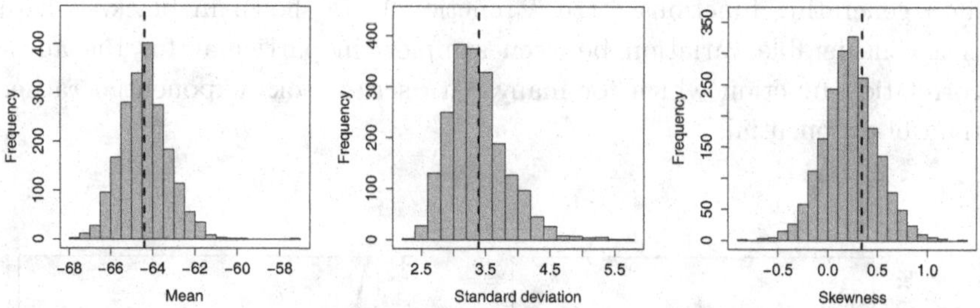

Fig. 14.3 Histograms for sample means, standard deviations and skewness for 2000 simulated data series. The values from the data generating model are shown as dashed lines.

computed the corresponding bootstrap confidence intervals, see the third line in Table 14.1. Obviously, the confidence intervals based on independence are much shorter, but they are not trustworthy due to the strong auto-correlation in the data. In particular, the statistical significance of the skewness suggested by the confidence interval $(0.249, 0.431)$ cannot be trusted.

Finally, I also computed confidence intervals from bootstrap samples of the variance-gamma model with exponentially decreasing auto-correlations, corresponding to the dotted curve in auto-correlation plot in Fig. 14.1. These are listed in the last line in Table 14.1, and are not too different from those based on the bi-exponential auto-correlation function. This agreement is not surprising considering the large variation in empirical auto-correlations from sample to sample, cf. the right panel of Fig. 14.2.

Table 14.1 Bootstrap confidence intervals for summaries for the invariant distribution, computed under three different assumptions about the dependence structure.

	Mean	SD	Skewness
Point estimate	-64.4	3.39	0.340
95% CI, bi-exponential	$(\mathbf{-67.7, -62.4})$	$(\mathbf{2.37, 4.11})$	$(\mathbf{-0.099, 0.972})$
95% CI, independence	$(-64.5, -64.3)$	$(3.32, 3.47)$	$(0.249, 0.431)$
95% CI, exponential	$(-66.3, -62.6)$	$(2.33, 4.29)$	$(-0.297, 1.128)$

Originally, my plan was to fit the variance-gamma distribution for each simulated data series. However, the estimate of λ often reached an artificial upper limit (8.446). This seems to be due to the implementation of the modified Bessel function of the third kind in R Core Team (2016), but is also related to identifiability problems with the variance-gamma distribution. Although the parametrization is indeed injective, several quite different parameter sets may in practice give rise to almost identical distributions, and even for fixed value of λ the variability was large for estimators of the remaining parameters. It could be interesting to see if estimation strategies employing the complete model structure, rather than just the marginal distribution, would have better properties.

14.3 Conclusion

Statistical inference for complicated SDE models is inherently difficult. Still, a complete model for the data generating mechanism is extremely useful and the paper by Bibby, Skovgaard and Sørensen is of great benefit in that respect. It is easy to simulate from SDE models and thereby study the behaviour of data, estimators and test statistics in order to carry out valid statistical inference for dynamical stochastic systems.

Acknowledgements

I am grateful to Rune W. Berg for making his experimental data available.

Bernoulli **11**(2), 2005, 191–220

Diffusion-type models with given marginal distribution and autocorrelation function

BO MARTIN BIBBY[1*], IB MICHAEL SKOVGAARD[1**] and MICHAEL SØRENSEN[2]

[1]*Department of Mathematics and Physics, The Royal Veterinary and Agricultural University, Thorvaldsensvej 40, DK-1871 Frederiksberg C, Denmark. E-mail: *bibby@dina.kvl.dk; **ims@kvl.dk*

[2]*Department of Applied Mathematics and Statistics, Institute of Mathematical Sciences, University of Copenhagen, Universitetsparken 5, DK-2100 København Ø, Denmark. E-mail: michael@math.ku.dk*

Flexible stationary diffusion-type models are developed that can fit both the marginal distribution and the correlation structure found in many time series from, for example, finance and turbulence. Diffusion models with linear drift and a known and prespecified marginal distribution are studied, and the diffusion coefficients corresponding to a large number of common probability distributions are found explicitly. An approximation to the diffusion coefficient based on saddlepoint techniques is developed for use in cases where there is no explicit expression for the diffusion coefficient. It is demonstrated theoretically as well as in a study of turbulence data that sums of diffusions with linear drift can fit complex correlation structures. Any infinitely divisible distribution satisfying a weak regularity condition can be obtained as a marginal distribution.

Keywords: ergodic diffusions; generalized hyperbolic distributions; long-range dependence; saddlepoint approximation; stochastic differential equation; turbulence

1. Introduction

We consider the problem of choosing a continuous-time model based on discrete-time observations $X_{t_1}, \ldots, X_{t_n}$. Ideally the choice of a model should be based on an understanding of the processes governing the system from which the data are obtained. Often such a description of a system is made using a number of ordinary differential equations – in the case of a single ordinary differential equation,

$$\frac{\mathrm{d}X_t}{\mathrm{d}t} = b(X_t), \qquad t \geqslant 0.$$

A natural extension of this model is to add a white noise term,

$$\mathrm{d}X_t = b(X_t)\mathrm{d}t + \sqrt{v(X_t)}\mathrm{d}W_t, \qquad t \geqslant 0,$$

where W is a standard Wiener process. This introduces an uncertainty in the description of the system behind the data and results in dependence between the observations. See Pedersen (2000) for an example of this approach in the modelling of nitrous oxide emissions from the

1350–7265

soil surface. In this paper we show how, with a given drift function b, any probability density satisfying weak regularity conditions can be obtained as a marginal distribution by choosing v suitably. This result is useful when choosing a parametrized class of diffusion coefficients v in the light of data. A linear specification of b is studied in detail.

In many cases the mechanisms driving the process of interest are not understood well enough or are too complicated to be described using a simple drift function, b, and a more data-driven approach must be taken. The main aim of this paper is to propose a method of choosing a model based on data also in such cases. Specifically, we show how to construct a model for X with a given marginal density f, $X_t \sim f$, and autocorrelation function $\rho(t) = \text{corr}(X_s, X_{s+t})$, $s, t \geqslant 0$, where f is infinitely divisible and satisfies a weak regularity condition, and where $\rho(t)$ belongs to a large and very flexible class of autocorrelation functions. The model is usually not Markovian. Expressions for f and ρ are typically chosen so that they fit a histogram of the data and the empirical autocorrelation function. Aït-Sahalia (1996) took the same approach as we do in the case of an exponentially decreasing autocorrelation function, but instead of a parametric model for the marginal density, he estimated this density nonparametrically. In Bibby and Sørensen (1997; 2001) a similar approach based only on the marginal density f was used in connection with financial data. The construction in this paper, which involves sums of diffusion processes, is related to the sums of Ornstein–Uhlenbeck processes driven by Lévy processes introduced by Barndorff-Nielsen *et al.* (1998) and provides a continuous-path alternative to these jump processes. Therefore, the models introduced in this paper can be used to construct stochastic volatility models without jumps that are alternatives to the models of Barndorff-Nielsen and Shephard (2001); see Bibby and Sørensen (2004). Constructions different from ours of Markovian martingales with prescribed marginal distributions have recently been considered by Madan and Yor (2002).

In Section 2 we introduce the method in the situation where X is a diffusion process with a linear drift and hence has an exponentially decreasing autocorrelation function. For a large number of commonly used probability distributions, explicit diffusion models are given with linear drift and with these distributions as marginal distributions. Moreover, general expressions for exponential families and normal variance mixtures are established. Also nonlinear drift functions are considered. Section 3 contains a result on an approximation of the squared diffusion coefficient that enlarges the class of possible marginal densities for which a diffusion model can be handled in practice. The approximation is based on saddlepoint techniques, and the marginal density of the resulting model is approximately proportional to the saddlepoint approximation of the original marginal density. In Section 4 models for X with more realistic autocorrelation functions are constructed based on the results in Sections 2 and 3. These models are finite sums of diffusion processes and hence not Markovian. Here the marginal distribution must be infinitely divisible. Relations to long-range dependence are investigated. Infinite sums of diffusions are also briefly considered. In Section 5 multivariate models are introduced. Finally, in Section 6 we study an example involving turbulence data.

2. Construction of diffusions

In this section we describe the construction of diffusion process models with an exponential autocorrelation function and a specified marginal distribution. The diffusion will be constructed such that the marginal distribution is concentrated on a set (l, u) $(-\infty \leqslant l < u \leqslant \infty)$, and has a prespecified density f with respect to the Lebesgue measure on the state space (l, u). The construction of a diffusion from its marginal distribution and an exponential autocorrelation function first appeared in Aït-Sahalia (1996), who instead of using a parametric model estimated the marginal density nonparametrically. In this way he obtained a nonparametric estimator of the diffusion coefficient. In particular, Aït-Sahalia (1996) also derived the basic equations (2.2) (2.4) and (2.9) below. In the rest of this section, let f be a probability density satisfying the following condition.

Condition 2.1. *The probability density f is continuous, bounded, and strictly positive on (l, u), zero outside (l, u), and has finite variance.*

Consider the stochastic differential equation

$$dX_t = -\theta(X_t - \mu)dt + \sqrt{v(X_t)}dW_t, \qquad t \geqslant 0, \tag{2.1}$$

where $\theta > 0$, $\mu \in (l, u)$ and v is a non-negative function defined on the set (l, u). Theorem 2.1 below shows that if

$$v(x) = \frac{2\theta\int_l^x (\mu - y)f(y)dy}{f(x)} = \frac{2\theta\mu F(x) - 2\theta\int_l^x yf(y)dy}{f(x)}, \qquad l < x < u, \tag{2.2}$$

where F is the distribution function associated with the density f, then the solution X is ergodic with invariant density f. The process X is mean-reverting, and if it is stationary the autocorrelation function is $e^{-\theta t}$.

Theorem 2.1. *Suppose the probability density f has expectation μ and satisfies Condition 2.1. Then the following holds.*

(i) *The stochastic differential equation given by (2.1) and (2.2) has a unique Markovian weak solution. The diffusion coefficient is strictly positive for all $l < x < u$.*

(ii) *The diffusion process X that solves (2.1) and (2.2) is ergodic with invariant density f.*

(iii) *The function $f(x)v(x)$ satisfies*

$$\int_l^u v(x)f(x)dx < \infty, \tag{2.3}$$

and $E(X_{s+t} \mid X_s = x) = xe^{-\theta t} + \mu(1 - e^{-\theta t})$. If $X_0 \sim f$, then X is stationary, and the autocorrelation function of X is given by

$$\mathrm{corr}(X_{s+t}, X_s) = e^{-\theta t}, \qquad s, t \geqslant 0. \tag{2.4}$$

(iv) *If $-\infty < l$ or $u < \infty$, then the diffusion given by (2.1) and (2.2) is the only ergodic diffusion with drift $-\theta(x - \mu)$ and invariant density f. If the state space is*

$\mathbb{R}$, *it is the only ergodic diffusion with drift* $-\theta(x-\mu)$ *and invariant density* f *for which the condition* (2.3) *is satisfied.*

Remark. When f has infinite second moment but finite first moment, the stochastic differential equation given by (2.1) and (2.2) also has a unique Markovian weak solution with invariant density f. In this case (2.3) is not satisfied. A finite first moment is obviously needed for the construction (2.2).

Remark. If the state space is the real line, the stochastic differential equation given by (2.1) and (2.8) with $C > 0$ has a unique Markovian weak solution with invariant density f.

Remark. When one (or both) of l or u is finite, densities that are unbounded at the finite boundary can be of interest, for instance a gamma distribution with shape parameter smaller than one. The scale measure will in all cases of practical interest be finite at the boundary where the density is unbounded, so that with positive probability the diffusion will hit this boundary at a finite time. A stationary diffusion with density f is also obtained in this case with v given by (2.2) provided that the diffusion is made instantly reflecting at the boundary, because f will satisfy the appropriate differential equation. This is a nice theoretical solution to the problem that is, however, less easy to implement when simulating the process.

The following lemma is used in the proof of Theorem 2.1.

Lemma 2.2. *Suppose the expectation of* f *is smaller than or equal to* μ, *and that* v *is given by* (2.2). *Then the function*

$$g(x) = f(x)v(x) = 2\theta \int_l^x (\mu - y)f(y)\mathrm{d}y \tag{2.5}$$

is strictly positive for all $l < x < u$, *and* $\lim_{x\to l} g(x) = 0$. *If* f *has expectation equal to* μ, *then* $\lim_{x\to u} g(x) = 0$.

Proof. Since $g(x) = 2\theta \int_l^x (\mu - y)f(y)\mathrm{d}y$, we see that g is strictly increasing on (l, μ) and strictly decreasing on (μ, u), and that $\lim_{x\to l} g(x) = 0$ and $\lim_{x\to u} g(x) \geqslant 0$. Hence $g(x) > 0$ for all $l < x < u$. □

Proof of Theorem 2.1. That $v(x) > 0$ for all $l < x < u$ follows from Lemma 2.2 and the fact that f is continuous. For $l < x < u$, define the scale density

$$s(x) = \exp\left(2\theta \int_{x^*}^x \frac{y-\mu}{v(y)}\mathrm{d}y\right) = \frac{g(x^*)}{g(x)}, \tag{2.6}$$

for some interior point $l < x^* < u$, and the scale function

$$S(x) = \int_{x^*}^x s(y)\mathrm{d}y = g(x^*)\int_{x^*}^x \frac{1}{g(y)}\mathrm{d}y.$$

The function g is given by (2.5), and we have used the fact that $(\log g(y))' = -2\theta(y-\mu)/v(y)$. The function S is strictly increasing, twice continuously differentiable and maps (l, u) onto $\mathbb{R}$. If $(l, u) = \mathbb{R}$, this follows immediately from Lemma 2.2. If u is finite, it follows from Condition 2.1 that there exists a $K > 0$ such that

$$g(x) = 2\theta \int_x^u (y-\mu) f(y) \mathrm{d}y \leqslant K(u-x),$$

which implies that $\lim_{x\to u} S(x) = \infty$. If l is finite, a similar argument shows that $\lim_{x\to l} S(x) = -\infty$.

The stochastic differential equation

$$\mathrm{d}Y_t = s(S^{-1}(Y_t))\sqrt{v(S^{-1}(Y_t))}\mathrm{d}W_t \tag{2.7}$$

satisfies the conditions of Engelbert and Schmidt's (1985) Theorem 2.2 because the function $s(S^{-1}(x))\sqrt{v(S^{-1}(x))}$ is continuous on $\mathbb{R}$. Hence it has a unique Markovian weak solution with state space $\mathbb{R}$. By Itô's formula, the process $S^{-1}(Y_t)$ solves (2.1). This is the only solution because if X is a solution of (2.1), then $S(X_t)$ solves (2.7), again by Itô's formula. We have thus proved (i).

Regarding (ii), we need only check that the scale measure diverges at both end-points and that the speed measure has a density proportional to f (and hence is finite); see, for example, Skorokhod (1989). The invariant density is proportional to the density of the speed measure; see Karlin and Taylor (1981). We have already proved the first assertion, and the second follows easily because the speed measure has density

$$\frac{1}{v(x)s(x)} = \frac{f(x)}{g(x^*)},$$

where we have used (2.5) and (2.6).

Now to (iii). Note that if we can show that (2.3) holds, then it easily follows from (2.1) that $\mathrm{E}(X_{s+t} \mid X_s = x) = x\mathrm{e}^{-\theta t} + \mu(1-\mathrm{e}^{-\theta t})$, which again implies (2.4). If $-\infty < l$ and $u < \infty$, (2.3) follows from Lemma 2.2. Otherwise it must be checked that $v(x)f(x)$ goes sufficiently fast to zero at infinite boundaries. The condition that f has finite variance is exactly enough to ensure this. If $u = \infty$,

$$\int_\mu^\infty g(x)\mathrm{d}x = 2\theta \int_\mu^\infty \int_x^\infty (y-\mu)f(y)\mathrm{d}y\,\mathrm{d}x = 2\theta \int_\mu^\infty \int_\mu^y \mathrm{d}x(y-\mu)f(y)\mathrm{d}y$$

$$= 2\theta \int_\mu^\infty (y-\mu)^2 f(y)\mathrm{d}y < \infty,$$

where we have used Tonelli's theorem. If $l = -\infty$, (2.3) is verified in a similar way.

Finally, to show (iv), note that for an ergodic diffusion of the form (2.1) with invariant density f, necessarily

$$f(x) = \frac{K}{v(x)} \exp\left(-2\theta \int_{x^*}^x \frac{y-\mu}{v(y)} \mathrm{d}y\right)$$

for some positive constant K. Here we have used the general expression for the speed measure. We see that the function $g = fv$ is differentiable and that $(\log g(x))' = -2\theta(x-\mu)/v(x)$ or $g'(x) = -2\theta(x-\mu)f(x)$. Thus

$$v(x) = \frac{2\theta \int_l^x (\mu - y) f(y) \mathrm{d}y + C}{f(x)}, \tag{2.8}$$

for some constant C. To ensure that $v(x) > 0$ for all $l < x < u$, it is necessary that $C \geqslant 0$, since by Lemma 2.2 the integral goes to zero at the boundaries. If one of the boundaries is finite, it is necessary that $C = 0$ for the scale measure $1/(fv)$ to diverge at that boundary, again because the integral in (2.8) goes to zero at the boundaries. If both boundaries are infinite, (2.8) defines an ergodic diffusion with invariant density f for all $C \geqslant 0$. However, (2.3) holds only when $C = 0$. □

By the arguments used to prove (2.3) for $u = \infty$ and $l = -\infty$, it follows that under the assumptions of Theorem 2.1,

$$\int_l^u v(x) f(x) \mathrm{d}x = 2\theta \int_l^u (y-\mu)^2 f(y) \mathrm{d}y = 2\theta \operatorname{var}(X_0).$$

The last equality holds, of course, only when $X_0 \sim f$.

The construction in Theorem 2.1 is a particular case of the following general result, the proof of which is analogous to the proof of Theorem 2.1.

Theorem 2.3. *Let b be a drift function with reversion defined on (l, u), that is, there exists a $\kappa \in (l, u)$ such that $b(x) > 0$ for $l < x < \kappa$ and $b(x) < 0$ for $\kappa < x < u$. Suppose f is a strictly positive, continuous probability density on (l, u) satisfying*

$$\int_l^u b(x) f(x) \mathrm{d}x = 0,$$

and that the function bf is continuous and bounded on (l, u) Then

$$v(x) = \frac{2\int_l^x b(y) f(y) \mathrm{d}y}{f(x)} > 0 \tag{2.9}$$

for all $l < x < u$, and the stochastic differential equation

$$\mathrm{d}X_t = b(X_t)\mathrm{d}t + \sqrt{v(X_t)}\mathrm{d}W_t, \qquad t \geqslant 0$$

has a unique Markovian weak solution which is ergodic with invariant density f.

The condition that b has reversion is only made for convenience. A sufficient condition is that the inequality (2.9) holds for all $l < x < u$.

In Bibby and Sørensen (2001) another method of constructing diffusion processes with a given marginal density was discussed. In that paper the squared diffusion coefficient was chosen proportional to the inverse of the marginal density raised to a power and an expression for the drift was then determined from the relationship between the drift, the diffusion coefficient and the invariant density. In Bibby and Sørensen (1997) a special case of this approach was considered, namely a diffusion process with no drift and diffusion coefficient proportional to $1/\sqrt{f}$.

When the invariant density belongs to an exponential family with a linear component in the canonical statistic the squared diffusion coefficient can be determined from the following theorem.

Theorem 2.4. *Consider an invariant density for a diffusion process which belongs to an exponential family of the form*

$$f(x;\xi) = a(\xi)b(x)e^{\xi_1 x + \alpha(\xi)\cdot t(x)}, \tag{2.10}$$

where $\xi = (\xi_1, \dots, \xi_p)$, *and where* α *and* t *may be vectors. Then the squared diffusion coefficient is given by*

$$v(x;\xi) = -\frac{2\theta}{f(x;\xi)}\frac{\partial}{\partial \xi_1}F(x;\xi), \qquad l < x < u. \tag{2.11}$$

Proof. Since the cumulant transform for f is given by

$$\kappa(t) = \log a(\xi) - \log a(\xi_1 + t, \xi_2, \dots, \xi_p),$$

we obtain that

$$\mu = -\frac{1}{a(\xi)}\frac{\partial}{\partial \xi_1}a(\xi),$$

and hence that

$$\begin{aligned}\frac{\partial}{\partial \xi_1}F(x;\xi) &= \frac{F(x;\xi)}{a(\xi)}\frac{\partial}{\partial \xi_1}a(\xi) + \int_l^x y f(y;\xi)\mathrm{d}y \\ &= -\mu F(x;\xi) + \int_l^x y f(y;\xi)\mathrm{d}y,\end{aligned}$$

yielding (2.11). □

The result of simple linear transformations is given in the following lemma, from which it follows that we need only consider centred and standardized distributions.

Lemma 2.5. *Let* X *be a stationary diffusion process with linear drift and invariant density* f. *Consider the linear transformation given by*

$$Y_t = \alpha + \sigma X_t, \qquad \sigma > 0, \alpha \in \mathbb{R}.$$

Then

$$v_g(y) = \sigma^2 \cdot v_f\left(\frac{y-\alpha}{\sigma}\right),$$

where g denotes the invariant density of Y, and v_f and v_g denote the squared diffusion coefficients obtained by (2.2) *from f and g, respectively.*

We shall now give an extensive list of examples to show that the diffusion coefficient can be found explicitly for many much-used probability distributions, so that our modelling approach is often easy to use in practice. We begin by considering in detail examples of diffusions with an invariant density on the whole real line, that is $-l = u = \infty$, on the half-line, and with compact support.

Example 2.1 *Student's t distribution.* In this example we consider a diffusion process with invariant density equal to a $t(\nu)$ distribution, that is,

$$f(x) = \frac{\Gamma((\nu+1)/2)}{\sqrt{\nu\pi}\Gamma(\nu/2)}\left(1+\frac{1}{\nu}x^2\right)^{-(\nu+1)/2}, \qquad x \in \mathbb{R}, \nu > 0.$$

Here we have taken $\mu = 0$. We only consider t distributions for which the variance exists, so we assume that $\nu > 2$. In this case

$$\int_{-\infty}^{x} yf(y)\mathrm{d}y = -\frac{\Gamma((\nu+1)/2)\nu^{\nu/2}}{(\nu-1)\sqrt{\nu}\Gamma(\nu/2)}(\nu+x^2)^{-(\nu-1)/2},$$

so we obtain that

$$v(x) = \frac{2\theta}{\nu-1}(\nu+x^2), \qquad x \in \mathbb{R}.$$

The function v is also well defined for $\nu = 2$, and we saw above that it defines an ergodic diffusion with the $t(2)$ distribution as invariant distribution.

In the following example we consider an invariant density on the half-axis (l, ∞), where $l > -\infty$. In this situation it may be more convenient to rewrite the expression in (2.2) as

$$v(x) = \frac{2\theta\left(\int_x^\infty (1-F(y))\mathrm{d}y - (\mu - x)(1-F(x))\right)}{f(x)}, \qquad x > l. \tag{2.12}$$

In the case of positive diffusions, where $l = 0$, the squared diffusion coefficient can be expressed in terms of the hazard function λ and the integrated hazard function Λ in the following way:

$$v(x) = \frac{2\theta\left(\mathrm{e}^{\Lambda(x)}\int_x^\infty \mathrm{e}^{-\Lambda(y)}\mathrm{d}y + x - \mu\right)}{\lambda(x)}, \qquad x > 0. \tag{2.13}$$

Example 2.2 *The gamma distribution.* Consider a diffusion process with an invariant density from the gamma distribution, that is,

$$f(x) = \frac{\lambda^\alpha}{\Gamma(\alpha)}x^{\alpha-1}\mathrm{e}^{-\lambda x} \qquad x > 0, \alpha > 0, \lambda > 0.$$

In order for the density to be bounded, we suppose that $\alpha \geqslant 1$. In this case the expectation is $\mu = \alpha/\lambda$. The distribution function is given by

$$F(x) = \frac{\Gamma(\lambda x;\, \alpha)}{\Gamma(\alpha)},$$

where

$$\Gamma(x;\, \alpha) = \int_0^x y^{\alpha-1} e^{-y}\, dy$$

is an incomplete gamma function. For the gamma invariant density we get that

$$\int_0^x y f(y) dy = \frac{\alpha}{\lambda} F(x) - \frac{x}{\lambda} f(x),$$

and therefore

$$v(x) = \frac{2\theta x}{\lambda}.$$

This process is well known and was proposed by Cox *et al.* (1985) as a model for the short-term interest rate.

The following is a simple example of an invariant density with compact support.

Example 2.3 *The beta distribution.* Consider a diffusion process with an invariant density corresponding to the beta distribution, that is,

$$f(x) = B(\alpha, \beta)^{-1} x^{\alpha-1}(1-x)^{\beta-1}, \qquad 0 < x < 1,\ \alpha > 0,\ \beta > 0,$$

where $B(a, b) = \Gamma(a)\Gamma(b)/\Gamma(a+b)$ is the beta function. In this case the distribution function is given by

$$F(x) = I_x(\alpha, \beta) = \frac{\int_0^x y^{\alpha-1}(1-y)^{\beta-1} dy}{B(\alpha, \beta)}, \qquad 0 < x < 1,$$

and the mean is $\mu = \alpha/(\alpha+\beta)$. Similarly, we obtain that

$$\int_0^x y f(y) dy = \frac{\alpha}{\alpha+\beta} I_x(\alpha+1, \beta), \qquad 0 < x < 1.$$

Since we have that

$$I_x(\alpha, \beta) - I_x(\alpha+1, \beta) = \frac{\Gamma(\alpha+\beta+1)}{\Gamma(\alpha+1)\Gamma(\beta+1)} x^{\alpha}(1-x)^{\beta},$$

the squared diffusion coefficient takes the form

$$v(x) = \frac{2\theta}{\alpha+\beta} x(1-x), \qquad 0 < x < 1.$$

Table 1. The squared diffusion coefficient for the most common distributions. For some parameter values the Student, Pareto, inverse gamma, and the F distributions do not have finite variance

Distribution	Density function $f(x)$	State space (l, u)	Mean μ	Parameter space	Squared diffusion $v(x)$
Normal	$\frac{1}{\sqrt{2\pi}} e^{-x^2/2}$	$(-\infty, \infty)$	0	–	2θ
Student	$\frac{\Gamma((\nu+1)/2)\nu^{\nu/2}}{\sqrt{\pi}\Gamma(\nu/2)}(\nu+x^2)^{-(\nu+1)/2}$	$(-\infty, \infty)$	0	$\nu > 1$	$\frac{2\theta}{\nu-1}(\nu+x^2)$
Laplace	$\frac{\alpha^2-\beta^2}{2\alpha} e^{\beta x-\alpha\|x\|}$	$(-\infty, \infty)$	$\frac{2\beta}{\alpha^2-\beta^2}$	$\alpha^2 > \beta^2$	$\frac{2\theta}{\alpha^2-\beta^2}(1+\alpha\|x\|+\beta x)$
Logistic	$\frac{e^x}{(1+e^x)^2}$	$(-\infty, \infty)$	0	–	$2\theta[(e^x+e^{-x}+2)\log(1+e^x) -x(1+e^x)]$
Extreme value	$e^{-x-e^{-x}}$	$(-\infty, \infty)$	γ	–	$2\theta e^x\left(\gamma - x + e^{e^{-x}} Ei(-e^{-x})\right)$
Pareto	$\alpha(1+x)^{-\alpha-1}$	$(0, \infty)$	$\frac{1}{\alpha-1}$	$\alpha > 1$	$2\theta\mu x(1+x)$
Exponential	$\lambda e^{-\lambda x}$	$(0, \infty)$	$\frac{1}{\lambda}$	$\lambda > 0$	$\frac{2\theta}{\lambda}x$
Gamma	$\frac{\lambda^\alpha}{\Gamma(\alpha)} x^{\alpha-1} e^{-\lambda x}$	$(0, \infty)$	$\frac{\alpha}{\lambda}$	$\alpha \geqslant 1,\ \lambda > 0$	$\frac{2\theta}{\lambda}x$
χ^2	$\frac{1}{2^{\nu/2}\Gamma(\nu/2)} x^{(\nu/2)-1} e^{-x/2}$	$(0, \infty)$	ν	$\nu \geqslant 2$	$4\theta x$

Inverse gamma	$\frac{\delta^{\lambda}}{\Gamma(\lambda)} x^{-\lambda-1} e^{-\delta/x}$	$(0, \infty)$	$\frac{\delta}{\lambda-1}$	$\delta > 0,\ \lambda > 1$	$\frac{2\theta}{\lambda-1} x^2$
Inverse Gaussian	$\sqrt{\frac{\lambda}{2\pi x^3}} e^{-\lambda(x-\delta)^2/(2\sigma^2 x)}$	$(0, \infty)$	δ	$\lambda > 0,\ \delta > 0$	$\frac{4\theta\delta}{f(x)} e^{2\lambda/\delta} \Phi\left(-\sqrt{\frac{\lambda}{x}}\left(\frac{x}{\delta}+1\right)\right)$
F	$\frac{\alpha^{\alpha/2}\beta^{\beta/2}}{B(\alpha/2, \beta/2)} \frac{x^{(\alpha/2)-1}}{(\beta+\alpha x)^{(\alpha+\beta)/2}}$	$(0, \infty)$	$\frac{\beta}{\beta-2}$	$\alpha \geqslant 2,\ \beta > 2$	$\frac{4\theta}{\alpha(\beta-2)} x(\beta+\alpha x)$
log-normal	$\frac{1}{\sqrt{2\pi\sigma^2}x} e^{-(\log x-\delta)^2/(2\sigma^2)}$	$(0, \infty)$	$e^{\delta+2\sigma^2}$	$\sigma^2 > 0$	$\frac{2\theta\mu}{f(x)}\left(\Phi\left(\frac{\log x-\delta}{\sigma}\right) - \Phi\left(\frac{\log x-\delta}{\sigma} - \sigma\right)\right)$
Weibull	$cx^{c-1} e^{-x^c}$	$(0, \infty)$	$\Gamma\left(\frac{1}{c}+1\right)$	$c > 0$	$\frac{2\theta}{f(x)}\left(\Gamma\left(\frac{1}{c}+1\right)(1-e^{-x^c}) - \Gamma\left(x^c; \frac{1}{c}+1\right)\right)$
Uniform	1	$(0, 1)$	$\frac{1}{2}$	–	$\theta x(1-x)$
Beta	$\frac{\Gamma(\alpha+\beta)}{\Gamma(\alpha)\Gamma(\beta)} x^{\alpha-1}(1-x)^{\beta-1}$	$(0, 1)$	$\frac{\alpha}{\alpha+\beta}$	$\alpha > 0,\ \beta > 0$	$\frac{2\theta}{\alpha+\beta} x(1-x)$

This process has been used to model the variation of exchange rates in a target zone by De Jong *et al.* (2001) (for $\alpha = \beta$) and Larsen and Sørensen (2003).

In Table 1 the squared diffusion coefficient is given for a large number of common distributions. In the table, Φ denotes the standard normal distribution function, $\Gamma(x; \alpha)$ the incomplete gamma function given by (2.14), and Ei is the exponential integral function given by

$$Ei(x) = -\int_{-x}^{\infty} y^{-1} e^{-y} \, dy, \qquad x < 0.$$

Furthermore, γ denotes Euler's constant, $\gamma \approx 0.577\,22$.

Example 2.4 *Normal variance mixtures.* Consider the normal variance mixture

$$f(x) = \int_0^{\infty} \frac{1}{\sqrt{2\pi u}} e^{-x^2/(2u)} h(u) du,$$

where the mixing distribution has density h on $\mathbb{R}_+$ and finite expectation μ_h. Let f^* be the normal variance mixture with mixing density $h^*(u) = uh(u)/\mu_h$. From the fact that a diffusion with marginals that are normally distributed with variance u is obtained when $v(x) = 2\theta u$ (the Ornstein–Uhlenbeck process), it follows easily that a diffusion with marginal density f emerges when

$$v(x) = \frac{2\theta\mu_h f^*(x)}{f(x)}.$$

If h belongs to a family of densities with a factor of the form x^{κ} ($\kappa > 0$), h^* belongs to the same class. An example is the class of generalized inverse Gaussian densities which contains, among many others, the inverse Gaussian densities, the gamma densities and the inverse gamma densities. When h is a generalized inverse Gaussian density, both f and f^* are explicitly known generalized hyperbolic densities. This result provides an alternative derivation in the case of the Student distribution, which is a normal variance mixture with an inverse gamma mixing distribution. As another example, a diffusion with a symmetric variance-gamma density, that is, (3.9) with $\beta = 0$, is obtained when

$$v(x) = |x - \mu| \frac{K_{\lambda+\frac{1}{2}}(\alpha|x - \mu|)}{K_{\lambda-\frac{1}{2}}(\alpha|x - \mu|)},$$

where K_λ is the modified Bessel function of the third kind with index λ. The variance-gamma distribution was introduced in the finance literature by Madam and Seneta (1990) and has often been connected with jump processes. For details of generalized inverse Gaussian and generalized hyperbolic distributions, see Bibby and Sørensen (2003).

3. Approximations

For some useful classes of distributions it is not possible to determine an explicit expression for the squared diffusion coefficient. However, for several such distributions the moment generating function exists and is known explicitly, so that the following approximation can be applied. Let M denote the moment generating function corresponding to the density f, that is,

$$M(t) = \int_l^u e^{tx} f(x) dx, \tag{3.1}$$

defined for t in the set

$$T = \left\{ t \in \mathbb{R} \middle| \int_l^u e^{tx} f(x) dx < \infty \right\}.$$

Similarly, we let κ denote the cumulant transform, $\kappa(t) = \log M(t)$, and note that it is twice differentiable for all $t \in \text{int}(T)$. Consider the approximation to v given by

$$\tilde{v}(x) = \frac{2\theta(x - \mu)}{\hat{t}_x}, \tag{3.2}$$

where $\hat{t}_x$ is the saddlepoint given by

$$\kappa'(\hat{t}_x) = x. \tag{3.3}$$

Clearly $\tilde{v}(x)$ is positive for $l < x < u$ since $x - \mu = \kappa'(\hat{t}_x) - \kappa'(0)$ and κ is a convex function. Since κ is analytic the singularity of $\tilde{v}(x)$ at $x = \mu$ is removable; in fact, the limiting value of $\tilde{v}$ is $2\theta\kappa''(0)$ and $\tilde{v}$ has derivatives of all orders.

The function $\tilde{v}$ emerges in a natural way when making a substitution in the expression for v in (2.2). Define

$$r_x = \text{sign}(\hat{t}_x)\sqrt{2(x\hat{t}_x - \kappa(\hat{t}_x))}. \tag{3.4}$$

Then the saddlepoint approximation to the density can be written

$$f(x) \approx (\kappa''(\hat{t}_x))^{-1/2} \varphi(r_x), \tag{3.5}$$

where φ is the standard normal density function. For the following computation, note that r_x is increasing in x, that $r_x \, dr_x = \hat{t}_x \, dx$ and that $(x - \mu)/r_x$ is a differentiable function when extended by continuity at $x = \mu$ where $r_x = 0$. Now define

$$I(x) = \int_x^u (y - \mu) f(y) \, dy = \int_{r_x}^{r_u} r_y \varphi(r_y) G(r_y) \, dr_y,$$

where

$$G(r_y) = \frac{(y - \mu)}{r_y} \frac{f(y)}{\varphi(r_y)} \frac{dy}{dr_y} = \frac{(y - \mu)}{\hat{t}_y} \frac{f(y)}{\varphi(r_y)}.$$

Integration by parts now yields

$$I(x) = \left[-\varphi(r_y)G(r_y)\right]_{r_x}^{r_u} + \int_{r_x}^{r_u} \varphi(r_y)G'(r_y)\,\mathrm{d}r_y$$

$$\approx \varphi(r_x)G(r_x) = f(x)\frac{x-\mu}{\hat{t}_x},$$

from which the approximation (3.2) is obtained using $v(x) = 2\theta I(x)/f(x)$. In this computation we have discarded two terms for the following reasons. First the upper limit r_u is usually infinity even if u is finite, but if r_u is finite, the term $-\varphi(r_u)G(r_u)$ is exponentially small in standard asymptotic analysis because of the factor $\varphi(r_u)$. Second, although there are no asymptotic considerations in the present setting, we may consider what happens when the density, f, corresponds to (a standardized version of) a convolution of n independent replications, thus approaching the normal. In particular, this fits in naturally with the infinitely divisible distributions. In that case the integral arising in the integration by parts above will be of low order compared to the leading term. More precisely, it is of order $O(n^{-1/2})$ relative to the leading term uniformly, improving to a relative error of order $O(n^{-1})$ for large deviations, that is, for arguments $x - \mu$ growing proportionally to $\sqrt{n}$ in the standardized scale. In view of Condition 4.1 in the next section, it may be noted that these asymptotic results are valid as $n \to \infty$ for a family of densities, f_n say, with characteristic functions

$$C_n(t) = \{C_0(t/\sqrt{n})\}^n \mathrm{e}^{\mathrm{i}t\mu}, \tag{3.6}$$

where C_0 is the characteristic function, some power of which must be integrable, of a centred distribution with finite Laplace transform in some neighbourhood of zero. For integer values of n this follows from asymptotic results for saddlepoint approximations. Using the method of contour integrals for the saddlepoint approximation (see Daniels 1954; 1987), the same techniques may be used to prove the validity for real (positive) values of n when C_0 corresponds to an infinitely divisible distribution. In summary, we may expect the approximation $\tilde{v}$ to work reasonably well near the mean and very well in the tails.

The approximation may be refined by inclusion of further terms according to the method outlined in Bleistein (1966). In asymptotic analysis as described above, the order of error would improve to $O(n^{-1})$ by the approximation

$$I(x) \approx \varphi(r_x)G(r_x) + G'(0)(1 - \Phi(r_x))$$

where Φ is the standard normal distribution function and

$$G'(0) = \frac{f(0)}{\varphi(0)}\left(-\frac{\kappa_3}{2\sqrt{\kappa_2}} + \frac{f'(0)}{f(0)}\kappa_2^{3/2}\right),$$

where κ_2 and κ_3 are the second and third cumulants of the distribution with density f.

Some properties of a diffusion process

$$\mathrm{d}X_t = \theta(\mu - X_t)\mathrm{d}t + \sqrt{\tilde{v}(X_t)}\mathrm{d}W_t, \qquad t \geqslant 0, \tag{3.7}$$

with $\tilde{v}$ as squared diffusion coefficient, are stated in the following theorem.

Theorem 3.1. *Let the density f have expectation μ and satisfy Condition 2.1. Assume that the function*

$$h(x) = x\hat{t}_x - \kappa(\hat{t}_x)$$

is such that $\int_\mu^x \exp\{h(y)\}\,\mathrm{d}y$ tends to ∞ as x tends to l and as x tends to u. Then, for a normalizing constant $c > 0$, the density

$$\tilde{f}(x) = \frac{c}{\tilde{v}(x)} \mathrm{e}^{-(x\hat{t}_x - \kappa(\hat{t}_x))}, \qquad x \in (l, u), \tag{3.8}$$

has mean μ, is the marginal density of a diffusion process given by the stochastic differential equation (3.7), and conclusions (i), (ii), *and* (iii) *of Theorem 2.1 hold with v and f replaced by $\tilde{v}$ and $\tilde{f}$.*

Remark. Note that $\tilde{f}$ in (3.8) is approximately proportional to the saddlepoint approximation to f. This is seen by observing that both $\tilde{v}(x)$ and $\sqrt{\kappa''(\hat{t}_x)}$ are approximately proportional to $\kappa''(0) + \frac{1}{2}\kappa^{(3)}(0)\hat{t}_x$ near the mean of the distribution, while the exponential is identical to that of the saddlepoint approximation. Moreover, $\tilde{v}$ is continuous, and hence so is $\tilde{f}$.

Remark. The condition on the function h is satisfied at the upper end if $u = \infty$ and also if u is finite and either the limiting value or any of the derivatives of f at u is non-zero. Similarly at the lower end, l. The proof of this assertion is trivial in the case $u = \infty$ because $h(x)$ tends to infinity; the other part is derived from Tauberian theorems on the Laplace transform of the density. It seems a reasonable conjecture that the theorem holds without the condition on h, but we have not been able to prove this. Incidentally, the inverse Gaussian distribution provides an example of a density with a (lower) end-point of support at which the density and all its derivatives vanish; the conclusion of the theorem is, however, also valid for this distribution.

Proof of Theorem 3.1. Notice first that h is strictly convex with derivative $h'(x) = \hat{t}_x$ and with minimum $h(\mu) = 0$. For later use we now prove that $h(x)$ tends to infinity as x tends to u. This is trivial if $u = \infty$; otherwise, assume without loss of generality that $u = 0$. Then $\kappa(t)$ is decreasing in t with $\kappa(t) \to -\infty$ as $t \to \infty$. For x satisfying $l < x < 0$ we have

$$tx - \kappa(t) < h(x),$$

for any $t \neq \hat{t}_x$, because κ is strictly convex and the derivative of the left-hand side vanishes at $t = \hat{t}_x$. For arbitrary but fixed $t > 0$, we see that $h(x) \geqslant -\kappa(t)$ for x sufficiently close to zero, because h is increasing and hence has a limit. Since this holds for any $t > 0$ and $-\kappa(t)$ tends to infinity, so does $h(x)$ as x approaches zero. Thus, h tends to infinity at the upper end-point, u, and the same result holds for the lower end-point, l, by the same argument.

Next we prove that the squared diffusion coefficient corresponding to $\tilde{f}$ derived from (2.2) is $\tilde{v}$ from (3.2). For $x > \mu$, consider the integral

$$\int_\mu^x (y - \mu)\tilde{f}(y)\,\mathrm{d}y = \int_0^{h(x)} \frac{c}{2\theta}\,\mathrm{e}^{-h}\,\mathrm{d}h = \frac{c}{2\theta}\left(1 - \mathrm{e}^{-h(x)}\right),$$

where we have used $h'(x) = \hat{t}_x$ to substitute h for y in the integral. Similarly, for $x < \mu$, we have

$$\int_x^{\mu} (y-\mu)\tilde{f}(y)\,\mathrm{d}y = -\frac{c}{2\theta}\left(1 - \mathrm{e}^{-h(x)}\right).$$

Thus, since h tends to infinity at both ends, the mean of $\tilde{f}$ is μ. Furthermore, substitution in (2.2) shows that $\tilde{v}$ is indeed the squared diffusion coefficient calculated from this equation when the density is $\tilde{f}$.

The proof that the pair consisting of $\tilde{v}$ and $\tilde{f}$ admits the remaining conclusions in (i)–(iii) of Theorem 2.1 now copies the arguments of the proof of that theorem, except that instead of providing Condition 2.1 for $\tilde{f}$ we have directly assumed that the scale function is unbounded at the two end-points. Furthermore, notice that the integral in (2.3) in the present case is proportional to $\int \exp\{-h(x)\}\,\mathrm{d}x$, so that the convexity of h directly implies that the integral is finite. □

Let us consider some examples. For background material and detail about the variance-gamma distribution, the normal-inverse Gaussian distribution, and other generalized hyperbolic distributions, see Bibby and Sørensen (2003).

Example 3.1 *The variance-gamma distribution.* The variance-gamma distribution is a special case of the generalized hyperbolic distribution that has proved useful in the modelling of turbulence and financial data. The density function is given by

$$f(x) = \frac{(\alpha^2-\beta^2)^{\lambda}}{\sqrt{\pi}\Gamma(\lambda)(2\alpha)^{\lambda-1/2}}|x-\delta|^{\lambda-1/2}K_{\lambda-1/2}(\alpha|x-\delta|)\mathrm{e}^{\beta(x-\delta)}, \qquad x \in \mathbb{R}, \tag{3.9}$$

where K_{λ} is the modified Bessel function of the third kind with index λ. The domain of the four parameters is $\lambda > 0$, $\alpha > |\beta|$ and $\delta \in \mathbb{R}$. The mean is of the form

$$\mu = \delta + \frac{2\beta\lambda}{\alpha^2 - \beta^2}. \tag{3.10}$$

Apart from the symmetric case ($\beta = 0$) treated in Example 2.4, it is not obvious how to determine an expression for the squared diffusion coefficient v. The moment generating function is, however, rather simple:

$$M(t) = \mathrm{e}^{\delta t}\left(\frac{\alpha^2-\beta^2}{\alpha^2-(\beta+t)^2}\right)^{\lambda}, \qquad |\beta + t| < \alpha. \tag{3.11}$$

The cumulant transform and its first derivative are given by

$$\kappa(t) = \delta t + \lambda(\log(\alpha^2-\beta^2) - \log(\alpha^2-(\beta+t)^2)),$$

$$\kappa'(t) = \delta + \frac{2\lambda(\beta+t)}{\alpha^2-(\beta+t)^2},$$

and so the saddlepoint is

$$\hat{t}_x = \begin{cases} -\beta, & x = \delta, \\ \dfrac{\sqrt{\lambda^2 + \alpha^2(x-\delta)^2} - \lambda}{x - \delta} - \beta, & x \neq \delta. \end{cases}$$

The approximate squared diffusion coefficient thus takes the form

$$\tilde{v}(x) = \begin{cases} \dfrac{4\theta\lambda}{\alpha^2 - \beta^2}, & x = \delta, \\ \dfrac{2\theta(x-\delta)(x - \delta - 2\beta\lambda/(\alpha^2 - \beta^2))}{\sqrt{\lambda^2 + \alpha^2(x-\delta)^2} - \lambda - \beta(x-\delta)}, & x \neq \delta. \end{cases} \tag{3.12}$$

Example 3.2 *The normal-inverse Gaussian distribution.* The normal-inverse Gaussian distribution is another member of the class of generalized hyperbolic distributions. According to Barndorff-Nielsen and Shephard (2001), it often fits the distribution of financial returns very well. Its density is given by

$$f(x) = \frac{\alpha\lambda K_1(\alpha\sqrt{\lambda^2 - (x+\delta)^2})}{\pi\sqrt{\lambda^2 + (x-\delta)^2}} \cdot e^{\lambda\sqrt{\alpha^2 - \beta^2} + \beta(x-\delta)}, \qquad x \in \mathbb{R}, \tag{3.13}$$

where we assume that $\lambda > 0$, $\alpha > |\beta|$, and $\delta \in \mathbb{R}$. The mean is

$$\mu = \delta + \frac{\beta\lambda}{\sqrt{\alpha^2 - \beta^2}}.$$

As in the previous example, the symmetric case ($\beta = 0$) can be handled by the result in Example 2.4, whereas it is hard to determine the squared diffusion coefficient explicitly in the general case. Again the approximation is readily obtained. The moment generating function of the normal-inverse Gaussian distribution is of the form

$$M(t) = e^{\delta t + \lambda(\sqrt{\alpha^2 - \beta^2} - \sqrt{\alpha^2 - (\beta + t)^2})}, \qquad |\beta + t| < \alpha, \tag{3.14}$$

giving the following expression for the derivative of the cumulant transform,

$$\kappa'(t) = \delta + \frac{\lambda(\beta + t)}{\sqrt{\alpha^2 - (\beta + t)^2}}.$$

This means that the saddlepoint is given by

$$\hat{t}_x = \frac{\alpha(x - \delta)}{\sqrt{\lambda^2 + (x-\delta)^2}} - \beta,$$

and therefore the following approximate squared diffusion coefficient emerges:

$$\tilde{v}(x) = \frac{2\theta\sqrt{\lambda^2 + (x-\delta)^2}(x - \delta - \beta\lambda/\sqrt{\alpha^2 - \beta^2})}{\alpha(x-\delta) - \beta\sqrt{\lambda^2 + (x-\delta)^2}}. \tag{3.15}$$

□

Example 3.3 *The inverse Gaussian distribution.* For the inverse Gaussian distribution we have that

$$v(x) = 4\theta\delta\sqrt{\frac{2\pi x^3}{\lambda}}\, e^{2\lambda/\delta} e^{\lambda(x-\delta)^2/2\delta^2 x}\Phi\left(-\sqrt{\frac{\lambda}{x}}\left(\frac{x}{\delta}+1\right)\right), \qquad x > 0; \qquad (3.16)$$

see Table 1. In this case the moment generating function is given by

$$M(t) = \exp\left(\frac{\lambda}{\delta}\left(1-\sqrt{1-\frac{2\delta^2 t}{\lambda}}\right)\right) \qquad t \leqslant \frac{\lambda}{2\delta^2}.$$

The cumulant transform and its first derivative take the form

$$\kappa(t) = \frac{\lambda}{\delta}\left(1-\sqrt{1-\frac{2\delta^2 t}{\lambda}}\right),$$

$$\kappa'(t) = \frac{\delta\sqrt{\lambda}}{\sqrt{\lambda - 2\delta^2 t}}.$$

This means that the saddlepoint is given by

$$\hat{t}_x = \frac{\lambda(x+\delta)(x-\delta)}{2\delta^2 x^2},$$

and so

$$\tilde{v}(x) = \frac{4\theta\delta^2 x^2}{\lambda(x+\delta)}. \qquad (3.17)$$

In Figure 1 the two versions of the squared diffusion coefficient, (3.16) and (3.17), corresponding to the parameter values $\theta = 1$, $\lambda = 5$ and 25, and $\delta = 5$ and 25 are drawn. Note that $M(t)$ is of the form $M_0(t)^\nu$ with $\nu = \lambda/\delta$ and $M_0(t) = \exp(1 - \sqrt{1 - 2\delta^2 t/\lambda})$, so from the remarks after (3.6) we expect the approximation to improve as λ/δ increases, which is in accordance with Figure 1.

Just as Theorem 2.1 could be generalized to diffusions with nonlinear drift function, as shown in Theorem 2.3, we may also generalize Theorem 3.1 to such cases. This may be viewed not only as an approximation but also as a result providing a class of diffusions with nonlinear drift and an (exact) analytic expression for the stationary density. The approximation is derived just as for the case with linear drift, and the proof follows that of Theorem 3.1.

Theorem 3.2. *Consider a probability density, f, satisfying the conditions of Theorem 3.1, and a drift function, $b(x)$, satisfying $b(x) > 0$ for $l < x < \mu$ and $b(x) < 0$ for $\mu < x < u$. Assume further that*

$$\int_l^u \frac{1}{\tilde{v}(x)} e^{-(x\hat{t}_x - \kappa(\hat{t}_x))}\, dx < \infty,$$

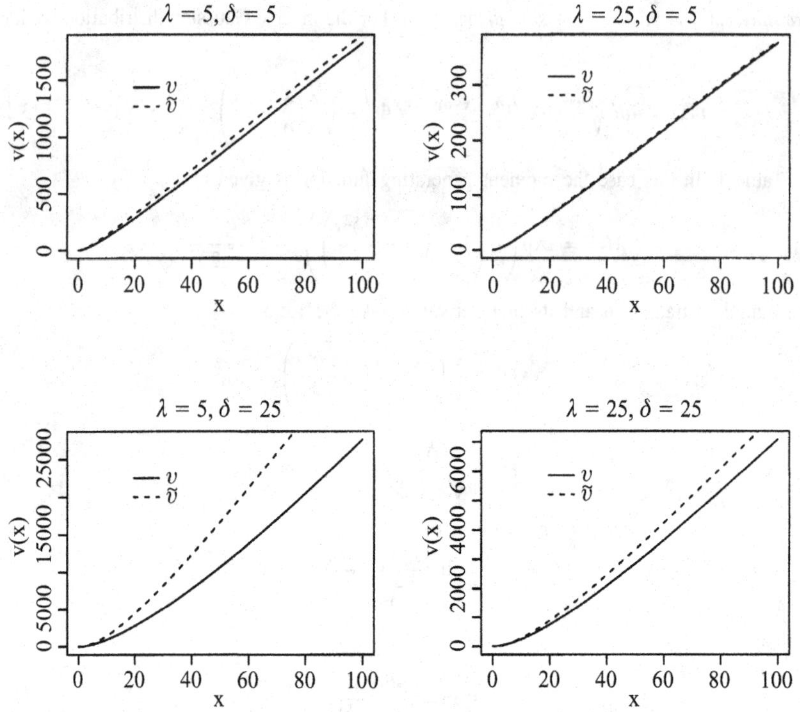

Figure 1. The two squared diffusion coefficients, (3.16) and (3.17), corresponding to the parameter values $\theta = 1$, $\lambda = 5$ and 25, and $\delta = 5$ and 25.

where

$$\tilde{v}(x) = \frac{-2b(x)}{\hat{t}_x}$$

(defined by continuity at x= *0) replaces* (3.2). *Then the differential equation*

$$\mathrm{d}X_t = b(X_t)\,\mathrm{d}t + \sqrt{\tilde{v}(X_t)}\,\mathrm{d}W_t, \qquad t \geqslant 0,$$

has a unique Markovian solution which is ergodic with invariant probability density $\tilde{f}$ given by (3.8).

Remark. Unlike the linear case, it is no longer true in general that μ is the mean of the distribution with density $\tilde{f}$. But the mean of the drift function, $b(X_t)$, is zero (provided X is stationary), so when $b(x)$ is anti-symmetric around μ, the mean is still μ. Similarly, an anti-symmetric drift guarantees that the same approximation results, relating $\tilde{f}$ to f, hold as in the case with linear drift, but in the general case $\tilde{f}$ may not comply with the saddlepoint approximation to f to the same degree of accuracy around $x = 0$; see the remark just below Theorem 3.1.

4. Sums of diffusions

Very often the correlation structure found in time series data is more complex than the exponentially decreasing autocorrelation of the models defined in Section 2. For diffusion models with a nonlinear drift the autocorrelation function is usually not known explicitly, but the autocorrelation function is bounded by a decreasing exponential function for all ρ-mixing diffusions. A stationary, ergodic diffusion is ρ-mixing under rather weak conditions; see Genon-Catalot *et al.* (2000). In order to obtain models with a more flexible correlation structure, we will therefore consider stochastic processes that are sums of processes of the type introduced in Section 2. Such processes have an explicit autocorrelation function of the form

$$\rho(t) = \phi_1 e^{-\theta_1 t} + \phi_2 e^{-\theta_2 t} + \ldots + \phi_m e^{-\theta_m t}, \qquad t \geqslant 0, \tag{4.1}$$

where $\phi_i > 0$, $i = 1, \ldots, m$ and $\phi_1 + \phi_2 + \ldots + \phi_m = 1$. This functional form is very flexible and can be fitted to a lot of empirical autocorrelation functions; see the discussion below. The construction considered in this section is closely related to the sums of Ornstein–Uhlenbeck processes driven by Lévy processes introduced in Barndorff-Nielsen *et al.* (1998) and used by Barndorff-Nielsen & Shephard (2001) to model financial volatility.

Our aim is to construct a stationary process X with a given marginal density f and with autocorrelation function given by (4.1) for some given integer m. We assume that f satisfies the following condition.

Condition 4.1. *The probability density f, with characteristic function C, is infinitely divisible, that is C^{ϕ} is a characteristic function for all positive ϕ. Assume, moreover, that there exists a $\phi_0 \geqslant 0$ such that for $\phi > \phi_0$ the probability distribution corresponding to C^{ϕ} has a density satisfying Condition 2.1.*

Note that Condition 4.1 excludes all distributions on a bounded interval since such distributions cannot be infinitely divisible. If f satisfies Condition 2.1, the only problem in the last part of Condition 4.1 is the boundedness of the density corresponding to C^{ϕ} because infinitely divisible densities are necessarily positive on (l, u). Properties of infinitely divisible distributions are reviewed in Steutel (1983). The reader is reminded that if C^{ϕ} is Lebesgue integrable on $\mathbb{R}$, then the corresponding distribution has a density with respect to the Lebesgue measure that is continuous and bounded as required in Condition 2.1. This is useful when the density cannot be found explicitly.

Let $f^{(i)}$ denote the density function corresponding to the characteristic function C^{ϕ_i} $(i = 1, \ldots, m)$. Since f satisfies Condition 4.1, we can, according to Theorem 2.1, define a stationary process $X^{(i)}$ of the type introduced in Section 2 with marginal density $f^{(i)}$, provided that $\phi_i > \phi_0$, $i = 1, \dots, m$. We will assume this to be the case. Specifically, let $W^{(1)}, W^{(2)}, \dots, W^{(m)}$ be m mutually independent Wiener processes, define

$$v_i(x) = \frac{2\theta_i \int_l^x (\phi_i \mu - y) f^{(i)}(y) \mathrm{d}y}{f^{(i)}(x)}, \qquad i = 1, \dots, m, \tag{4.2}$$

where μ is the expectation of f, and let $X^{(i)}$ be the solution of the stochastic differential equation

$$\mathrm{d}X_t^{(i)} = \theta_i(\phi_i \mu - X_t^{(i)})\mathrm{d}t + \sqrt{v_i(X_t^{(i)})}\mathrm{d}W_t^{(i)}, \qquad i = 1, \dots, m. \tag{4.3}$$

Then the processes $X_t^{(1)}, \dots, X_t^{(m)}$ are independent, and $X_t^{(i)} \sim f^{(i)}$, $i = 1, \dots, m$, that is, the distribution of $X_t^{(i)}$ has characteristic function C^{ϕ_i}. Hence the process X constructed as the sum

$$X_t = X_t^{(1)} + X_t^{(2)} + \dots + X_t^{(m)}, \tag{4.4}$$

has marginal density f, and since

$$\mathrm{corr}(X_{s+t}^{(i)}, X_s^{(i)}) = \mathrm{e}^{-\theta_i t}, \qquad i = 1, \dots, m, \tag{4.5}$$

the autocorrelation function of X is given by (4.1). It is not difficult to see that

$$\phi_i = \frac{\mathrm{var}(X_t^{(i)})}{\mathrm{var}(X_t)}, \qquad i = 1, 2, \dots, m. \tag{4.6}$$

The spectral density of the process X is given by

$$e(\omega) = \frac{2}{\pi}\left(\frac{\phi_1\theta_1}{\theta_1^2 + \omega^2} + \dots + \frac{\phi_m\theta_m}{\theta_m^2 + \omega^2}\right), \tag{4.7}$$

which follows immediately from the fact that a process with autocorrelation function $\mathrm{e}^{-\theta t}$ has spectral density $2\theta/(\pi(\theta^2 + \omega^2))$.

The motivation for models of the type (4.4) is that the random variation quite frequently is a compound of processes with different time scales. An example is the velocity fluctuations in a turbulent wind that are caused by eddies with different time scales. The process $X^{(i)}$ represents random variation with a time scale θ_i^{-1}.

The construction of the process X is particularly simple if the marginal distribution of X belongs to a class of distributions which is closed under convolution. The following two examples illustrate this.

Example 4.1 *The gamma distribution.* Here we construct a stationary stochastic process X for which the marginal density is a gamma distribution, $X_t \sim \Gamma(\alpha, \lambda)$, and the autocorrelation function is of the form (4.1). This process can be obtained as the sum of m independent diffusion processes (4.4), where $X_t^{(i)}$ is the solution of

$$\mathrm{d}X_t^{(i)} = \theta_i(\phi_i \alpha\lambda^{-1} - X_t^{(i)})\mathrm{d}t + \sqrt{2\theta_i\lambda^{-1}X_t^{(i)}}\mathrm{d}W_t^{(i)}, \qquad i = 1, \dots, m. \tag{4.8}$$

According to Example 2.2, $X_t^{(i)} \sim \Gamma(\phi_i\alpha, \lambda)$, $i = 1, \ldots, m$, and $X_t^{(i)}$ satisfies (4.5). Here the ϕ_0 of Condition 4.1 equals α^{-1}, so the construction is only possible when $\phi_i \geqslant \alpha^{-1}$, $i = 1, \ldots, m$. If there exists a $\phi_i < \alpha^{-1}$, then the process $X^{(i)}$ is not ergodic and can hit the boundary zero in finite time with positive probability.

Example 4.2 *The variance-gamma distribution.* In this example we construct a stochastic process X whose marginal density is a variance-gamma distribution, $X_t \sim VG(\lambda, \alpha, \beta, \delta)$ – see Example 3.1 – and whose autocorrelation function is of the form (4.1). Let $X^{(1)}, \ldots, X^{(m)}$ be independent diffusions constructed according to (4.3) and (4.2) with μ given by (3.10). Then

$$X_t^{(i)} \sim VG(\phi_i\lambda, \alpha, \beta, \phi_i\delta), \qquad i = 1, \ldots, m,$$

and X given by the sum (4.4) has the right distribution and autocorrelation function. In practice v_i has to be replaced by the approximation $\tilde{v}_i$; see Example 3.1 and Section 6.

Finally, a more difficult example.

Example 4.3 *The hyperbolic distribution.* The moment generating function of the centred symmetric hyperbolic distribution is

$$M(t) = \frac{\alpha \cdot K_1(\delta(\alpha^2 - t^2))}{\sqrt{\alpha^2 - t^2} \cdot K_1(\delta\alpha)}, \qquad |t| < \alpha.$$

The hyperbolic distribution is infinitely divisible, so $M(t)^{\phi_i}$ is again a moment generating function, but there seems to be no way of inverting it to get an expression for $f^{(i)}$. If one wishes to simulate the process of type (4.4) with centred symmetric hyperbolic marginal distribution, it is therefore necessary to use the approximation introduced in Section 3. This can clearly only be done numerically.

The following theorem states exactly which autocorrelation functions can be approximated by an autocorrelation function of the form (4.1).

Theorem 4.1. *The class of functions obtained as limits, as $m \to \infty$, of pointwise convergent sequences $\rho_m(t)$ of autocorrelation functions given by* (4.1) *equals the class of all Laplace transforms for distributions on* $(0, \infty)$, *that is, the class of functions given by*

$$r(u) = \int_0^\infty e^{-uv} dP(v), \qquad u \geqslant 0,$$

for some probability measure P on $(0, \infty)$.

Proof. An autocorrelation function

$$\rho_m(t) = \phi_1^{(m)} e^{-\theta_1^{(m)} t} + \ldots + \phi_m^{(m)} e^{-\theta_m^{(m)} t}$$

is equal to the Laplace transform of the distribution concentrated in $\theta_1^{(m)}, \ldots, \theta_m^{(m)}$ with probabilities $\phi_1^{(m)}, .\phi_m^{(m)}$. If the sequence $\rho_m(t)$ is convergent, the sequence of distributions converges weakly to a probability distribution on $(0, \infty)$ and the limit function is the Laplace transform of this distribution. On the other hand, any probability distribution on $(0, \infty)$ can be obtained as the limit of probability distributions concentrated on a finite set. To see this consider a suitable sequence of discretizations of the distribution in question. □

We see, in particular, that we can only approximate autocorrelation functions that are decreasing and convex. Moreover, the logarithm of the autocorrelation function must also be convex. In fact, it is well known that the class of all Laplace transforms of distributions on $(0, \infty)$ equals the class of *completely monotone functions* r with $r(0) = 1$; see Feller 1971, p. 439). A function r on $[0, \infty)$ is called completely monotone if

$$(-1)^n r^{(n)}(u) \geqslant 0, \qquad u > 0$$

for all $n \in \mathbb{N}$, where $r^{(n)}$ is the nth derivative of r.

One motivation for using models with autocorrelations of the type (4.1) is to be able to fit a relatively simple model to data to which some might think it necessary to fit a model with *long-range dependence*. Let us therefore briefly discuss the fact that an autocorrelation function of the type (4.1) can be close to an autocorrelation function of a process with long memory. Suppose the series

$$r(u) = \sum_{j=1}^{\infty} \phi_j e^{-\theta_j u} \tag{4.9}$$

is convergent. If we, moreover, assume that

$$\sum_{j=1}^{\infty} \phi_j/\theta_j = \infty, \tag{4.10}$$

then

$$\int_0^{\infty} r(u)\mathrm{d}u = \infty,$$

so $r(u)$ is the autocorrelation function of a process with long memory that can be approximated as well as we wish by an autocorrelation function of the type (4.1). To give a specific example, we choose

$$\phi_j \sim j^{-1-2(1-H)}, \qquad \theta_j \sim j^{-1}$$

with $0 < H < 1$. Then

$$\phi_j/\theta_j \sim j^{-2(1-H)},$$

and when $\frac{1}{2} \leqslant H < 1$,

$$r(u) \sim L(u)u^{-2(1-H)},$$

where L is a slowly varying function (for a definition, see Feller 1971, p. 276), so a process with autocorrelation function r has long memory with Hurst exponent H.

The convergence of the sum (4.9) implies mean-square and hence almost sure convergence of the sum

$$X_t = \sum_{i=1}^{\infty} X_t^{(i)}, \tag{4.11}$$

where the random variables $X_t^{(i)}$ are given by (4.2) and (4.3) with $f^{(i)}$ denoting the density function corresponding to $C(t)^{\phi_i}$. It is again assumed that $f^{(i)}$, $i = 1, \ldots$, are continuous and bounded on their support. The limit process X is stationary with marginal density f and has autocorrelation function $r(t)$. It is thus possible to define an infinite version of the sum (4.4). Usually this is, however, an unnecessary complication because a good fit to data can be obtained for a small value of m in (4.4). When the long-memory condition (4.10) is satisfied, the limit process (4.11) is closely related to the long-range dependent processes constructed in Cox (1984), Barndorff-Nielsen *et al.* (1990), and Barndorff-Nielsen (1998).

5. Multivariate models

In this section we shall briefly show how to construct multivariate processes where each coordinate is a process of the type introduced in Section 4.

As in Section 4, we consider a probability density f with characteristic function C satisfying Condition 4.1. For given $\phi_i > \phi_0$ $(i = 1, \ldots, m)$ satisfying $\phi_1 + \ldots + \phi_m = 1$, define $v_i(x)$ by (4.2). Let the processes $X_t^{(k,i)}$, $i = 1, \ldots, m$, $k = 1, \ldots, d$, be given by

$$\mathrm{d}X_t^{(k,i)} = -\theta_i\left(X_t^{(k,i)} - \phi_i\mu\right)\mathrm{d}t + \sqrt{v_i\left(X_t^{(k,i)}\right)}\,\mathrm{d}W_t^{(k,i)},$$

where $W^{(1,1)}, \ldots, W^{(d,m)}$ are independent standard Wiener processes. Then we can define a d-dimensional process X by

$$X_t = (X_{1t}, \ldots, X_{dt})$$

with

$$X_{jt} = X_t^{(\nu_{j1},1)} + \ldots + X_t^{(\nu_{jm},m)},$$

where $\nu_{ji} \in \{1, \ldots, d\}$, $i = 1, \ldots, m$. The point is that some of the ν_{ji} can be identical for different j so that the same process appears in different coordinates. As previously, we can interpret the process $X_t^{(k,i)}$ as random variation with time scale θ_i^{-1}. Dependence between two coordinates is thus caused by the random variation on certain time scales being the same for the two coordinates. Extra flexibility in the modelling of dependence can be obtained by taking some of the θ_i to be identical, so that it is possible for only a part of the random variation on a certain time scale to be the same in two coordinates.

The density of X_{jt} is f, and the autocorrelation function of X_j is given by (4.1). Moreover,

$$\mathrm{corr}(X_{j_1 t}, X_{j_2 t}) = \sum_{i \in M_{j_1 j_2}} \phi_i,$$

where $M_{j_1 j_2} = \{i \mid \nu_{j_1 i} = \nu_{j_2 i}\}$. The final process is obtained by making location–scale transformations of the marginals X_{jt}.

6. Case study

In this section we consider a data set consisting of 5415 measurements of the streamwise wind velocity component measured at Ferring on the Danish west coast in an experiment carried out in September 1985. The data were recorded using a sonic anemometer on a 30 metre mast erected on the shore around 60 m from the shoreline, with a 10 Hz frequency. The experiment is described in further detail in Mikkelsen (1988; 1989); the data were analysed in Barndorff-Nielsen *et al.* (1993) using a sum of independent autoregressions and in Bibby and Sørensen (2001) based on a hyperbolic diffusion model. The data are presented in Figure 2. See also Barndorff-Nielsen *et al.* (1990).

In Figures 3 and 4 a histogram and a log histogram of the wind velocity data are given, along with fitted curves corresponding to a variance-gamma density function; see Example 3.1. The fitted curves are determined by maximum likelihood based on a multinomial likelihood function where the groups are defined by the points (mid-points) in Figures 3 and 4.

In Figures 5 and 6 the empirical autocorrelation function and its logarithm are drawn for lag values up to 500. The following three exponential functions (and their logarithms) are included in the figures:

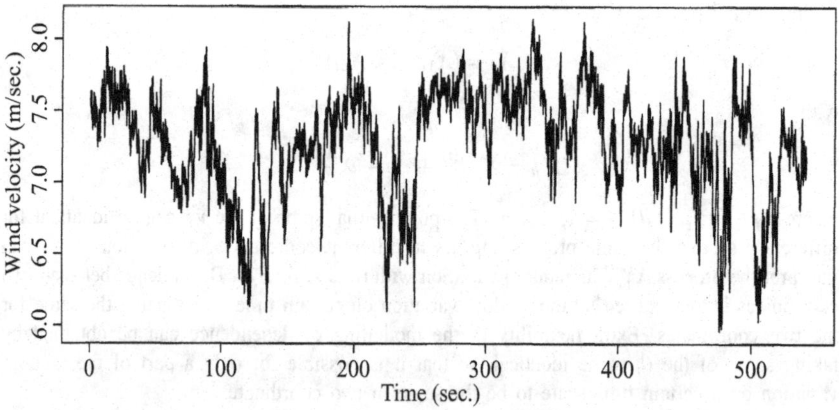

Figure 2. Streamwise wind velocity component in metres per second plotted against time in seconds.

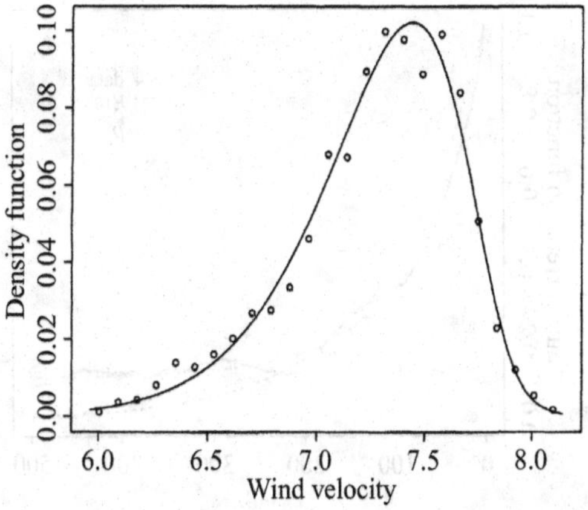

Figure 3. Histogram of the wind velocity data with a fitted variance-gamma density function

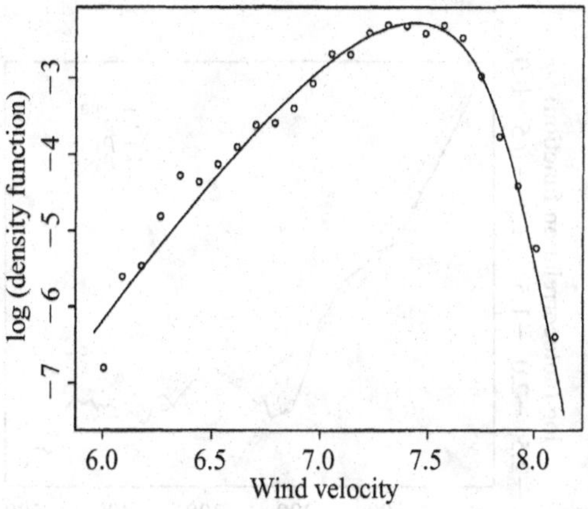

Figure 4. Log histogram of the wind velocity data with a fitted variance-gamma log-density function.

$$\rho_1(t) = e^{-0.0093t},$$

$$\rho_2(t) = 0.83e^{-0.0154t} + 0.17e^{-0.0009t},$$

$$\rho_3(t) = 0.80e^{-0.0125t} + 0.09e^{-0.0986t} + 0.11e^{-0.0001t}.$$

Based on Figures 3–6, we wish to consider a stochastic process with a variance-gamma marginal distribution,

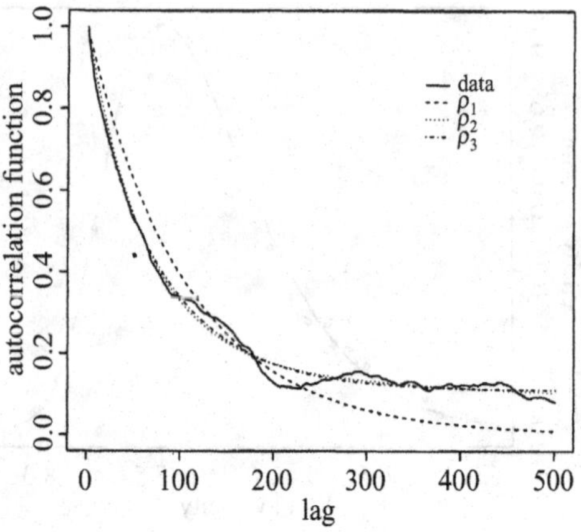

Figure 5. The empirical autocorrelation function of the wind velocity data with fitted curves corresponding to one, two, and three exponential functions.

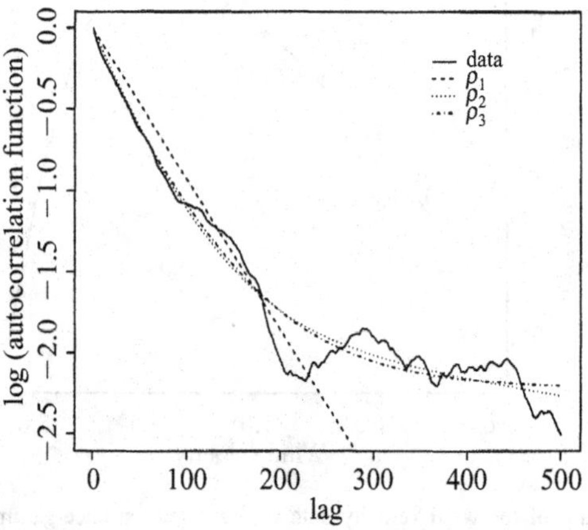

Figure 6. The log empirical autocorrelation function of the wind velocity data with fitted curves corresponding to the logarithm of one, two, and three exponential functions.

$$X_t \sim VG(\lambda, \alpha, \beta, \delta),$$

given as the sum of two diffusion processes,

$$X_t = X_t^{(1)} + X_t^{(2)},$$

where

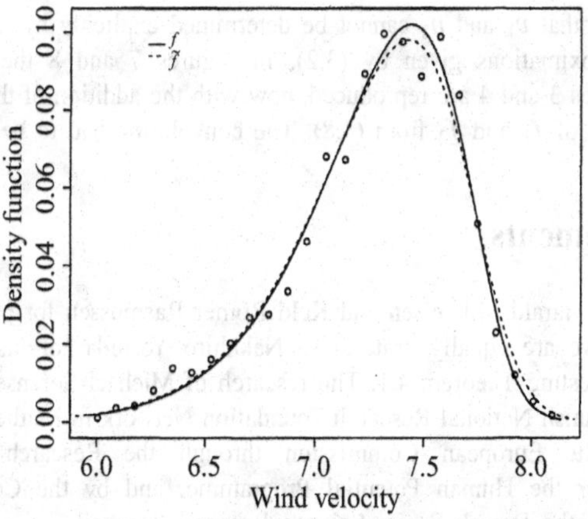

Figure 7. Histogram of the wind velocity data with fitted corresponding to a variance-gamma density (f) and an approximate variance-gamma density ($\tilde{f}$).

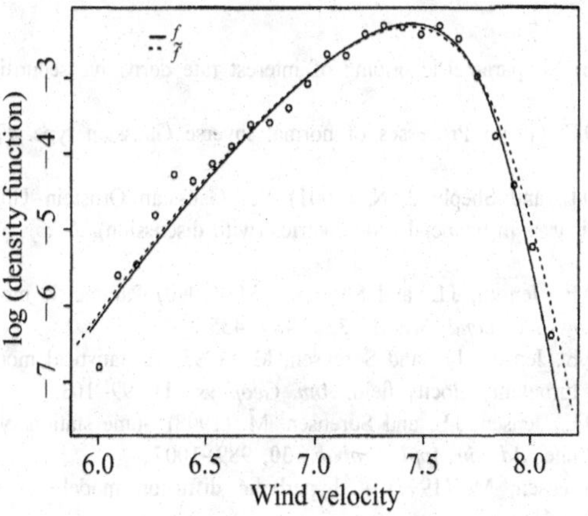

Figure 8. Log histogram of the wind velocity data with fitted curves corresponding to a variance-gamma log-density (f) and an approximate variance-gamme log-density ($\tilde{f}$).

$$\operatorname{corr}(X_{s+t}^{(i)}, X_s^{(i)}) = \mathrm{e}^{-\theta_i t}, \quad i = 1, 2.$$

This can be done using the construction given in Example 4.2.

From the fit of the histogram and the empirical autocorrelation function we obtain

$$\hat{\lambda} = 3.9134, \quad \hat{\alpha} = 13.1760, \quad \hat{\beta} = -7.8232, \quad \hat{\delta} = 7.8128,$$

$$\hat{\phi}_1 = 0.8346, \quad \hat{\phi}_2 = 0.1654, \quad \hat{\theta}_1 = 10.0154, \quad \hat{\theta}_2 = 0.0009.$$

A problem here is that v_1 and v_2 cannot be determined explicitly by (2.2). Instead we can consider the approximations given by (3.2). In Figures 7 and 8 the histogram and log histogram in Figures 3 and 4 are reproduced, now with the addition of the estimated $\tilde{f}$ given by the convolution of $\tilde{f}_1$ and $\tilde{f}_2$ from (3.8). The convolution had to be done numerically.

Acknowledgements

We are grateful to Harald Mikkelsen and Keld Rømer Rasmussen for putting the wind data at our disposal. We are equally grateful to Nakahiro Yoshida for his comments, and in particular for suggesting Theorem 4.1. The research of Michael Sørensen was supported by MaPhySto, The Danish National Research Foundation Network in Mathematical Physics and Stochastics, by the European Commission through the Research Training Network DYNSTOCH under the Human Potential Programme, and by the Centre for Analytical Finance funded by the Danish Social Science Research Council.

References

Aït-Sahalia, Y. (1996) Nonparametric pricing of interest rate derivative securities. *Econometrica*, **64**, 527–560.

Barndorff-Nielsen, O.E. (1998) Processes of normal inverse Gaussian type. *Finance Stochastics*, **2**, 41–68.

Barndorff-Nielsen, O.E. and Shephard, N. (2001) Non-Gaussian Ornstein–Uhlenbeck-based models and some of their uses in financial econometrics (with discussion). *J. Roy. Statist. Soc. Ser. B.*, **63**, 167–241.

Barndorff-Nielsen, O.E., Jensen, J.L. and Sørensen, M. (1990) Parametric modelling of turbulence. *Philos. Trans. Roy. Soc. Lond. Ser. A.*, **332**, 439–455.

Barndorff-Nielsen, O.E., Jensen, J.L. and Sørensen, M. (1993) A statistical model for the streamwise component of a turbulent velocity field. *Ann. Geophys.*, **11**, 99–103.

Barndorff-Nielsen, O.E., Jensen, J.L. and Sørensen, M. (1998) Some stationary processes in discrete and continuous time. *Adv. in Appl. Probab.*, **30**, 989–1007.

Bibby, B.M. and Sørensen, M. (1997) A hyperbolic diffusion model for stock prices. *Finance Stochastics*, **1**, 25–41.

Bibby, B.M. and Sørensen, M. (2001) Simplified estimating functions for diffusion models with a high-dimensional parameter. *Scand. J. Statist.*, **28**, 99–112.

Bibby, B.M. and Sørensen, M. (2003) Hyperbolic processes in finance. In S. Rachev (ed.), *Handbook of Heavy Tailed Distributions in Finance*, pp. 211–248. Amsterdam: Elsevier Science.

Bibby, B.M. and Sørensen, M. (2004) Flexible stochastic volatility models of the diffusion type. In preparation

Bleistein, N. (1966) Uniform asymptotic expansions of integrals with stationary point near algebraic singularity. *Comm. Pure Appl. Math.*, **19**, 353–370.

Cox, D.R. (1984) Long-range dependence: A review. In H.A. David and H.T. David (eds), *Statistics: An Appraisal.* Ames: Iowa State University Press.

Cox, J.C., Ingersoll Jr., J.E. and Ross, S.A. (1985) A theory of the term structure of interest rates. *Econometrica*, **53**, 385–407.

Daniels, H.E. (1954) Saddlepoint approximations in statistics. *Ann. Math. Statist.*, **25**, 631–650.

Daniels, H.E. (1987) Tail probability approximations. *Internat. Statist. Rev.*, **55**, 37–48.

De Jong, F., Drost, F.C. and Werker, B.J.M. (2001) A jump-diffusion model for exchange rates in a target zone. *Statist. Neerlandica*, **55**, 270–300.

Engelbert, H.J. and Schmidt, W. (1985) On solutions of one-dimensional stochastic differential equations without drift. *Z. Wahrscheinlichkeitstheorie Verw. Geb.*, **68**, 287–314.

Feller, W. (1971) *An Introduction to Probability Theory and its Applications, Vol. II.* New York: Wiley.

Genon-Catalot, V., Jeantheau, T. and Larédo, C. (2000) Stochastic volatility models as hidden Markov models and statistical applications. *Bernoulli*, **6**, 1051–1079.

Karlin, S. and Taylor, M. (1981) *A Second Course in Stochastic Processes*. Orlando, FL: Academic Press.

Larsen, K.S. and Sørensen, M. (2003) A diffusion model for exchange rates in a target zone. Preprint no. 6, Department of Applied Mathematics and Statistics, University of Copenhagen.

Madan, D.B. and Seneta, E. (1990) The variance gamma (V.G.) model for share market returns. *J. Business*, **63**, 511–524.

Madan, D.B. and Yor, M. (2002) Making Markov martingales meet marginals: with explicit constructions. *Bernoulli*, **8**, 509–536.

Mikkelsen, H.E. (1988) *Turbulence in the Wake Zone of a Vegetated 2-Dimensional Dune*, Geoskrifter 30. Geology Institute, University of Aarhus.

Mikkelsen, H.E. (1989) *Wind Flow and Sediment Transport over a Low Coastal Dune*, Geoskrifter 32. Geology Institute, University of Aarhus.

Pedersen, A.R. (2000) Estimating the nitrous oxide emission rate from the soil surface by means of a diffusion model. *Scand. J. Statist.*, **27**, 385–404.

Skorokhod, A.V. (1989) *Asymptotic Methods in the Theory of Stochastic Differential Equation*. Providence, RI: Americal Mathematical Society.

Steutel, F.W. (1983) Infinite divisibility. In N.L. Johnson, S. Kotz and C.B. Read (eds), *Encyclopedia of Statistical Sciences, Vol. 4*, pp. 114–116. New York: Wiley.

Received April 2003 and revised July 2004

Chapter 15

Likelihood ratio tests in curved exponential families with nuisance parameters present only under the alternative

Introduction by Nina Munkholt Jakobsen and Michael Sørensen

University of Copenhagen

15.1 Introduction to the paper

The standard asymptotic theory does not apply to the likelihood ratio test when there are nuisance parameters which are present only under the alternative hypothesis. There are several instances of this for models of practical interest. One example is mixture models, where the parameters determining a mixture component are not present under the hypothesis that the component has probability zero. The asymptotic theory of even simple mixture models is highly complex; see Hartigan (1985), Bickel and Chernoff (1993), Liu and Shao (2004), and Hall and Stewart (2005).

Another example is the exponential family setting considered by Ritz and Skovgaard (2005). Their framework allows a relatively direct and transparent derivation of a non-standard asymptotic result. It is, however, sufficiently general to cover several important models, including their motivating longitudinal model. In order to emphasize the main ideas of the proof, we sketch the derivation in a slightly simpler setting.

Consider a basic model for n independent observations $x_1, \ldots, x_n$ with log-likelihood function $l_n(\alpha)$, $\alpha \in A \subseteq \mathbb{R}^p$. A submodel $\mathcal{M}$ is parametrized by $(\phi, \rho) \in \mathbb{R} \times P$, where $P \subseteq \mathbb{R}^q$, $q \leq p-1$, and it is given by $\alpha = \psi(\phi, \rho)$ for some function $\psi : \mathbb{R} \times P \mapsto A$. Suppose we wish to test the hypothesis $H : \phi = \phi_0$. If the nuisance parameter ρ is present under the hypothesis, the likelihood ratio test statistic is $-2 \log Q = 2 \left[l_n\{\psi(\hat{\phi}, \hat{\rho})\} - l_n\{\psi(\phi_0, \hat{\rho}_{\phi_0})\} \right]$, where $(\hat{\phi}, \hat{\rho})$ is the maximum likelihood estimator under the model $\mathcal{M}$, and

$\hat{\rho}_{\phi_0}$ is the maximum likelihood estimator under the hypothesis H. By the well-known Wilks' theorem, $-2 \log Q$ is asymptotically chi-squared distributed with one degree of freedom as $n \to \infty$ under standard regularity conditions. The proof of this classical result relies on the fact that in the limit, the effect of $\hat{\rho}$ on the first term of the test statistic is cancelled by the effect of the estimator $\hat{\rho}_{\phi_0}$ on the second term. When the parameter ρ is not present under the hypothesis, this cannot happen, and a different limit distribution is to be expected.

Let us now see what happens if the nuisance parameter ρ is not present under the hypothesis. Specifically, let us assume that there exists an α_0 such that $\psi(\phi_0, \rho) = \alpha_0$ for all $\rho \in P$. Ritz and Skovgaard show that when deriving the asymptotic distribution, the log-likelihood function of the basic model may be replaced by the quadratic approximation

$$l_n^*(\alpha) = l_n(\alpha_0) + \partial_\alpha l_n(\alpha_0)^\mathsf{T}(\alpha - \alpha_0) - \frac{n}{2}(\alpha - \alpha_0)^\mathsf{T} I(\alpha_0)(\alpha - \alpha_0),$$

where $nI(\alpha) = -E_{\alpha_0}\{\partial_\alpha^2 l_n(\alpha)\}$ is the Fisher information matrix of the basic model. Here, vectors are column vectors, T denotes transposition, and ∂_α denotes partial differentiation with respect to the coordinates of α. Under Ritz and Skovgaard's regularity conditions,

$$\left| \sup_{\phi \in \mathbb{R}, \rho \in P} l_n\{\psi(\phi, \rho)\} - \sup_{\phi \in \mathbb{R}, \rho \in P} l_n^*\{\psi(\phi, \rho)\} \right| \xrightarrow{\mathcal{P}} 0, \tag{15.1}$$

where $\xrightarrow{\mathcal{P}}$ denotes convergence in probability under the hypothesis. If ψ is an affine function of ϕ, $\partial_\phi \psi(\phi, \rho)$ does not depend on ϕ, and we may write $\psi(\phi, \rho) = \psi(\phi_0, \rho) + \partial_\phi \psi(\rho)(\phi - \phi_0)$. Under this assumption

$$l_n^*\{\psi(\phi, \rho)\} = l_n(\alpha_0) + \partial_\phi l_n\{\psi(\phi_0, \rho)\}(\phi - \phi_0) - \frac{n}{2} I^{\mathcal{M}}(\phi_0, \rho)(\phi - \phi_0)^2,$$

with $\partial_\phi l_n\{\psi(\phi_0, \rho)\} = \partial_\alpha l_n(\alpha_0)^\mathsf{T} \partial_\phi \psi(\rho)$, and where $I^{\mathcal{M}}(\phi_0, \rho) = \partial_\phi \psi(\rho)^\mathsf{T} I(\alpha_0) \partial_\phi \psi(\rho)$ is the Fisher information for the parameter ϕ evaluated at (ϕ_0, ρ). By standard calculations, the quadratic function $\phi \mapsto l_n^*\{\psi(\phi, \rho)\}$ attains its unique maximum at $\phi = \phi_0 + \partial_\phi l_n\{\psi(\phi_0, \rho)\}/\{nI^{\mathcal{M}}(\phi_0, \rho)\}$, so

$$\sup_{\phi \in \mathbb{R}} l_n^*\{\psi(\phi, \rho)\} - l_n(\alpha_0) = \frac{\{\partial_\alpha l_n(\alpha_0)^\mathsf{T} \partial_\phi \psi(\rho)/\sqrt{n}\}^2}{2I^{\mathcal{M}}(\phi_0, \rho)}.$$

Note that the approximate test statistic $\sup_{\phi \in \mathbb{R}, \rho \in P} l_n^*\{\psi(\phi, \rho)\} - l_n(\alpha_0)$ is the largest score test for the hypothesis H among all models with a fixed $\rho \in P$. By the central limit theorem, $\partial_\alpha l_n(\alpha_0)/\sqrt{n} \xrightarrow{w} W$

with $W \sim N_p\{0, I(\alpha_0)\}$, where $\xrightarrow{w}$ denotes weak convergence under the hypothesis. Since the mapping from $\mathbb{R}^p$ into $\mathbb{R}$ given by $x \mapsto \sup_{\rho\in P} \frac{1}{2}\{x^\top \partial_\phi \psi(\rho)\}^2/I^{\mathcal{M}}(\phi_0, \rho)$ is continuous, it follows that under the hypothesis H

$$\sup_{\phi\in\mathbb{R}, \rho\in P} l_n^*\{\psi(\phi,\rho)\} - l_n(\alpha_0) \xrightarrow{w} Z/2.$$

Here

$$Z = \sup_{\rho\in P}\{V^\top \xi(\rho)\}^2,$$

where the distribution of V in $\mathbb{R}^p$ is standard normal, and $\xi(\rho) = I(\alpha_0)^{\frac{1}{2}}\partial_\phi\psi(\rho)/\|I(\alpha_0)^{\frac{1}{2}}\partial_\phi\psi(\rho)\|$ is a unit vector. Thus, $V^\top\xi(\rho)$ is the projection of V onto the subspace spanned by $\xi(\rho)$, and for fixed ρ, $V^\top\xi(\rho)$ has a standard normal distribution. By (15.1) it follows that $-2\log Q \xrightarrow{w} Z$.

15.2 The limit distribution

In the special case $P = \{\rho_0\}$, Z is the square of a standard normally distributed random variable, and we recover the classical limit distribution. However, for general P, the limit distribution can be quite different from a chi-squared distribution.

In Ritz and Skovgaard's motivating longitudinal example, the alternative hypothesis is one-sided, which is not the case in our simplified setting. In connection with their longitudinal model, Ritz and Skovgaard mention that in applications, when ρ is one-dimensional, the limit distribution of $-2\log Q$ is sometimes assumed to be chi-squared with one or two degrees of freedom, or a mixture of these two distributions. They also note the absence of a theoretical foundation for this assumption. Munkholt (2012) investigated the limit distributions in some examples within the simplified framework presented here, and compared the distributions to the two chi-squared distributions and their mixtures. Some of the results of this investigation are presented in the following.

15.2.1 *Example I*

Consider the case where $p = 2$ and $\psi(\phi,\rho) = (\phi, c\phi\rho)^\top$ for some given $c > 0$, with $\rho \in P = [-1, 1]$. Let us first assume that $I(\alpha_0)$ equals the 2×2-identity matrix, such that $\xi_c(\rho) = \partial_\phi\psi(\rho)/\|\partial_\phi\psi(\rho)\| = (1, c\rho)^\top/(1 + c^2\rho^2)^{1/2}$. In the parameter set of the basic model, the submodel $\mathcal{M}$ is the double cone $\mathcal{C}_c = \cup_{\rho\in P} \operatorname{span} \xi_c(\rho)$.

By geometrical considerations

$$\sup_{\rho\in P}\{V^{\mathsf{T}}\xi_c(\rho)\}^2 = \begin{cases} \|V\|^2 & \text{if } V \in \mathcal{C}_c \\ \{V^{\mathsf{T}}\xi_c(1)\}^2 & \text{if } V_2 > cV_1 > 0 \text{ or } V_2 < cV_1 < 0 \\ \{V^{\mathsf{T}}\xi_c(-1)\}^2 & \text{if } -V_2 > cV_1 > 0 \text{ or } -V_2 < cV_1 < 0, \end{cases}$$

i.e., $\sup_{\rho\in P}\{V^T\xi_c(\rho)\}^2$ equals the squared length of the projection of V onto the double cone $\mathcal{C}_c$. Since $\|V\|^2$ is chi-squared distributed with two degrees of freedom, and both $\{V^{\mathsf{T}}\xi_c(1)\}^2$ are $\{V^T\xi_c(-1)\}^2$ are chi-squared distributed with one degree of freedom, one might expect the asymptotic distribution of $-2\log Q$ to lie in-between these two distributions, in some sense. Note also that $\sup_{\rho\in P}\{V^{\mathsf{T}}\xi_c(\rho)\}^2 \to \|V\|^2$ as $c \to \infty$, while $\sup_{\rho\in P}\{V^{\mathsf{T}}\xi_c(\rho)\}^2 \to V_1^2$ as $c \to 0$.

Let $\theta_c \in (0, \pi/2)$ denote the angle between the positive x-axis and the vector $\xi_c(1)$, one of the vectors spanning the boundary of $\mathcal{C}_c$. Then $c = \tan(\theta_c)$, with $\theta_c \to 0$ as $c \to 0$, and $\theta_c \to \pi/2$ as $c \to \infty$. The limit distribution, henceforth denoted $\mathbf{C}(\theta_c)$ in this example, depends on the angle θ_c.

It was shown by Munkholt (2012) that the distributions in the family $\{\mathbf{C}(\theta) \mid \theta \in (0, \pi/2)\}$ are stochastically larger than the chi-squared distribution with one degree of freedom, but stochastically smaller than the chi-squared distribution with two degrees of freedom. This is in accordance with the above intuition. However, calculations of the means and variances of the $\mathbf{C}(\theta)$ distributions, as illustrated in Figure 15.1, reveal that these distributions are not mixtures of the two chi-squared distributions. In Figure 15.1 the solid curve indicates the means and variances of the $\mathbf{C}(\theta)$ distributions for $\theta \in (0, \pi/2)$, while the dashed curve represents those of the $p : (1-p)$ mixtures of the chi-squared distributions with one and two degrees of freedom for $p \in (0, 1)$. The $\times$'s indicate the means and variances of the chi-squared distributions with one, two, and three degrees of freedom.

Due to the stochastic ordering of the $\mathbf{C}(\theta)$ distributions relative to the two chi-squared distributions, any quantile of a $\mathbf{C}(\theta)$ distribution will always be smaller than the corresponding quantile of the chi-squared distribution with two degrees of freedom. Consequently, using the $(1-\alpha)$-quantile of the chi-squared distribution with two degrees of freedom, in place of that of the relevant $\mathbf{C}(\theta)$ distribution in a likelihood ratio test of level $\alpha \in (0, 1)$, would imply a conservative test.

If the Fisher information matrix $I(\alpha_0)$ is not equal to the identity matrix, the vectors $\xi_c(\rho)$ are defined differently, but $\sup_{\rho\in P}\{V^{\mathsf{T}}\xi_c(\rho)\}^2$ still equals the squared length of the projection of V onto a (possibly rotated)

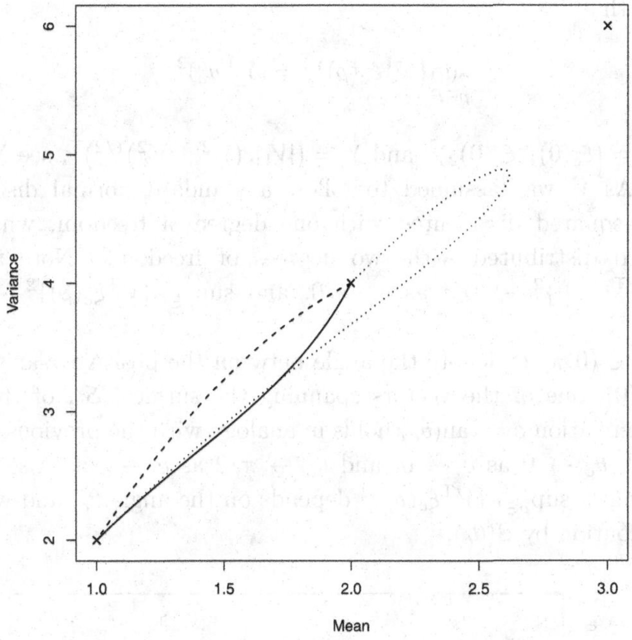

Fig. 15.1 (Mean, variance) plot for the $\mathbf{C}(\theta)$ distributions (solid curve), the $\mathbf{S}(\theta)$ distributions (dotted curve), and mixtures of the chi-squared distributions with one and two degrees of freedom (dashed curve). The $\times$'es indicate the means and variances of the chi-squared distributions with one, two, and three degrees of freedom.

double cone. Since the distribution of V is invariant under rotation, the distribution of $\sup_{\rho\in P}\{V^\top \xi_c(\rho)\}^2$ still belongs to the family of $\mathbf{C}(\theta)$ distributions with θ related to the double cone in a similar way.

15.2.2 *Example II*

Now consider the case where $p = 3$ and $\psi(\phi,\rho) = (\phi, c\phi\cos\rho, c\phi\sin\rho)^\top$ for some given $c > 0$, with $\rho \in P = [-\pi,\pi]$. We assume that $I(\alpha_0)$ equals the 3×3 identity matrix, such that $\xi_c(\rho) = \partial_\phi\psi(\rho)/\|\partial_\phi\psi(\rho)\| = (1, c\cos\rho, c\sin\rho)^\top/(1+c^2)^{1/2}$. In the parameter set of the basic model, the submodel $\mathcal{M}$ is the surface $\mathcal{S}_c = \cup_{\rho\in P}\operatorname{span}\xi_c(\rho)$ of a double cone, where c has an interpretation similar to that in the previous example.

By expressing V in cylindrical coordinates $(V_1, (V_2^2+V_3^2)^{1/2}, \Gamma_V)$, where Γ_V is the (counterclockwise) angle between $(0,1,0)$ and $(0,V_2,V_3)$, it can be shown that

$$\sup_{\rho \in P}\{V^{\mathsf{T}}\xi_c(\rho)\}^2 = (Y^{\mathsf{T}}u_c)^2,$$

where $u_c = (\xi_c(0)_1, \xi_c(0)_2)^{\mathsf{T}}$ and $Y = (|V_1|, (V_2^2 + V_3^2)^{1/2})^{\mathsf{T}}$, see Munkholt (2012). As V was assumed to follow a standard normal distribution, Y_1^2 is chi-squared distributed with one degree of freedom, while Y_2^2 is chi-squared distributed with two degrees of freedom. Note also that $\sup_{\rho\in P}\{V^{\mathsf{T}}\xi_c(\rho)\}^2 \to Y_1^2$ as $c \to 0$, and $\sup_{\rho\in P}\{V^{\mathsf{T}}\xi_c(\rho)\}^2 \to Y_2^2$ as $c \to \infty$.

Let $\theta_c \in (0, \pi/2)$ denote the angle between the positive x-axis and the vector $\xi_c(0)$, one of the vectors spanning the surface, $\mathcal{S}_c$, of the double cone. The relation $c = \tan(\theta_c)$ holds in analogy with the previous example, again with $\theta_c \to 0$ as $c \to 0$, and $\theta_c \to \pi/2$ as $c \to \infty$. Also here the distribution of $\sup_{\rho\in P}\{V^{\mathsf{T}}\xi_c(\rho)\}^2$ depends on the angle θ_c, and we denote this distribution by $\mathbf{S}(\theta_c)$.

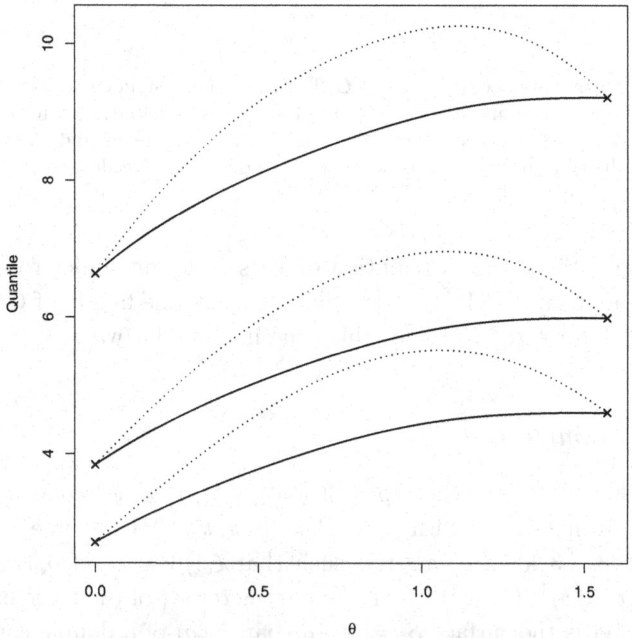

Fig. 15.2 0.90 (bottom), 0.95 (middle), and 0.99 (top) quantiles of the $\mathbf{S}(\theta)$ distributions (dotted curves) and $\mathbf{C}(\theta)$ distributions (solid curves). The $\times$'es indicate the quantiles of the chi-squared distributions with one and two degrees of freedom.

Since $(Y^\top u_c)^2 \leq \|Y\|^2\|u_c\|^2 \leq Y_1^2 + Y_2^2$, the distributions $\mathbf{S}(\theta)$, $\theta \in (0, \pi/2)$, are dominated by the chi-squared distribution with three degrees of freedom. Means and variances of the distributions $\mathbf{S}(\theta)$ were calculated by Munkholt (2012) and are plotted in Figure 15.1 (the dotted curve). It is seen from this figure that the expectations of many $\mathbf{S}(\theta)$ distributions exceed the expectation of the chi-squared distribution with two degrees of freedom. Such distributions cannot be stochastically smaller than the chi-squared distribution with two degrees of freedom, as stochastic ordering implies the same ordering of expectations. This finding is not in accordance with the standard intuition that a hypothesis which essentially reduces the number of parameters by two, should not result in the loss of more than two degrees of freedom.

Figure 15.2 shows the 0.90 (bottom), 0.95 (middle), and 0.99 (top) quantiles of the $\mathbf{S}(\theta)$ distributions (dotted curves), together with the corresponding quantiles of the $\mathbf{C}(\theta)$ distributions from the previous example for comparison (solid curves). The quantiles of the chi-squared distributions with one and two degrees of freedom, respectively, are indicated with $\times$'es. Here we see that using, e.g., the 0.95 quantile of the chi-squared distribution with two degrees of freedom in place of the quantile of the relevant $\mathbf{S}(\theta)$ distribution would, for most values of θ, not result in a conservative test of level $\alpha = 0.05$. If instead the quantile of the chi-squared distribution with three degrees of freedom is used, a conservative test is obtained. This follows from the stochastic domination result above.

The results for these simple examples demonstrate the importance of Ritz and Skovgaard's general distributional result. Their work is an interesting and illuminating contribution to statistical science concerning models of relevance to applied statistics for which the failure of standard regularity conditions can have significant practical consequences.

Acknowledgements

The authors are grateful to Ernst Hansen for fruitful discussions.

Biometrika (2005), **92**, 3, *pp*. 507–517

Printed in Great Britain

Likelihood ratio tests in curved exponential families with nuisance parameters present only under the alternative

BY CHRISTIAN RITZ AND IB M. SKOVGAARD

Department of Natural Sciences, The Royal Veterinary and Agricultural University, Thorvaldsensvej 40, 1871 Frederiksberg C, Denmark

ritz@dina.kvl.dk ib.m.skovgaard@imf.kvl.dk

SUMMARY

For submodels of an exponential family, we consider likelihood ratio tests for hypotheses that render some parameters nonidentifiable. First, we establish the asymptotic equivalence between the likelihood ratio test and the score test. Secondly, the score-test representation is used to derive the asymptotic distribution of the likelihood ratio test. These results are derived for general submodels of an exponential family without assuming compactness of the parameter space. We then exemplify the results on a class of multivariate normal models, where null hypotheses concerning the covariance structure lead to loss of identifiability of a parameter. Our motivating problem throughout the paper is to test a random intercepts model against an alternative covariance structure allowing for serial correlation.

Some key words: Covariance structure; Exponential family; Likelihood ratio test; Multivariate normal distribution; Nonidentifiability; Repeated measurements model.

1. INTRODUCTION

The large number of covariance structures available for longitudinal models or repeated measurements models allow one to accommodate various patterns of dependence (Diggle et al., 1994; Verbeke & Molenberghs, 1997). In deciding on the appropriate covariance structure we can often start with a relatively general covariance structure and then use hypothesis testing to assess whether or not this structure can be simplified. A null hypothesis arising from such a simplification in the covariance structure often renders a parameter nonidentifiable under the null hypothesis, thus invalidating standard asymptotic results such as the chi-squared approximation to the likelihood ratio test.

To fix ideas consider a multivariate normal model having covariance matrices of the general form

$$(\phi - \phi_0)M(\rho) + \gamma_1 M_1 + \ldots + \gamma_q M_q \tag{1}$$

with parameters $\phi, \rho, \gamma_1, \ldots, \gamma_q$ and matrices $M(\rho), M_1, \ldots, M_q$. Many spatial correlation models fit into formulation (1), such as all models given in Little et al. (1996, pp. 304–6). We consider the likelihood ratio test, or the score test, for the null hypothesis $\phi = \phi_0$ versus the one-sided alternative $\phi > \phi_0$. This null hypothesis reduces the models to random coefficients models, but renders the parameter ρ nonidentifiable at the same time. In applications it is sometimes assumed that the asymptotic distribution of the likelihood ratio statistic is a χ^2 distribution with one or two degrees of freedom or a mixture of these two distributions when ρ is one-dimensional, but this assumption lacks theoretical foundation. We consider a one-sided alternative here in accordance with usual applications of the model, but for the derivation of the asymptotic distribution of the test statistics the boundary problem thus introduced is minor compared to the vanishing parameter.

In the present paper we derive the asymptotic distribution for the likelihood ratio test as well as for the asymptotically equivalent score test in a general setting with submodels of an exponential family. The resulting asymptotic distribution agrees with that of the maximum of a given Gaussian process as proposed by Davies (1977). However, it has been pointed out by Hartigan (1985) that Davies's result does not generally apply, even in seemingly well-behaved models, when the parameter space is not compact. Furthermore it is not a trivial task in general, even with a compact parameter space, to establish the validity of Davies's results; see for example Lemdani & Pons (1995, 1997) dealing with mixtures of binomials, Dacunha-Castelle & Gassiat (1999) dealing with a test of the order of autoregressive moving average processes and also with tests in mixture models, and Cierco (1998), who derives the asymptotic distribution of the likelihood ratio test for testing the absence of a quantitative trait locus.

Lindsay (1995) discusses likelihood ratio tests for general mixture models, and he distinguishes between three types of likelihood ratio test. In his terminology the likelihood ratio tests in the presence of a nonidentifiable parameter fall into the category of 'type III likelihood ratio tests', which do not have χ^2 distributions or mixtures of χ^2 distributions as asymptotic distributions. One way to think of these likelihood ratio tests is in terms of score statistics: for each value of the parameter which is nonidentifiable we have a score test. If one were to combine all these score tests into a simultaneous test, a simple approach would be to take the supremum over all score tests as the relevant test statistic; this is referred to as the generalised score test and agrees with the statistic considered by Davies (1977, 1987) and Hartigan (1985).

2. A MODEL FOR REPEATED MEASUREMENTS

Let $y_1, \ldots, y_n$ be independent vector-valued responses of length $l_1, \ldots, l_n$. Assume that y_i follows a multivariate normal distribution with mean $x_i\mu$, depending on a design matrix x_i and a parameter vector μ, and variance matrix V_i also depending on a parameter vector. The vectors y_i could be perceived as measurements on subjects taken at different instances on a continuous scale, recorded in the vector t_i, say. It could for example be measurements on persons in a dietary study at different time points.

Now we consider the structure of the variance matrix. In order to take account of possible correlation between measurements on the same subject, a random intercepts model (Longford, 1993) could be used. The random intercepts model has a single level of random effects, one random effect for each y vector, imposing constant correlation between different measurements in the same y vector. The resulting covariance structure is often referred to as the compound symmetry structure, in which the corresponding variance-covariance matrix has diagonal elements $\sigma_e^2 + \sigma_u^2$ and off-diagonal elements σ_u^2, conveniently expressed as

$$\sigma_u^2 M_1 + \sigma_e^2 M_2,$$

where the matrix M_1 has the value 1 in all entries, and M_2 is the identity matrix. Often such a model will appropriately accommodate the dependence between measurements, but as a starting point it may be more natural to consider models allowing for a more flexible correlation structure (Verbeke & Molenberghs, 1997, p. 126).

Diggle (1988) introduces a model which provides, by means of two additional parameters, a covariance structure that introduces an additional decreasing correlation as measurements move apart on the continuous scale. We will refer to this correlation as serial correlation.

The model for measurement y_{ij} from subject i at time t_{ij} $(j=1, \ldots, l_i)$ can be written

$$y_{ij} = x_{ij}^{\mathrm{T}}\mu + u_i + w_i(t_j) + e_{ij},$$

where x_{ij}^{T} is the jth row in the design matrix x_i and e_{ij}, u_i and $w_i(t_j)$ are random variables. The e_{ij}'s are distributed as $N(0, \sigma_e^2)$, the u_i's are distributed as $N(0, \sigma_u^2)$ and $\{w_i(t)\}_{t\in\mathbb{R}}$ is a stationary Gaussian process with mean 0 and covariance function $\operatorname{cov}\{w_i(s), w_i(t)\} = \phi \exp(-\rho|s-t|^c)$ for a fixed c, usually equal to 1 or 2, and depending on two parameters $\phi \geqslant 0$ and $\rho \geqslant 0$. The e_{ij}'s, u_i's and $\{w_i(t)\}_{t\in\mathbb{R}}$ are mutually independent. The variance-covariance matrix has components

$$\operatorname{var}(y_{ij}) = \sigma_e^2 + \sigma_u^2 + \phi, \quad \operatorname{cov}(y_{ij}, y_{ij'}) = \sigma_u^2 + \phi \exp(-\rho|t_{ij} - t_{ij'}|^c) \quad (j \neq j').$$

The covariance structure is linear in the variance parameters, except for ρ, and it can be seen that this model falls within the general form given in (1) in § 1 as the variance-covariance matrix can be written as

$$\phi M_i(\rho) + \sigma_u^2 M_1 + \sigma_e^2 M_2,$$

where the jj' element of $M_i(\rho)$ is $\exp(-\rho|t_{ij} - t_{ij'}|^c)$. We will exploit this linearity in § 3. Note that the covariance structure can be continuously extended at $\rho = \infty$. For $\rho = 0$ and $\rho = \infty$ the model reduces to random intercepts models, but with different parameterisations. Thus there is no need to include these values in the alternative. Note that we allow the full open parameter set $]0, \infty[$ for ρ, contrary to other approaches which require compactness; see for example Lemdani & Pons (1995).

For $\phi = 0$ the model reduces to a random intercepts model. Thus, in applications it could be of interest to test the null hypothesis $\phi = 0$ versus the alternative $\phi > 0$ to see whether or not serial correlation is present. The null hypothesis renders the parameter ρ nonidentifiable, and therefore the likelihood ratio statistic for testing the null hypothesis does not asymptotically follow a χ^2 distribution with one or two degrees of freedom or a mixture of such χ^2 distributions. We will return to this model in § 4.

3. Likelihood ratio tests

3·1. *Quadratic approximation to the loglikelihood*

Consider a general framework where we have a submodel of an exponential family (Berk, 1972). In this subsection, in Theorem 1, we establish approximations to the loglikelihood and to the likelihood ratio statistic. This essentially amounts to showing that the log likelihood ratio test statistic is asymptotically equivalent to the score statistic. In § 3·2 an explicit approximation to the likelihood ratio test statistic based on the quadratic expression is obtained and this approximation yields the asymptotic representation.

The development in this subsection is subject to Assumption 1, which invokes the standard conditions for exponential families.

Assumption 1. The model considered is a submodel of an exponential family, $\{P_\eta\}_{\eta \in E}$, with densities $a(y)\exp\{t(y)^{\mathrm{T}}\eta - \kappa(\eta)\}$, where $a(y)$ and $t(y)$ are measurable functions, the latter being the canonical sufficient vector statistic, η is the canonical vector parameter with domain E, say, and $\exp\{\kappa(\eta)\}$ is the normalising constant. Furthermore we assume that $\mathrm{var}\{t(y)\}$ is positive definite.

Let $y_1, \ldots, y_n$ be independent and with distribution P_{η_0}, where η_0 belongs to the interior, int(E), of E. We shall need to consider a smoothly parameterised submodel, so, with some abuse of notation, let also η be a map from an open set A in some normed space to int(E), and assume that η is a homeomorphism, twice continuously differentiable and having a full-rank matrix of derivatives for every element α in A. Assume that there is an α_0 such that $\eta_0 = \eta(\alpha_0)$. The map η changes the canonical parameterisation into an arbitrary smooth parameterisation. The canonical parameterisation is used to establish the asymptotic approximations, whereas the second parameterisation allows us to consider models in a form convenient for the formulation of submodels.

Denote by $l_n(\alpha)$ the loglikelihood of $y_1, \ldots, y_n$ expressed in terms of the α parameter. Consider the quadratic approximation

$$l_n^*(\alpha) = l_n(\alpha_0) + A_n(\alpha_0)^{\mathrm{T}}(\alpha - \alpha_0) + \tfrac{1}{2}(\alpha - \alpha_0)^{\mathrm{T}}\{-nI(\alpha_0)\}(\alpha - \alpha_0), \qquad (2)$$

where $A_n(\alpha)^{\mathrm{T}} = \partial l_n(\alpha)/\partial\alpha$ is the score statistic and $nI(\alpha_0) = -E(\partial^2 l_n(\alpha)/\partial\alpha^2)$ is the expected Fisher information matrix evaluated at α_0.

Now consider any set of submodels indexed by a parameter ρ. For each ρ the submodel is given by the subset $D_\rho \subseteq A$ with the constraint that $\alpha_0 \in D_\rho$, meaning that the submodel is not misspecified. The following theorem is used to show that the asymptotic likelihood ratio distributions for the submodels may be derived from the quadratic approximation, l_n^*, instead of from l_n.

THEOREM 1. *Define*

$$G_n(\rho) = \sup_{\alpha \in D_\rho} l_n(\alpha), \quad G_n^*(\rho) = \sup_{\alpha \in D_\rho} l_n^*(\alpha).$$

The difference $\sup_\rho G_n(\rho) - \sup_\rho G_n^*(\rho)$ *converges in probability to* 0 *as n tends to infinity.*

The proof is given in the Appendix.

Remark 1. Theorem 1 implies that the likelihood ratio statistic is asymptotically equivalent to a generalised score test statistic (Lindsay, 1995, p. 107). This equivalence trivially extends to composite hypotheses in models of the type considered, since the log likelihood ratio is then just the difference between two simple log likelihood ratios.

3·2. *Asymptotic distribution of the likelihood ratio test statistic*

Let Θ have the form $\Phi \times \Gamma \times R$ with elements (ϕ, γ, ρ), and assume that Φ is one-dimensional, that R is p-dimensional and that Γ is open and q-dimensional. Where convenient the pair (ϕ, γ) is denoted by β and $\Phi \times \Gamma$ is denoted by B. The parameter β is the identifiable parameter, whereas ρ is the nonidentifiable parameter. Let $A_n(\rho)^{\mathrm{T}}$ denote the derivative $\partial l_n/\partial\beta$ evaluated at (β_0, ρ). Furthermore define $I(\rho) = (\partial\alpha/\partial\beta)^{\mathrm{T}} I(\alpha_0)(\partial\alpha/\partial\beta)$

with $\partial\alpha/\partial\beta$ evaluated at β_0 and ρ. Denote by $A_{1n}(\rho)$ and $A_{2n}(\rho)^{\mathrm{T}}$ the partition of $A_n(\rho)^{\mathrm{T}}$ according to the components ϕ and γ of β. Similarly $I_1(\rho)$ and $I_2(\rho)$ are the square submatrices of $I(\rho)$ corresponding to ϕ and γ, respectively.

We need the following assumptions on the maps η and α.

Assumption 2. Let A and Γ be open sets. Assume that the map $\eta: A \mapsto \mathrm{int}(E)$ is a homeomorphism, twice differentiable with a full-rank matrix of derivatives for every element α in A. Assume also that there is α_0 such that $\eta_0 = \eta(\alpha_0)$. The map $\alpha: \Theta \mapsto A$ is assumed differentiable in β with a continuous derivative, and $\alpha(\theta) = \alpha(\beta, \rho)$ is linear in the parameter β. There exists $\beta_0 = (\phi_0, \gamma_0) \in B$ such that the equality $\alpha(\beta_0, \rho) = \alpha_0$ holds for $\rho \in R$, that is nonidentifiability.

The assumption that Γ is open implies that the value γ_0 is not on the boundary. In the setting of a multivariate normal distribution, Assumption 2 is satisfied, for example, whenever the mean structure is linear in the parameters and the covariance matrix has the form (1); that is

$$(\phi - \phi_0)M(\rho) + \gamma_1 M_1 + \ldots + \gamma_q M_q,$$

where $M(\rho)$ is a covariance matrix depending on ρ and $M_1, \ldots, M_q$ are known covariance matrices.

In order to handle boundary cases in the nonidentifiable parameter, we introduce the following assumption.

Assumption 3. Assume that $I(\rho)$ is invertible for all ρ in R.

Consider the negative log likelihood ratio test statistic for testing the null hypothesis $H_0: \phi = \phi_0$ versus the alternative $\phi > \phi_0$:

$$\Lambda_n = 2 \sup_{\rho \in R} l_n[\{\hat{\phi}_n(\rho), \hat{\gamma}_n(\rho)\}, \rho] - 2l_n(\phi_0, \hat{\gamma}_{0n}),$$

where $\{\hat{\phi}_n(\rho), \hat{\gamma}_n(\rho)\}$ is the maximum likelihood estimator under the alternative hypothesis for a fixed ρ, and $\hat{\gamma}_{0n}$ is the maximum likelihood estimator under the null hypothesis. The statistic Λ_n is split into two terms,

$$2\left(\sup_{\rho \in R} l_n[\{\hat{\phi}_n(\rho), \hat{\gamma}_n(\rho)\}, \rho] - l_n(\phi_0, \gamma_0)\right) - 2\{l_n(\phi_0, \hat{\gamma}_{0n}) - l_n(\phi_0, \gamma_0)\}. \tag{3}$$

We deal with the two terms in (3), Λ_{1n} and Λ_{2n} say, separately, establishing the weak convergence for each of them which extends to their simultaneous weak convergence. In particular Lemma 1, proved in the Appendix, deals with Λ_{1n}.

LEMMA 1. *Assume that Assumptions* 2 *and* 3 *hold. Then, for any fixed* $\rho \in R$, *the approximation* $l_n^*(\alpha) - l_n(\alpha_0)$ *attains the maximum value*

$$G_n^*(\rho) - l_n(\alpha_0) = \tfrac{1}{2} A_n(\rho)^{\mathrm{T}} \{nI(\rho)\}^{-1} A_n(\rho) 1_{[\{I(\rho)^{-1}A_n(\rho)\}_1 \geqslant 0]} + \tfrac{1}{2} A_{2n}^{\mathrm{T}} (nI_2)^{-1} A_{2n} 1_{[\{I(\rho)^{-1}A_n(\rho)\}_1 < 0]}, \tag{4}$$

where $\{\,\}_1$ *around a vector denotes the first component in the vector. Furthermore, for all* $n \in \mathbb{N}$, *the stochastic process* $\{G_n^*(\rho) - l_n(\alpha_0)\}$ *with index set* R *has bounded sample paths and converges weakly to a tight process.*

The main result in this paper is the following theorem.

THEOREM 2. *Under Assumptions 1–3 the likelihood ratio test statistic for testing the null hypothesis $\phi=\phi_0$ versus $\phi>\phi_0$ converges weakly under the null hypothesis to the tight random variable*

$$\sup_{\rho\in R}\left[\left\{\frac{\{I(\rho)^{-1}A(\rho)\}_1}{(\mathrm{var}[\{I(\rho)^{-1}A(\rho)\}_1])^{\frac{1}{2}}}\right\}^2 1_{[\{I(\rho)^{-1}A(\rho)\}_1\geqslant 0]}\right],$$

which also can be written as $(\sup_{\rho\in R}X_\rho^2)1_{[X_\rho\geqslant 0]}$, *where* X_ρ *is a zero-mean Gaussian process with* $\mathrm{var}(X_\rho)=1$ *for all* $\rho\in R$ *and covariance function*

$$\mathrm{cov}(X_{\rho_1},X_{\rho_2})=u(\rho_1,\rho_2)/\{u(\rho_1,\rho_1)u(\rho_2,\rho_2)\}^{\frac{1}{2}}, \tag{5}$$

where

$$u(\rho_1,\rho_2)=\mathrm{cov}\{A_1(\rho_1),A_1(\rho_2)\}-\mathrm{cov}\{A_1(\rho_1),A_2\}\{\mathrm{var}(A_2)\}^{-1}\,\mathrm{cov}\{A_1(\rho_2),A_2\}^{\mathrm{T}}$$

for all $\rho_1,\rho_2\in R$.

Proof. Theorem 1 implies that the term Λ_{1n} is asymptotically equivalent to

$$2\sup_{\rho\in R}\{G_n^*(\rho)\}-l_n(\alpha_0),$$

which by Lemma 1 equals

$$\sup_{\rho\in R}[A_n(\rho)^{\mathrm{T}}\{nI(\rho)\}^{-1}A_n(\rho)1_{[\{I(\rho)^{-1}A_n(\rho)\}_1\geqslant 0]}+A_{2n}^{\mathrm{T}}(nI_2)^{-1}A_{2n}1_{[\{I(\rho)^{-1}A_n(\rho)\}_1<0]}].$$

Note that Λ_{2n} is the likelihood ratio test statistic for testing the simple null hypothesis $\gamma=\gamma_0$ under the assumption that $\phi=\phi_0$. The score statistic for testing $\gamma=\gamma_0$ is asymptotically equivalent to the likelihood ratio test (van der Vaart, 1998, p. 231). Thus for asymptotic considerations we can replace Λ_{2n} by $A_{2n}^{\mathrm{T}}(nI_2)^{-1}A_{2n}$, which converges weakly to the random variable $A_2^{\mathrm{T}}I_2^{-1}A_2$.

Then the difference $\Lambda_{1n}-\Lambda_{2n}$ converges weakly to the limit

$$\sup_{\rho\in R}[\{A(\rho)^{\mathrm{T}}I(\rho)^{-1}A(\rho)-A_2^{\mathrm{T}}I_2^{-1}A_2\}1_{[\{I(\rho)^{-1}A(\rho)\}_1\geqslant 0]}]. \tag{6}$$

After a little algebra we can write the difference as

$$\sup_{\rho\in R}\left[\left\{\frac{\{I(\rho)^{-1}A(\rho)\}_1}{(\mathrm{var}[\{I(\rho)^{-1}A(\rho)\}_1])^{\frac{1}{2}}}\right\}^2 1_{[\{I(\rho)^{-1}A(\rho)\}_1\geqslant 0]}\right],$$

which also can be written as $(\sup_{\rho\in R}X_\rho^2)1_{[X_\rho\geqslant 0]}$, with X_ρ a Gaussian process because $\{A(\rho)\}_{\rho\in R}$ is a Gaussian process; see Lemma 1. □

Remark 2. The result in Theorem 2 also holds for inference based on restricted maximum likelihood (Patterson & Thompson, 1971). This is seen by writing down the restricted loglikelihood (McCulloch & Searle, 2001, Ch. 6) and noting that the extra term vanishes in the loglikelihood difference and that the difference between the maximum likelihood estimator and the restricted maximum likelihood estimator converges in probability to zero (Jiang, 1996).

Remark 3. The covariance function depends on the parameter γ_0 and thus, in order to apply the result in Theorem 2, an estimate has to be inserted. Since the estimator is consistent in the exponential family, this does not alter the asymptotic validity.

Remark 4. In the case where ρ is known, so that R equals a one-point set, the likelihood ratio test in Theorem 2 reduces to the usual test for a single parameter at the boundary, possibly in the presence of nuisance parameters. Self & Liang (1987) show that the asymptotic distribution of this likelihood ratio statistic is a 50:50 mixture of a χ^2 distribution with zero degrees of freedom, degenerate at 0, and a χ_1^2 distribution, that is $\frac{1}{2}\chi_0^2+\frac{1}{2}\chi_1^2$. This result is recovered from the expression in Theorem 2. As the distribution of the likelihood ratio test in Theorem 2 is given in the form of a supremum of such mixtures, the distribution $\frac{1}{2}\chi_0^2+\frac{1}{2}\chi_1^2$ forms a lower bound. An immediate consequence is that at most 50% of the probability mass is placed at 0, a result also obtained by Lindsay (1995, p. 76) in a mixture model setting.

4. Results for the motivating example

Using Theorem 2 we are now able to formulate the following theorem for models of the form (1).

THEOREM 3. *The likelihood ratio test statistic for testing the null hypothesis* $H_0:\phi=\phi_0$ *versus* $\phi>\phi_0$ *converges weakly under the null hypothesis to* $(\sup_{\rho\in]0,\infty[} X_\rho^2)1_{[X_\rho\geqslant 0]}$, *where* X_ρ *is a Gaussian process with covariance function of the general form* (5) *with explicit expressions for the associated variance-covariance matrices given below. Let the* V_{i0}*'s be compound symmetry matrices with dependence on* $(\sigma_{u0}^2,\sigma_{e0}^2)$ *suppressed in the notation. The variance matrix* $\mathrm{var}(A_2)$ *has components*

$$\{\mathrm{var}(A_2)\}_{k_1,k_2}=n^{-1}\sum_{i=1}^{n}\tfrac{1}{2}\mathrm{tr}(V_{i0}^{-1}M_{k_1}V_{i0}^{-1}M_{k_2}),$$

for $k_1,k_2=1,\ldots,q$. *The covariance matrix* $\mathrm{cov}\{A_1(\rho),A_2\}$ *has components*

$$[\mathrm{cov}\{A_1(\rho),A_2\}]_{1,k}=n^{-1}\sum_{i=1}^{n}\tfrac{1}{2}\mathrm{tr}\{V_{i0}^{-1}M(\rho)V_{i0}^{-1}M_k\},$$

for $k=1,\ldots,q$ *and* $\rho\in]0,\infty[$. *The covariance* $\mathrm{cov}\{A_1(\rho_1),A_1(\rho_2)\}$ *is equal to*

$$n^{-1}\sum_{i=1}^{n}\tfrac{1}{2}\mathrm{tr}\{V_{i0}^{-1}M(\rho_1)V_{i0}^{-1}M(\rho_2)\}$$

with $\rho_1,\rho_2\in]0,\infty[$.

Proof. Assumptions 1–3 can easily be seen to hold and the general result in Theorem 2 can be applied. The expressions for the variance-covariance matrices are obtained by computation, using formulae from McCulloch & Searle (2001, pp. 179–81). □

Remark 5. The repeated measurements model discussed in § 2 is a special case of (1) with $\phi_0=0$ and $q=2$, and therefore Theorem 3 is directly applicable and allows additional specialisation by insertion of the specific matrices M_1, M_2 and $M(\rho)$ into the expressions for the variances and covariances. In this case the covariance function in Theorem 3 can be seen to depend only on the quotient $\sigma_{u0}^2/\sigma_{e0}^2$; by Assumption 2 both σ_{u0}^2 and σ_{e0}^2 are positive. This follows from the form of the matrices V_{i0}^{-1}.

As the process X_ρ in Theorem 3 is Gaussian with a covariance function that can be estimated consistently, discretised realisations of the limiting process over a grid of ρ points can be simulated. Then the last step in approximating the asymptotic distribution of the likelihood ratio statistic is to find the maximum values of the realisations. This is relatively easy because the underlying Gaussian process X_ρ does not fluctuate much, ρ only entering the variance function in terms of the form $\exp(-\rho|t_{ij}-t_{ij'}|^c)$; compare with the more difficult situation in Lemdani & Pons (1995).

As a small simulation study we construct 200 datasets of size m, corresponding to m independent subjects, from a null hypothesis model with zero mean and a compound symmetry covariance structure. For each of the m subjects there are five measurements at five equidistant time points. In vector form the model is

$$y_i = 0 + zu_i + e_i \quad (i = 1, \ldots, m),$$

where z is a vector of length 5 with 1 in all entries, u_i and e_i are independent, $u_i \sim N(0, 2)$ and $e_i \sim N_5(0, 2)$. We study the agreement between the asymptotic distribution and the empirical distribution of the likelihood ratio statistic for testing $\phi = 0$ by means of Q–Q plots, shown in Fig. 1 for m equal to 10, 20, 40 and 80.

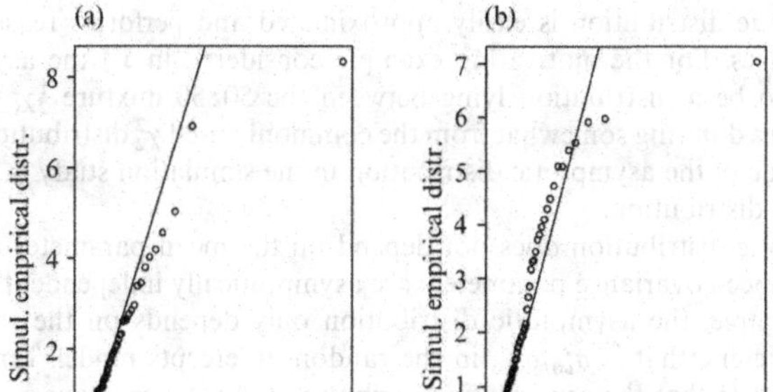

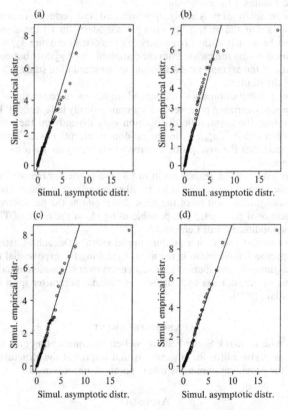

Fig. 1. Q–Q plots comparing the asymptotic distribution of the likelihood ratio statistic to the empirical distribution of the likelihood ratio statistic based on 200 simulated datasets, for (a) $m = 10$, (b) $m = 20$, (c) $m = 40$ and (d) $m = 80$ subjects.

As expected the asymptotic approximation improves as the number of subjects m increases, and even for the smallest subject size the substantial deviations appear beyond the 90% quantile, indicating that the asymptotic distribution may also be used for small or moderate sample sizes.

5. Discussion

Although we have focused on multivariate normal models and one-sided alternative hypotheses concerning the covariance structures, the result also applies to other exponential families, and two-sided alternatives can also be handled similarly.

The result in Theorem 2 is given for the case with a common collection $(l, t, x, M_1, \ldots, M_q)$ for all y_i, which covers balanced designs. It can be extended to the situation where the length of the response vector l, the time points t, the design matrix x and the M matrices vary with i in such a way that the collection $(l_i, t_i, x_i, M_{i1}, \ldots, M_{iq})$ only takes finitely many values, according to some distribution. The approach is to use Theorem 1 and Lemma 1 for each value.

The asymptotic distribution is easily approximated and performs reasonably well for small sample sizes. For the motivating example considered in § 4 the asymptotic distribution seems to be a distribution lying between the 50:50 mixture $\frac{1}{2}\chi_0^2 + \frac{1}{2}\chi_1^2$ and a χ_1^2 distribution, thus differing somewhat from the commonly used χ_2^2 distribution. For example the 95% quantile of the asymptotic distribution in the simulation study is 3·46, compared to 5·99 for a χ_2^2 distribution.

The asymptotic distribution does not depend on the mean parameter if estimators of mean and variance-covariance parameters are asymptotically independent. In fact, for the motivating example, the asymptotic distribution only depends on the ratio of the two variance parameters, that is $\sigma_{u0}^2/\sigma_{e0}^2$, in the random intercepts model. Empirical studies, not shown, indicate that the asymptotic distribution is nearly constant over a wide range of values of the ratio.

We have only explored the general result in § 3 for a particular repeated measurements model, but application of the results in § 3 to other covariance structures is possible. The resulting Gaussian process will be of the same dimension as the parameter ρ. Extension to a multi-dimensional parameter ϕ is possible using (6) in the proof of Theorem 2, but it requires some modification of Lemma 1.

In contrast to similar results in a mixture model setting (Dacunha-Castelle & Gassiat, 1999), no compactness assumption is required. Exploiting the exponential family theory, we derive the asymptotic distribution as a supremum over the nonidentifiable parameter only, in contrast to the suprema over higher-dimensional parameter spaces in Dacunha-Castelle & Gassiat (1999).

Acknowledgement

The authors wish to thank Søren Feodor Nielsen for commenting on a first version of the paper, the associate editor for suggestions that improved the structure of the paper and the editor for providing numerous minor points of improvement.

Appendix

Proofs

Proof of Theorem 1. We prove the slightly stronger result that $\sup_\rho |G_n(\rho) - G_n^*(\rho)|$ converges in probability to 0 as n tends to infinity, under the distribution given by α_0.

Recall that by a sequence of $n^{-1/2}$ neighbourhoods of α_0 we mean a sequence $V_n = \alpha_0 + n^{-1/2}V$, where $\alpha_0 + V$ is a bounded neighbourhood of α_0. Standard likelihood considerations, as in Lehmann (1983, Theorem 4.1), yield that, for any sequence V_n of $n^{-1/2}$-neighbourhoods of α_0, the sequence of suprema $\sup_{\alpha \in V_n} |l_n(\alpha) - l_n^*(\alpha)|$ converges to 0 in probability.

From considerations of local existence and uniqueness of the maximum likelihood estimator in curved exponential families it follows that for any $\varepsilon > 0$ there exists a sequence V_n of $n^{-1/2}$-neighbourhoods of α_0 and a number $n_0 \in \mathbb{N}$ such that $\sup_{\alpha \in V_n^c} l_n(\alpha) < l_n(\alpha_0)$ holds with probability at least $1 - \varepsilon$ for all $n \geqslant n_0$. Here V_n^c denotes the complement of V_n.

We obtain a similar result for l_n^* using the inequality $(\alpha - \alpha_0)^T I(\alpha_0)(\alpha - \alpha_0) \geqslant \lambda_{\min} \|\alpha - \alpha_0\|^2$, where $\lambda_{\min}$ denotes the smallest eigenvalue of $I(\alpha_0)$. The result is that for any positive ε there exists a sequence $\tilde{V}_n$ of $n^{-1/2}$-neighbourhoods of α_0 and a number $n_0 \in \mathbb{N}$ such that $\sup_{\alpha \in \tilde{V}_n^c} l_n^*(\alpha) < l_n^*(\alpha_0)$ holds with probability at least $1 - \varepsilon$ for all $n \geqslant n_0$.

In the following argument let V_n denote a sequence of $n^{-1/2}$-neighbourhoods satisfying both the inequality for l_n and that for l_n^*. Consider a map α from the set Θ with elements θ to A. By definition the maximum likelihood estimator $\hat{\theta}_n$ of $l_n\{\alpha(\theta)\}$ satisfies the inequality $l_n(\alpha_0) \leqslant l_n\{\alpha(\hat{\theta}_n)\}$, which shows that $\alpha(\hat{\theta}_n)$ must lie in V_n. Similarly it follows that the maximiser $\hat{\theta}_n^*$ of $l_n^*\{\alpha(\theta)\}$ also lies in V_n. For given $\delta > 0$ and $\varepsilon > 0$ the previous inequalities can be used to establish that

$$G_n(\rho) \leqslant G_n^*(\rho) + 2\delta, \quad G_n^*(\rho) \leqslant G_n(\rho) + 2\delta.$$

Consequently $|G_n(\rho) - G_n^*(\rho)| \leqslant 2\delta$ holds for any ρ, for all $n \geqslant n_0$ with probability at least $1 - \varepsilon$. The convergence in probability is proved. □

Proof of Lemma 1. A first-order Taylor expansion of the map α in β around β_0 for fixed ρ gives the relationship

$$\alpha(\beta, \rho) - \alpha_0 = \frac{\partial \alpha}{\partial \beta}(\beta - \beta_0)$$

by the second, linearity, part of Assumption 2. From the chain rule we obtain $A_n(\rho)^T = A_n(\alpha_0)^T \partial\alpha/\partial\beta$. These two results yield

$$l_n^*\{\alpha(\beta, \rho)\} - l_n(\alpha_0) = A_n(\rho)^T(\beta - \beta_0) + \tfrac{1}{2}(\beta - \beta_0)^T\{-nI(\rho)\}(\beta - \beta_0) \tag{A1}$$

for all $\beta \in B$ and $\rho \in R$.

We seek the maximum of the quadratic approximation (A1) as a function of β for fixed ρ. Since $\alpha(\phi_0, \gamma, \rho)$ does not depend on ρ, $A_{2n}(\rho)^T$ and $I_2(\rho)$ do not depend on ρ and we can simply write A_{2n}^T and I_2. For $\rho \in R$, the unconstrained maximum of $l_n^*\{\alpha(\beta, \rho)\} - l_n(\alpha_0)$ is attained at $\beta_0 + \{nI(\rho)\}^{-1}A_n(\rho)$. Under the restriction that ϕ is greater than or equal to ϕ_0, the constrained maximum is attained at $\beta_0 + (0, (nI_2)^{-1}A_{2n})^T$ if the first component, $\phi_0 + \{\{nI(\rho)\}^{-1}A_n(\rho)\}_1$, is smaller than ϕ_0 and attained at $\beta_0 + \{nI(\rho)\}^{-1}A_n(\rho)$ otherwise. By insertion into (A1) the maximum value (4) is obtained.

In view of the definition of G_n^* in Theorem 1, the inequality

$$G_n^*(\rho) - l_n(\alpha_0) \leqslant \sup_{\alpha \in A} l_n^*(\alpha) - l_n(\alpha_0)$$

holds uniformly in $\rho \in R$ for all $n \in \mathbb{N}$, implying that the sample paths of G_n^* are bounded almost surely. Both the left-hand side and the right-hand side converge weakly. Note that the weak convergence of $n^{-1/2}A_n(\alpha_0)$ to a normal distribution implies that $n^{-1/2}A_n(\rho)$ converges to a Gaussian process, and then the continuous mapping theorem implies that the left-hand side converges weakly. The right-hand side is the score test for the simple hypothesis $\alpha = \alpha_0$, and hence converges weakly to a χ^2 distribution (van der Vaart, 1998, pp. 230–1). Therefore, the inequality also holds in the limit and we conclude that the limit of the left-hand side is dominated and tight. □

References

BERK, R. H. (1972). Consistency and asymptotic normality of MLE's for exponential models. *Ann. Math. Statist.* **43**, 193–204.

CIERCO, C. (1998). Asymptotic distribution of the maximum likelihood ratio test for gene detection. *Statistics* **31**, 261–85.

DACUNHA-CASTELLE, D. & GASSIAT, E. (1999). Testing the order of a model using locally conic parametrization: population mixtures and stationary ARMA processes. *Ann. Statist* **27**, 1178–209.

DAVIES, R. B. (1977). Hypothesis testing when a nuisance parameter is present only under the alternative. *Biometrika* **64**, 247–54.

DAVIES, R. B. (1987). Hypothesis testing when a nuisance parameter is present only under the alternative. *Biometrika* **74**, 33–43.

DIGGLE, P. J. (1988). An approach to the analysis of repeated measurements. *Biometrics* **44**, 959–71.

DIGGLE, P. J., LIANG, K.-Y. & ZEGER, S. L. (1994). *Analysis of Longitudinal Data.* Oxford: Oxford University Press.

HARTIGAN, J. A. (1985). A failure of likelihood for the mixture model. In *Proceedings of the Berkeley Symposium in Honor of Jerzy Neyman and Jack Kiefer*, Ed. L. LeCam and R. Olshen, pp. 807–10. New York: Wadsworth.

JIANG, J. (1996). REML estimation: asymptotic behavior and related topics. *Ann. Statist.* **24**, 255–86.

LEHMANN, E. L. (1983). *Theory of Point Estimation.* New York: Wiley.

LEMDANI, M. & PONS, O. (1995). Tests for genetic linkage and homogeneity. *Biometrics* **51**, 1033–41.

LEMDANI, M. & PONS, O. (1997). Likelihood ratio tests for genetic linkage. *Statist. Prob. Lett.* **33**, 15–22.

LINDSAY, B. G. (1995). *Mixture Models: Theory, Geometry and Applications.* Hayward, CA: Institute of Mathematical Statistics.

LITTLE, R. C., MILLIKEN, G. A., STROUP, W. W. & WOLFINGER, R. D. (1996). *SAS System for Mixed Models.* Cary: SAS Institute Inc.

LONGFORD, N. T. (1993). *Random Coefficient Models.* Oxford: Oxford University Press.

MCCULLOCH, C. E. & SEARLE, S. R. (2001). *Generalized, Linear and Mixed Models.* New York: Wiley.

PATTERSON, H. D. & THOMPSON, R. (1971). Recovery of interblock information when block sizes are unequal. *Biometrika* **58**, 545–54.

SELF, S. G. & LIANG, K.-Y. (1987). Asymptotic properties of maximum likelihood estimators and likelihood ratio tests under nonstandard conditions. *J. Am. Statist. Assoc.* **82**, 605–10.

VAN DER VAART, A. W. (1998). *Asymptotic Statistics.* Cambridge: Cambridge University Press.

VERBEKE, G. & MOLENBERGHS, G. (Ed.). (1997). *Linear Mixed Models in Practice. A SAS-oriented Approach.* New York: Springer-Verlag.

[*Received October* 2004. *Revised February* 2005]

Chapter 16

Nonparametric maximum likelihood estimation of randomly time-transformed curves

Introduction by Judith Rousseau

Université Paris Dauphine and CREST-ENSAE

It is with great pleasure that I write this introduction for this book on Ib's work. I met Ib at the end of my PhD studies, and he has been extremely encouraging and kind with me ever since. I have always been impressed by the depth of his understanding in higher order expansions, his rigour and his amazing modesty. He encouraged me shortly after my PhD. to combine theoretical statistics with applications and this paper is a great illustration of this.

This paper deals with an important problem in functional data analysis, namely the estimation of the *template curve* when the data consist of a set of curves which are noisy observations from curves which themselves vary around a common pattern (the template). The approach deals with the problem of pre-processing the data prior to most statistical analyses. This pre-processing of the data consists generally in alignment of the curves, also called data registration or time warping, followed by smoothing and data reduction. In this paper the authors consider the problem of smoothing and data registration simultanuously, using a modelling of the misalignment. These probabilistic deformable template models have wide applications in computer vision, computational anatomy, biology, neurosciences and more generally image analysis. There is now a huge literature on deformable models and curve registration, see for instance Ramsay and Silverman (2005) or James (2007).

The observations are thus curves $Y_i(\cdot)$ observed at values $t_{i,j}$ and satisfying

$$Y_i(t_{i,j}) = m\{g_{\theta_i}^{-1}(t_{i,j})\} + \epsilon_{i,j}, \quad \epsilon_{i,j} \overset{i.i.d}{\sim} \mathcal{N}(0,\sigma^2) \quad i \leq n, \quad j \leq N, \quad (16.1)$$

where g_{θ_i} is the deformation model or the warping function, $m(\cdot)$ is the template curve and θ_i are random parameters driving the transformation. In many applications, for instance, the transformation is reasonably modelled by a shift $g_\theta(t) = t + \theta$. To properly identify the structural mean parameter m, the transformation parameters are assumed to be finite dimensional and random with given distribution f_a. Hence the approach can be considered either as a random effects model or as a partially Bayesian method with a prior distribution f_a on the θ_i's.

The main aim of this paper is to estimate m, but predicting the deformation parameters θ_i is also considered. Following Gervini and Gasser (2005) and Rønn (2001), the maximum likelihood estimator satisfies the implicit equation:

$$\hat{m}(t) = \frac{\sum_{i=1}^{n}\sum_{j=1}^{N} Y_{i,j}\hat{f}_{Z_{i,j}|Y_i}(t)}{\sum_{i=1}^{n}\sum_{j=1}^{N} \hat{f}_{Z_{i,j}|Y_i}(t;\hat{m})}, \tag{16.2}$$

where $Z_{i,j} = g_\theta^{-1}(t_{i,j})$ is viewed as a random variable as a measurable transformation of θ, having posterior density $f_{Z_{i,j}|Y_i}$, $\hat{f}_{Z_{i,j}|Y_i}(\cdot;\hat{m})$ means that the posterior density is computed with $\hat{m}$ instead of m. Recall that

$$f_{a|Y_i}(\theta;m) = \frac{e^{-\|Y_i - m(g_{a=\theta}(\mathbf{t}_i))\|^2/(2\sigma^2)}(2\pi\sigma^2)^{-N/2}f_a(\theta)}{\int_\Theta e^{-\|Y_i - m(g_{a=\theta}(\mathbf{t}_i))\|^2/(2\sigma^2)}(2\pi\sigma^2)^{-N/2}f_a(\theta)d\theta}.$$

The maximum likelihood estimator is then calculated using an iterated procedure, akin to an EM algorithm which at each iteration consists in computing on a grid $\hat{f}^{\ell-1}_{Z_{i,j}|Y_i}$ using the value of $\hat{m}^{\ell-1}$ computed at the previous iteration and computing

$$\hat{m}^\ell(t) = \frac{\sum_{i=1}^{n}\sum_{j=1}^{N} Y_{i,j}\hat{f}_{Z_{i,j}|Y_i}(t;\hat{m}^{\ell-1})}{\sum_{i=1}^{n}\sum_{j=1}^{N} \hat{f}_{Z_{i,j}|Y_i}(t;\hat{m}^{\ell-1})}.$$

The main innovation of the paper is in proposing an approximate algorithm which is computationally feasible, in particular when the number of observed time points per individual, N, is large. The idea is to approximate the density $f_{Z_{i,j}|Y_i}(t)$ using the Laplace approximation of Tierney *et al.* (1989). For the approximation to be valid or precise, one needs that $-\{\log f_{Y_i|\theta}(Y_i) + \log f_a(\theta)\}$ behaves like $-Tb_T(\theta)$ for some quantity T going to infinity. Here typically one could imagine to have $T = N$ the number of observations for each curve. The authors also propose a cruder approximation based on a first order Gaussian approximation, like a Bernstein–von Mises approximation of the posterior density, see for instance Johnson (1970). Under regularity conditions Tierney *et al.* (1989) prove that

$$f_{Z_{i,j}|Y_i}(t) = \tilde{f}_{Z_{i,j}|Y_i}(t)\{1 + O(1/T^{3/2})\}, \quad \text{if} \quad |t - \hat{t}| \le \delta,$$

where $\delta > 0$ is fixed and $\tilde{f}_{Z_{i,j}|Y_i}$ is the resulting approximation, see the Corollary of Tierney *et al.* (1989).

Obviously the error is controlled only at each iteration and locally around the mode, for each time point $t_{i,j}$ (i.e. $Z_{i,j}$) and it is not clear that the global resulting error of the approximation remains small and does not accumulate. This type of approximation at each iteration of an algorithm based on many iterations is commonly used; see for instance the renowned algorithm, or family of algorithms, INLA, Rue *et al.* (2009). In Guihenneuc-Jouyaux and Rousseau (2005) we had, in a similar type of framework, but using a Laplace approximation within a Gibbs sampler, obtained some theoretical control in terms of N of the global error of the approximation. It is however typically difficult to obtain, in general, a control on the global resulting error under reasonable conditions, although these approximations have proved to be quite powerful in practice.

Brazilian Journal of Probability and Statistics
2009, Vol. 23, No. 1, 1–17
DOI: 10.1214/08-BJPS004

Nonparametric maximum likelihood estimation of randomly time-transformed curves

Birgitte B. Rønn[a] and Ib M. Skovgaard[b]
[a] *Genmab a/s*
[b] *University of Copenhagen*

Abstract. Alignment of curves by nonparametric maximum likelihood estimation can be done when the individual transformations of the time axis is assumed to be of a parametric form, known up to some individual unobserved random parameters. We suggest a fast algorithm, based on a Laplace approximation, to find the nonparametric maximum likelihood estimator (NPMLE) for the shape function. We find smooth estimates for the shape functions without choosing any smoothing parameters or kernel function and we estimate realizations of the unobserved transformation parameters that align the curves to satisfy the eye. The method is applied to two data examples of electrophoretic spectra on feta cheese samples and on wheat samples, respectively. A small simulation study indicates reasonable robustness against assumptions regarding the error covariance function.

1 Introduction

When data from a process evolving in time is recorded the variation between replications is both in amplitude and phase. Here amplitude refers to the (vertical) variation in the response variable, whereas the variation in phase is meant to cover the differences between the individual timing of the curves. The individual transformed time may correspond to biological time, physical time or some artificial time depending on circumstances of the experiment that might vary beyond the control of the researcher. An example of replicated process data is shown in the left panel of Figure 1, where nine electrophoretic spectra of the same feta cheese are plotted versus time. The technique used is capillary electrophoresis. Thus, the x-axis represents the time of migration through the capillary and the y-axis represents intensity, reflecting the concentration of the substance with the given migration time. There are 412 observations for each samples. The data are seen in the left panel of Figure 1.

Since the 9 samples were highly variable in their level of intensity, background intensity was initially removed before plotting and further analysis. This was done using the same method as in Glasbey, Vali and Gustafsson (1995). The data were

Key words and phrases. Curve alignment, Laplace approximation, nonparametric maximum likelihood estimation, self-modeling regression, semiparametric model, warping.
Received October 2007; accepted February 2008.

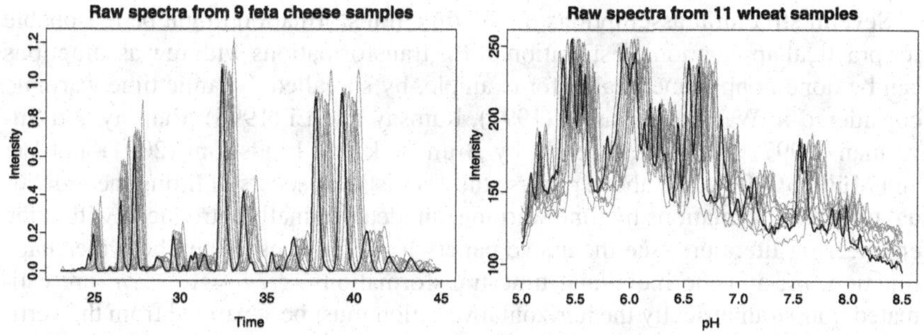

Figure 1 *Two examples of observed electrophoretic spectra: capillary electrophoresis with nine samples of feta cheese (left) and isoelectric focusing with 11 samples of wheat (right). One sample is enhanced in each plot.*

collected to obtain information on the protein profile of feta cheese and replications were made to determine which features were real and which were artifacts occurring only in some replicates due to disturbances from other sources than the cheese. Here horizontal variation between replications is unavoidable due to the nature of experiment. The variation in phase seems to be by far the most substantial and the cross-sectional mean clearly is a poor estimator of the shape function of the protein profile, since it will be less peaky than any of the nine individual curves. The feta cheese experiment is described in detail in Wium, Kristiansen and Qvist (1998).

Another sample of curves was recorded to characterize variety of wheat. Electrophoretic spectra of ten different varieties of wheat were made by isoelectric focusing on eleven different plates. The eleven spectra of variety number one are shown in the right panel of Figure 1. Here the x-axis represents the isoelectric pH-value for the substance, not time.

It is difficult to see much in the figure due to the substantial variation in phase between the individual curves. The features of the eleven curves differ more than the features of the nine feta cheese curves, and the variation in amplitude, the "measurement" error on the signal, is of significant size in the wheat data. Both types of variation must be accounted for in order to obtain a meaningful estimator for the shape function of interest. For details on the wheat experiment and data see Jensen et al. (1995).

Data of the type above can be modeled by a smooth function evolving in time, subject to individual transformations of the time axis, and measured with noise,

$$Y_{ij} = m\{g_i^{-1}(t_{ij})\} + e_{ij}, \tag{1.1}$$

where the jth observation of curve i at time point t_{ij} is denoted Y_{ij}, the common shape function is denoted m, the individual time transformation is denoted g_i and the error term is denoted e_{ij}.

Several structural assumptions on the time transformation might be reasonable for practical applications. Estimation of the transformations with few assumptions can be done nonparametrically, for example, by so-called dynamic time warping, considered in Wang and Gasser (1997), Ramsay and Li (1998), Ramsay and Silverman (1997) and more recently by Brumback and Lindstrom (2004) and Liu and Müller (2004). The above papers study statistical aspects of fitting the individual time transformations by time warping, an idea originally introduced within the engineering literature; see the above papers for references. When both the shape function, m, and the individual time transformations, g_i, $i = 1, \ldots, n$, are estimated nonparametrically the horizontal variation must be separated from the vertical. Otherwise, features of the shape function can be absorbed in the time transformations and visa versa. In some applications the focus might be on transforming observed curves to match a reference curve, hence nonparametric estimation of the transformation is clearly a method allowing for a flexible class of transformations. In other applications, as the two examples introduced above, the focus is on estimation of the common shape function, hence the transformations of the time axes are merely experimental noise and might therefore be modeled parametrically with the parameters considered as nuisance parameters. This can be done within the framework of self-modeling regression, which spans a wide class of regression models. The models similar to model (1.1), when the time transformation is assumed to be parametric, $g_i(t) = g_{\theta_i}(t)$, are models of this type. Here the time transformation is assumed to be the same, up to some parameter θ_i, for all individuals. Several suggestions have been made on how to estimate the parameters in self-modeling regression; see, for example, Lawton, Sylvestre and Maggio (1972), Stützle et al. (1980), Kneip and Gasser (1992) and Kneip and Engel (1995). The very simple special case of model (1.1), where the individual time axes are assumed to be the observed time axis subject to individual rigid shifts, $g_i^{-1}(t) = t - \theta_i$, has been studied in Rønn (2001). The interpretation of the time transformations as experimental variation has been adopted and the shift parameters have been modeled as unobserved nuisance parameters with some known distribution. A nonparametric maximum likelihood approach has been followed, inspired by the applications of this approach within event history analysis; see, for example, Gill (1989), Fernholz (1983), Groeneboom and Wellner (1992), Murphy (1995) and Parner (1998). This approach has been generalized to transformations involving more parameters in Decker, Rønn and Jørgensen (2000) and Gervini and Gasser (2005) where the nonparametric score equation for the shape function has been shown to lead to a fairly simple equation, immediately suggesting an iterative algorithm for estimation. The calculations involved are, however, substantial and while Gervini and Gasser (2005) suggest a simulation based estimation procedure, we show in the present paper that simple and accurate approximations lead to a relatively fast and reliable estimation procedure.

The approach from the latter three above-mentioned references is based on a nonparametric maximum likelihood estimation (NPMLE) for model (1.1) with a parametric time transformation, $g_i(t) = g_{\theta_i}(t)$, where g is a known function of the p-dimensional transformation parameter θ_i. The transformation parameters are considered to be unobserved random variables, and their distribution enters the likelihood as a penalty on extreme transformations. To approximate the critical terms in the algorithm, a Laplace approximation to the crucial integral is suggested and the expressions leading to an approximate NPMLE of the shape function are worked out. We then apply the algorithm to the feta cheese data and the wheat data and show in a small simulation study that the method also works well in situations where the assumptions from the working model are not met. The present paper deals with estimation of the shape function, but the proposed algorithm also provides estimates for the parameters of the time transformation and the set of smooth, aligned curves. Hence, the method can serve as pre-processing of data prior to functional analysis by principal component analysis or following approaches described in, for example, Brumback and Rice (1998), Ramsay and Silverman (2002) or Anselmo, Dias and Garcia (2005). The latter uses methods from Ramsay and Li (1998) as a step in the functional data analysis process. The wheat profiles presented represent 1 out of 10 varieties and the aim of the study was to classify the profiles. Good alignment of the profiles was essential to achieve successful classification. Similarly, protein profiles of plant seed oil have been successfully classified by the method proposed in the present paper; see Decker, Rønn and Jørgensen (2000). Note that, as in Rønn (2001), an advantage of the method is that the degree of smoothing is controlled by the data through the probability that a given data point contributes to the function value in question. The present work was introduced in the unpublished thesis Rønn (1998).

2 A model for randomly time-transformed curves

The shape invariant model corresponding to randomly time-transformed curves is given by

$$m_i(s) = m\{g_{\theta_i}^{-1}(s)\},$$

where m_i is the ith curve, θ_i is the ith transformation parameter, g_{θ_i} is the transformation function and m is the shared shape function, defined on some interval, J, on which also the transformations, g_θ, are defined. The transformation parameter θ_i is assumed to follow a distribution with continuous density function, f_θ, and compact support $\text{supp}(f_\theta) = \Theta \subset \mathbb{R}^p$. The mean of the distribution is assumed to correspond to no transformation, $g_\xi(t) = t$ for all $t \in \mathbb{R}$, where $\xi = \mathrm{E}(\theta_i)$. Each transformation, g_θ, is assumed to be strictly increasing in t,

$$\frac{\partial}{\partial s} g_\theta(s) > 0 \qquad \text{for all } s \in J \subset \mathbb{R}, \theta \in \Theta,$$

and hence a strictly increasing inverse, g_θ^{-1}, exists. In some settings other constraints on the transformation may be natural. If, for example, each curve represents the growth curve of an individual, the observed time axis represents the age of the individual. The biological age, which corresponds to the transformed observed age, must be 0 in the same point as the observed age for biological age to be meaningful. Thus, depending on the situation, boundary conditions on g, such as $g_\theta(t_0) = t_0$ for all θ and some fixed point t_0, may be natural.

Observations from the above model obtained at deterministic time points may be subject to further random variation modeled as the error term e_{ij} in the equation

$$Y_{ij} = m\{g_{\theta_i}^{-1}(t_{ij})\} + e_{ij}, \tag{2.1}$$

where the observation of the ith individual at the jth point in time denoted Y_{ij} and the corresponding time point is denoted t_{ij}, $i \in \{1,\dots,n\}$ and $j \in \{1,\dots,N\}$. The error terms, e_{ij}, are assumed to be independent, normally distributed with mean 0 and variance σ^2. This assumption is rarely realistic, but since it is used to derive the estimator of the mean, essentially by introducing a penalty on the residual vertical deviations, it is not as prohibitive as it may seem.

3 Nonparametric maximum likelihood estimation

The infinite-dimensional parameter we want to estimate is the shared shape function. The NPMLE for the shape function is the shape function that maximizes the likelihood function. The log-likelihood function, as a function of the parameter of interest, m, is

$$\begin{aligned} l(m) &= \sum_{i=1}^{n} \log\left\{\int_\Theta f_{Y_i|\theta}(u) f_\theta(u)\,du\right\} \\ &= \sum_{i=1}^{n} \log\left(\int_\Theta (2\pi\sigma^2)^{-N/2} \exp\left[-\frac{1}{2\sigma^2}\|Y_i - m\{g_u^{-1}(t_i)\}\|^2\right] \cdot f_\theta(u)\,du\right), \end{aligned}$$

where $f_\theta(u)$ and $f_{Y_i|\theta}(u)$ denote the density functions for the distribution of the transformation parameter and the density function of the observations from the ith curve given the transformation parameter, both evaluated in the p-dimensional vector u. The N-dimensional vectors Y_i, t_i, $g_u^{-1}(t_i)$ and $m\{g_u^{-1}(t_i)\}$ have elements Y_{ij}, t_{ij}, $g_u^{-1}(t_{ij})$ and $m\{g_u^{-1}(t_{ij})\}$, respectively. Furthermore, the usual norm of an N-dimensional vector $v \in \mathbb{R}^N$ is denoted $\|v\| = (\sum_j v_j^2)^{1/2}$, and the integration is with respect to the usual Lebesgue measure on $\mathbb{R}^p$.

In Decker, Rønn and Jørgensen (2000) and Gervini and Gasser (2005) the infinite-dimensional score function of m has been derived and shown to be zero at $m = \hat{m}$ if and only if

$$\hat{m}(t) = \frac{\sum_{i=1}^{n}\sum_{j=1}^{N} Y_{ij}\hat{f}_{Z_{ij}|Y_i}(t)}{\sum_{i=1}^{n}\sum_{j=1}^{N} \hat{f}_{Z_{ij}|Y_i}(t)}, \tag{3.1}$$

where

$$Z_{ij} = g_U^{-1}(t_{ij})$$

denotes the time-point corresponding to t_{ij} by the back-transformation given by U. Existence of a solution, $\hat{m}$, to the equation above has been shown only for compact parameter space, essentially ruling out an infinite-dimensional model. However, an approximate solution may well exist and be found by the algorithm.

Noting that the posterior densities on the right-hand side of the equation, $\hat{f}_{Z_{ij}|Y_i}$, are estimates depending on the shape function $\hat{m}$ itself, an obvious choice is to iterate the computation of the right-hand side of the equation, plugging in the current estimate of m. This is what is done in Rønn (2001) for the case of rigid shifts of the curves, and in Decker, Rønn and Jørgensen (2000) and Gervini and Gasser (2005) in the present case. More precisely the algorithm consists of the following:

A. *Initialization:*

- Decide on a grid of time points on which the score equation shall be fulfilled, for example, the grid of observation points, $t_1, \ldots, t_N$.
- Calculate an initial estimate, $\hat{m}_0$, of the smooth function m, for example, as the cross-sectional mean.

B. *Iteration:*

- Calculate estimates for the weight functions, $\hat{w}_{ij}(\hat{m}_{k-1})(t)$, by plugging in the estimate of the smooth function from the previous step $\hat{m}_{k-1}$.
- Find values of a new estimate, $\hat{m}_k$, of the smooth function on the grid by the weighted mean of observations obtained from the score equation,

$$\hat{m}_k(t) = \frac{\sum_{i=1}^{n} \sum_{j=1}^{N} Y_{ij} \hat{w}_{ij}(\hat{m}_{k-1})(t)}{\sum_{i=1}^{n} \sum_{j=1}^{N} \hat{w}_{ij}(\hat{m}_{k-1})(t)}. \tag{3.2}$$

  Values of the new estimate in any time point can then be found by for example cubic spline interpolation between the grid points.

The estimator $\hat{m}(t)$ is a Nadaraya–Watson type kernel estimator, where any observation Y_{ij} contributes according to the likelihood of the corresponding time point, t_{ij}, to be transformed into t, measured in the empirical posterior distribution of the transformation parameters given data. Hence the width of the kernel is determined by certainty in data of the time transformations.

4 Approximation to the empirical posterior transformation density

In the present section a Laplace-type approximation is introduced to approximate the crucial quantities needed in equation (3.1) such that its solution becomes computationally feasible. An even simpler approximation is also given, based on a normal approximation to the empirical posterior distribution of the transformation parameter for each curve. The two approximations are given in equations (4.1) and (4.3), respectively.

In each step of the algorithm we need to estimate the weight functions for a given shape function, given as

$$w_{ij}(t) = f_{Z_{ij}|Y_i}(t).$$

However, the calculation of this empirical posterior density of the transformed time given data is highly demanding. The empirical posterior density of the transformation parameter given data is

$$\begin{aligned} f_{\theta|Y_i}(u) &= \frac{f_{Y_i|\theta=u} f_\theta(u)}{\int_\Theta f_{Y_i|\theta=v} f_\theta(v)\,dv} = \frac{(2\pi\sigma^2)^{-N/2}\exp\{b_i(u)\}}{\int_\Theta (2\pi\sigma^2)^{-N/2}\exp\{b_i(v)\}\,dv} \\ &= c\exp(-b_i(u)), \end{aligned}$$

where c is the normalizing constant and

$$b_i(u) = -\frac{1}{2\sigma^2}\sum_{k=1}^{N}[Y_{ik} - m\{g_u^{-1}(t_{ik})\}]^2 + \log\{f_\theta(u)\}.$$

For this situation, with a posterior parameter-density of the form above, Tierney, Kass and Kadane (1989) give a Laplace-type approximation to a posterior density of a (smooth) function, g say, of the parameter; in our case

$$g_{ij}(u) = g_u^{-1}(t_{ij}).$$

Their approximation, calculated at the estimated function $\hat{m}$, reads

$$\hat{f}_{Z_{ij}|Y_i}(t) \approx (2\pi)^{-1/2}\beta_{ij}(t)\exp\{-b_i(\tilde{u}_{ij}(t)) + b_i(\hat{u}_i)\}, \tag{4.1}$$

where $\tilde{u}_{ij}(t)$ minimizes $b_i(u)$ subject to the condition $g_{ij}(u) = t$ while $\hat{u}_i$ is the unconstrained minimum, and where β_{ij}, G_{ij} and λ_{ij} are defined by

$$\begin{aligned} \beta_{ij}(t) &= |G_{ij}(\tilde{u}_{ij}(t))|^{-1/2}|g'_{ij}(\tilde{u}_{ij}(t))G_{ij}(\tilde{u}_{ij}(t))^{-1}g'_{ij}(\tilde{u}_{ij}(t))^T|^{-1/2} \\ &\quad\times |b''_i(\hat{u}_i)|^{1/2}, \\ G_{ij}(\tilde{u}_{ij}(t)) &= b''_i(\tilde{u}_{ij}(t)) - \lambda(t)g''_{ij}(\tilde{u}_{ij}(t)), \\ \lambda_{ij}(t) &= \frac{g'_{ij}(\tilde{u}_{ij}(t))b'_i(\tilde{u}_{ij}(t))^T}{g'_{ij}(\tilde{u}_{ij}(t))g'_{ij}(\tilde{u}_{ij}(t))^T}. \end{aligned}$$

Here and throughout this section we use a prime to indicate a (vector) derivative with respect to the argument of the function.

Notice the interpretation that $\hat{u}_i$ is the empirical posterior mode estimate of the transformation parameter for curve i while $\tilde{u}_{ij}(t)$ similarly estimates this parameter given that t_{ij} is transformed to t.

Tierney, Kass and Kadane (1989) prove their formula to give an asymptotic approximation of high accuracy when the exponent, $b_i(u)$, is proportional to some n going to infinity. The crucial point here is that the exponent is well approximated by a quadratic in the neighborhood of the minimum where the contribution to the integral matters. In our experience from examples like the ones presented in this paper, this is a very good approximation.

There is one remaining computational problem: the minimum must be found for each curve, i, for each combination of estimation point, t, and observation point, t_{ij}. However, for the large majority of these terms the contribution to the sum will be negligible because the empirical posterior probability may essentially rule out a transformation mapping t to the vicinity t_{ij}. Thus, as a first part, we truncate the sum over j to those values of t_{ij} that are in the neighborhood of

$$\hat{t}_{ij} = g_{ij}(\hat{u}_i),$$

the unconstrained estimate of the transformed value of t.

Second, we may for each j use a Taylor series expansion of $\tilde{u}_{ij}(t)$ as a function of t around $\hat{t}_{ij}$. The resulting Taylor series expansion to first order is

$$\tilde{u}_{ij}(t) \sim \hat{u}_i + \frac{\partial}{\partial t}\tilde{u}_{ij}(t)\bigg|_{t=\hat{t}_{ij}} (t - \hat{t}_{ij}),$$

where the derivative may be shown to be

$$\frac{\partial \tilde{u}_{ij}(t)}{\partial t}\bigg|_{t=\hat{t}_{ij}} = \frac{\{b''(\hat{u}_i)\}^{-1} g'_{ij}(\hat{u}_i)^T}{g'_{ij}(\hat{u}_i)\{b''(\hat{u}_i)\}^{-1} g'_{ij}(\hat{u}_i)^T}. \tag{4.2}$$

The only minimizations required for this method are one for each curve, leading to the estimated transformation $\hat{u}_i$.

Although the Laplace approximation is generally fast and accurate, the large number of data points often available in the warping problems calls for an even simpler and faster candidate approximation. Approximating the empirical posterior density of Z_{ij} given Y_i by a normal distribution with mean $\hat{u}_i$ and inverse variance $b''_i(\hat{u}_i)$, we get

$$\tilde{f}_{Z_{ij}|Y_i} = (2\pi)^{-1/2}|D|^{1/2} \exp\left\{-\frac{1}{2}D(t - \hat{t}_{ij})^2\right\}, \tag{4.3}$$

where

$$D = \{g'_{ij}(\hat{u}_i)(b''(\hat{u}_i))^{-1} g'_{ij}(\hat{u}_i)^T\}^{-1}.$$

We recommend that in each particular problem both approximations are tried for speed and accuracy, although the latter (simpler) approximation has been excellent in the examples we have tried.

Linear back-transformations including a shift

Simplifications for $\beta(t)$ arise for certain models. Consider back-transforms of the type

$$g_\theta^{-1}(t) = \theta_1 + \theta_2 a_2(t) + \cdots + \theta_p a_p(t),$$

where $a_k(t)$ for $k = 1, \ldots, p$ are known functions. Polynomial back-transformations with $a_k(t) = t^{k-1}$, as we shall use in the examples, are of this form.

Define the p-vector

$$a(t) = (1, a_2(t), \ldots, a_p(t)).$$

Then $g_{ij}(u)$ is linear in u with derivative $a(t_{ij})$ so that

$$\beta(t) = (a(t_{ij})\{b''(\tilde{u}_{ij})\}^{-1}a(t_{ij})^T)^{-1/2}|b''(\tilde{u}_{ij})|^{-1/2}|b''(\hat{u}_i)|^{1/2}, \quad (4.4)$$

and $G_{ij} = b''$.

Estimation of the variance and covariance parameters

We have, so far, neglected the estimation of the transformation parameters. In many applications we might only fix the distributions up to the residual variance, σ^2, and the covariance matrix of the transformation parameters, Σ. Approximate maximum likelihood estimates for these variance parameters can be calculated in each step of the above algorithm and used in the calculations of the weight functions. Differentiation of the log-likelihood function with respect to the error variance in combination with a Laplace approximation similar to those used in the previous section leads to the usual estimate of the error variance as the mean squared residual

$$\hat{\sigma}^2 = \frac{1}{nN}\sum_{i=1}^{n} \|Y_i - \hat{m}\{g_{\hat{\theta}_i}(t_i)\}\|^2.$$

By a similar argument we arrive at the plug-in type estimate of the covariance matrix for the transformation parameters,

$$\hat{\Sigma} = \frac{1}{n}\sum_{i=1}^{n}(\hat{u}_i - \xi)(\hat{u}_i - \xi)^T.$$

Whether the sum should rather be divided by $n-1$ seems less important with data as the two examples, where the many features of the individual curves ensure a very peaky empirical posterior distribution.

Prediction of the transformation parameters

The unobserved random variables, corresponding to the individuals considered in the study, are estimated by their conditional mean given data,

$$\hat{\theta}_i = \mathrm{E}(\theta | Y_i).$$

In order to calculate an explicit expression for the above conditional mean we will have to calculate integrals similar to the integrals in the weight function. Hence, again we need Laplace-type approximations, and we find that

$$\hat{\theta}_i \approx \arg\min_u b_i(u),$$

or, in other words, we use the conditional (or empirical posterior) mode. Thus, the realization of the unobserved random transformation variables for the curve i, is estimated by $\hat{u}_i$, a quantity we need to calculate anyway to obtain an estimator for the shape function.

5 Examples

Various parametric time transformations may be considered in practical data analysis. The nine curves in the feta cheese example introduced above can be modeled, with the simplest possible individual transformation, namely individual shifts,

$$g_\theta^{-1}(s) = s - \theta.$$

In Rønn (2001), the shape function for the protein profile was estimated, when the shifts were assumed to follow a slightly truncated normal distribution. The analysis indicated that the transformation with individual shifts is too simple to fit the data perfectly. The resulting alignment is seen in the left panel of Figure 2.

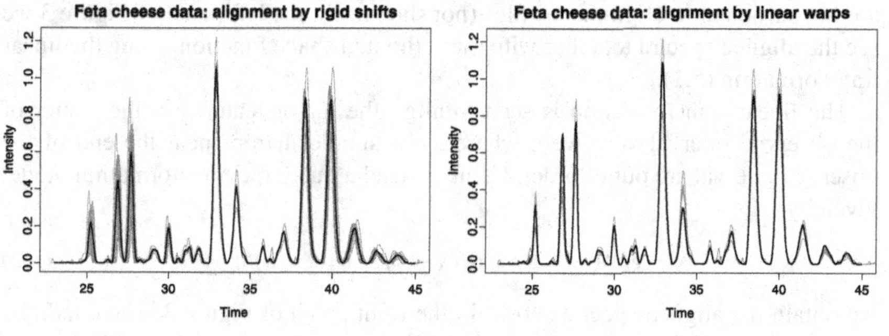

Figure 2 *The nine aligned profiles for the feta cheese example (grey) together with the estimated shape function (enhanced). The transformation alignment used is a rigid shift (left) and linear (right).*

A transformation that allows for stretching as well as shifting might be preferred. This two-dimensional transformation can be parameterized as follows:

$$g_\theta^{-1}(s) = \theta_1 + \theta_2(s - \bar{s}), \tag{5.1}$$

where the time scale has been centered to avoid too strong dependence between the two transformation parameters. The transformation parameters are assumed to follow a two-dimensional normal distribution

$$\theta_i \sim N_2\left\{\begin{pmatrix}0\\0\end{pmatrix}, \Sigma\right\}$$

in principle truncated to a (large) set giving monotone transformations. The estimates of the residual standard deviation and the standard deviations of θ_1 and θ_2 are $\hat{\sigma} = 0.046$, $\sqrt{\hat{\Sigma}_{11}} = 0.030$, $\sqrt{\hat{\Sigma}_{22}} = 0.016$, while the estimate of the correlation between the two transformation parameters is 0.0988. These results were obtained using the Laplace approximation (4.1), but virtually identical results were obtained using the even simpler and faster normal approximation (4.3). Thus, the deviations in the standard relative deviations given above were all less than 1 percent, and the gain in speed was about a factor five.

The estimated protein profile is shown, together with the aligned data profiles, in the right panel of Figure 2. Comparison of the two plots shows that the alignment as well as the estimated profile are clearly sharper with the linear alignment. The estimated individual linear time transformation is seen to align each of the nine protein profiles almost perfectly. Thus, the linear transformation model seems to fit the feta cheese data extremely well. It is also seen that the estimator is a smooth function that contains all the features present in all the nine curves. False peaks, occurring only in one curve each, are not visible in the estimator. Furthermore, the height of the peaks is close to the average height of the nine individual peaks, which is also a desirable property of an estimator.

For the wheat data, inspection of the profiles quickly reveals that rigid shifts are not sufficient to align the profiles (not shown). In the left panel of Figure 3 we see the aligned spectra together with the estimated shape function, using the linear transformation (5.1).

The linear transformation is seen to align the major features in the center of the observed interval very well, whereas the smaller bumps near the end of the observed interval are out of order. Using instead a quadratic transformation model given by

$$g_\theta^{-1}(s) = \theta_1 + \theta_2(s - \bar{s}) + \theta_3(s - \bar{s})^2, \tag{5.2}$$

we obtain the aligned spectra shown in the right panel of Figure 3. The quadratic transformation gives an almost perfect alignment of the data, although local fine-tuning on a very small scale might be desirable in a few instances.

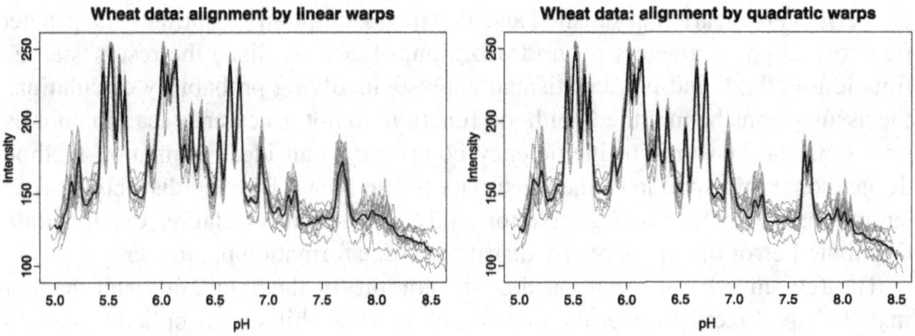

Figure 3 *The 11 aligned profiles for the wheat example (grey) together with the estimated shape function (enhanced). The transformation alignment used is linear (left) and quadratic (right).*

Thus, it is seen that in these realistic examples the model and the algorithm together provide a very satisfactory solution to the alignment of the curves and to the estimation of the common function.

The time per iteration for the two examples were between 2 and 4 seconds per iteration, using the (faster) normal approximation (4.3), and the convergence criteria were met after 33 iterations for the wheat data and 30 iterations for the feta cheese data. The algorithm was implemented in the statistical software R on a standard laptop. Visually, convergence was obtained after less than 10 iterations, but the stopping criteria were rather strict, demanding a relative change less than 0.0001 for all parameters including the transformation parameters and the function, m, evaluated on a fine grid. Finally, the hardest criterion to meet was an absolute change less than 0.01 in twice the negative log-likelihood, which involves the sum of squared errors for the entire set of observations and had a magnitude in the order 10000.

For comparison the we tried the function `register.fd` from the R-package `fda`, Ramsay (2007), which is based on methods from Ramsay and Li (1998) and used by Anselmo, Dias and Garcia (2005). Memory problems arose when a fine-scale representation matching the number of observations were attempted. With a representation of 100 base-vectors, alignment of one curve against another took 10 minutes and did not align even the major peaks correctly in the cases we tried. A reason is probably the large flexibility in the nonparametric class of time transformations. More knowledge of the function might well have helped as other representations and user choices may be made, but the method seems to be more directed towards smoother functions than the examples presented here.

6 Simulation study

The parametric transformation of the x-axis seems to align the profiles from the above examples very well. However, the working model assumptions of independent errors are clearly not satisfied and also the assumption on normally distributed transformation parameters is an idealization, of course. Since the result is an estimation method, and not a statistical analysis involving probability calculations, the assumptions behind the likelihood function do not necessarily lead to poor estimation, but, at worst, to inefficiency compared to an ideal estimation method. To get some information of the sensitivity of the NPMLE to the distributional assumptions we made a small simulation study, with serial correlation, exponentially distributed error or exponentially distributed transformation parameters.

The feta cheese data were used as inspiration for the simulation and the estimated shape function under the model with random shifts was applied as the true shape function m_0. Data from the following four models for the errors and the random shifts were simulated:

1. The working model: independent normally distributed errors with standard deviation $\sigma = 0.031$ and independent normally distributed random shifts with standard deviation $\omega = 0.26$.
2. Autoregressive process of order 1 with autocorrelation $\rho = 0.89$ and marginal standard deviation $\sigma = 0.031$ for the errors and independent normally distributed random shifts with standard deviation $\omega = 0.26$.
3. Independent exponentially distributed errors with standard deviation 0.031 and independent normally distributed random shifts with standard deviation $\omega = 0.26$.
4. Independent normally distributed errors with standard deviation $\sigma = 0.031$ and independent exponentially distributed random shifts with standard deviation 0.26.

These cases were chosen to reflect various kinds of discrepancies from the working model without altering the signal-to-noise ratio.

For each model 100 data sets consisting of 10 shifted profiles were simulated. The NPMLE assuming independent normally distributed errors and shifts were calculated for each simulated data set as described in the present paper. The estimation procedure converged for all data sets and the 100 estimated NPMLEs and their mean are plotted together with the true profile m_0 in Figure 4. As the NPMLEs are only well defined up to a shift, the 100 NPMLEs were aligned by rigid shifts before plotting.

It is seen that the profile is remarkably well estimated in all cases; in particular the negligible bias in the estimation of the peaks is noteworthy. This latter conclusion follows since the dashed white line, representing the true curve, virtually falls on top of the mean of the 100 estimates, represented by the black line.

Plots of the 100 estimated shifts were made against the 100 true shifts together with identity lines for all four simulation scenarios in Figure 5. The estimated and the true shifts showed almost perfect agrement, suggesting that the algorithm is robust towards misspecification of the distribution of the transformation parameters.

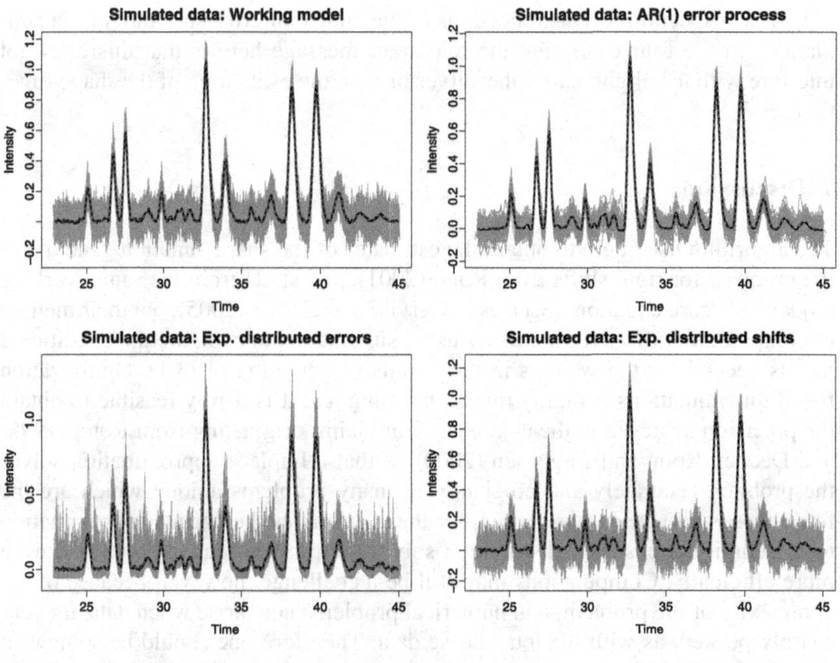

Figure 4 *The 100 estimated functions (grey) together with their mean (thick black line) and the true function (dashed white line) for each of the four simulation models.*

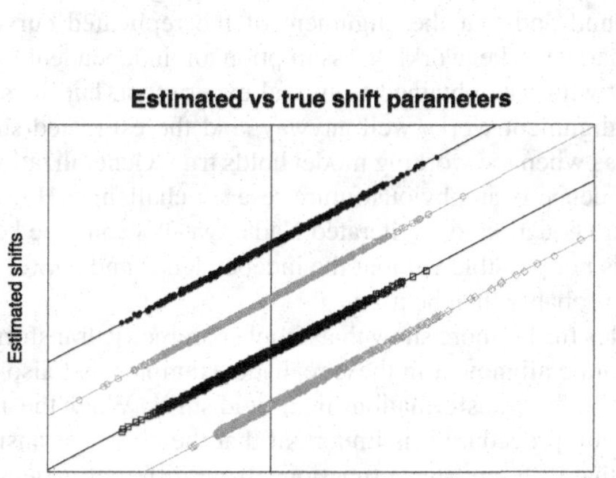

Figure 5 *The 100 estimated shifts versus the true shifts together with the identity line for four simulation scenarios. The vertical line is placed at zero, and the four cases have been shifted vertically together with the identity line drawn for each case.*

Closer inspection reveals reflections of the choice of error structure of the simulations in the four cases, but the important message here is that this does not interfere with the alignment of the curves or with the estimation of the shape function.

7 Discussion

The algorithm used here to obtain the estimator of the shape function generalizes the proposal for rigid shifts from Rønn (2001), and starts from the same working model and score equation, (3.1), as in Gervini and Gasser (2005), but their method of computation differs from ours by using simulations to compute the conditional means needed for the weights in the iterative calculation of (3.1). Optimization based on simulations is highly time consuming and it is hardly feasible to obtain the precision achieved in the examples. Our claim, originating from Rønn (1998) and Decker, Rønn and Jørgensen (2000), is that a Laplace approximation solves the problem accurately and efficiently in many problems among which are the feta cheese and wheat data. In fact, for these examples the normal approximation to the empirical posterior distribution solves the problem equally well and even more efficiently. Computations may still be a challenge, however, because of the complexity of the problem, and numerical problems may arise when data are very sharply peeked, as with the feta cheese data. Therefore one should be somewhat liberal with the convergence criteria, at the same time allowing for the possibility that an exact solution may not exist.

The two successful applications indicate that a good estimator for the shape function is found and that the alignment of the replicated curves is done in a satisfactory manner. The working assumption of independent normal errors is still in conflict with data obtained from real experiment, but the simulations suggest that the alignment works well anyway, and the estimated shape function is equally sharp as when the working model holds true. Generalization to more models with dependence is an obvious future reseach challenge. However, the reduction of the score equation to the iterated Nadaraya–Watson type kernel smoothing scheme (3.2) is not possible without the independence and entirely different algorithms should probably then be used.

The examples furthermore show that simple parametric transformations are sufficient for accurate alignment in these realistic examples, but also that it is important to consider other transformations than rigid shifts. When the alignment is part of a classification procedure it is important that the class of transformations used is not so rich that different shape functions, from different varieties, for example, easily align to look similar. This is avoided by use of low-dimensional parametric transformations.

The aligned profiles look even better if the profiles were preprocessed by removal of background. This was done in the feta cheese example but was avoided in the wheat example to show that the alignment method is largely insensitive to such pre-processing. Thus, both examples were tried with and without background removal, with similar results.

Gervini and Gasser (2005) show that when the space of shape functions is compact, the estimate of the shape function converges at the rate of $1/\sqrt{n}$ towards the true shape function, when the sup-norm is used. The compactness assumption, however, essentially rules out a nonparametric model, because a compact set is either of finite dimension or is very fragmented by being everywhere thin. That is, for no function in the model is any neighborhood of this function contained in the model. We conjecture, on the basis of results in Stone (1980), that when the parameter space is infinite-dimensional, the asymptotic rate will be slower in the same way as for nonparametric regression.

Acknowledgments

We are grateful to Ib Søndergaard and to Kirsten Jensen for providing the wheat data and to Karsten B. Qvist for providing the feta cheese data. Also, we thank two anonymous reviewers for their constructive and useful comments. The research was supported by The Danish Research and Development Programme for Food technology through the Center for Advanced Food Studies.

References

Anselmo, C. A. F., Dias, R. and Garcia, N. L. (2005). Adaptive basis selection for functional data analysis via stochastic penalization. *Computational & Applied Mathematics* **24** 209–229. MR2186845

Brumback, B. A. and Rice, J. (1998). Smoothing spline models for the analysis of nested and crossed samples of curves. *Journal of the American Statistical Association* **93** 961–976. MR1649194

Brumback, L. C. and Lindstrom, M. J. (2004). Self modelling with flexible, random time transformations. *Biometrics* **60** 461–470. MR2066281

Decker, M., Rønn, B. B. and Jørgensen, S. S. (2000). Thermally assisted in-line methylation and gas chromatography with statistical data analysis for determination of fatty acid distribution and fingerprinting of plant seeds and oils. *European Food Research and Technology* **211** 366–373.

Fernholz, L. T. (1983). *von Mises Calculus for Statistical Functionals*. Springer, New York. MR0713611

Gervini, D. and Gasser, T. (2005). Nonparametric maximum likelihood estimation of the structural mean of a sample of curves. *Biometrika* **92** 801–820. MR2234187

Gill, R. D. (1989). Non- and semi-parametric maximum likelihood estimators and the von Mises method (part 1). *Scandinavian Journal of Statistics. Theory and Applications* **16** 97–128. MR1028971

Glasbey, C., Vali, L. and Gustafsson, J. (1995). A statistical model for unwarping of 1-D electrophoresis gels. *Electrophoresis* **26** 4237–4242.

Groeneboom, P. and Wellner, J. A. (1992). *Information Bounds and Nonparamteric Maximum Likelihood Estimation*. Birkhauser, Basel. MR1180321

Jensen, K., Søndergaard, I., Skovgaard, I. and Nielsen, H. (1995). From image processing to classification: I. Modelling disturbance of isoelectric focusing patterns. *Electrophoresis* **16** 921–926.

Kneip, A. and Engel, J. (1995). Model estimation in nonlinear regression under shape invariance. *The Annals of Statistics* **23** 551–570. MR1332581

Kneip, A. and Gasser, T. (1992). Statistical tools to analyze data representing a sample of curves. *The Annals of Statistics* **20** 1266–1305. MR1186250

Lawton, W., Sylvestre, E. and Maggio, M. (1972). Self-modeling nonlinear regression. *Technometrics* **14** 513–532.

Liu, X. and Müller, H.-G. (2004). Functional convex averaging and synchronization for time-warped random curves. *Journal of the American Statistical Association* **99** 687–699. MR2090903

Murphy, S. (1995). Asymptotic theory for the frailty model. *The Annals of Statistics* **23** 182–198. MR1331663

Parner, E. (1998). Asymptotic theory for the correlated gamma-frailty model. *The Annals of Statistics* **26** 183–214. MR1611788

Ramsay, J. (2007). R-package `fda`: Functional data analysis. R Development Core Team. Available at http://cran.r-project.org/.

Ramsay, J. and Li, X. (1998). Curve registration. *Journal of the Royal Statistical Society. Series B* **60** 351–363. MR1616045

Ramsay, J. and Silverman, B. (1997). *Functional Data Analysis*. Springer, New York. MR2168993

Ramsay, J. and Silverman, B. (2002). *Applied Functional Data Analysis: Methods and Case Studies*. Springer, New York. MR1910407

Rønn, B. B. (1998). Analyses of functional data. Ph.D. thesis, Department of Mathematics and Physics, The Royal Veterinary and Agricultural University.

Rønn, B. B. (2001). Non-parametric maximum likelihood estimation for shifted curves. *Journal of the Royal Statistical Society. Series B* **63** 243–259. MR1841413

Stone, C. J. (1980). Optimal rate of convergence for nonparametric estimators. *The Annals of Statistics* **8** 1348–1360. MR0594650

Stützle, W., Gasser, T., Molinari, L., Largo, R., Prader, A. and Huber, P. (1980). Shape-invariant modeling of human growth. *Annals of Human Biology* **5** 1–24.

Tierney, L., Kass, R. E. and Kadane, J. B. (1989). Approximate marginal densities of nonlinear functions (Corrigenda **78** 233–234). *Biometrika* **76** 425–433. MR1040637

Wang, K. and Gasser, T. (1997). Alignment of curves by dynamic time warping. *The Annals of Statistics* **25** 1251–1276. MR1447750

Wium, H., Kristiansen, K. R. and Qvist, K. B. (1998). Proteolysis and its role in relation to texture of Feta cheese made from ultrafiltered milk with different amounts of rennet. *Journal of Dairy Research* **65** 665–674.

Biometrics, Genmab a/s
Copenhagen
Denmark
E-mail: brn@genmab.com

Department of Natural Sciences
Faculty of Life Sciences
University of Copenhagen
Denmark
E-mail: ims@life.ku.dk

Bibliography

Aalen, O. O. and Johansen, S. (1978). An empirical transition matrix for non-homogeneous Markov chains based on censored observations. *Scandinavian Journal of Statistics* **5**, pp. 141–150.

Albers, W., Bickel, P. J., and van Zwet, W. R. (1976). Asymptotic expansions for the power of distribution free tests in the one-sample problem. *Annals of Statistics* **4**, pp. 108–156.

Almudevar, A. (2016). Higher order density approximations for solutions to estimating equations. *Journal of Multivariate Analysis* **143**, pp. 424–439.

Almudevar, A., Field, C., and Robinson, J. (2000). The density of multivariate M-estimates. *Annals of Statistics* **28**, pp. 275–297.

Amari, S. (1982). Geometrical theory of asymptotic ancillarity and conditional inference. *Biometrika* **69**, pp. 1–18.

Anderson, T. W. (1951). Estimating linear restrictions on regression coefficients for multivarate normal distributions. *Annals of Mathematical Statistics (Correction, Annals of Statistics 8 (1980))* **22**, pp. 327–351.

Anderson, T. W. (1984). Estimating linear statistical relationships. *Annals of Statistics* **12**,pp. 1–45.

Andrews, D. W. K., Lieberman, O., and Marmer, V. (2006). Higher-order improvements of the parametric bootstrap for long-memory Gaussian processes. *Journal of Econometrics* **133**, pp. 673–702.

Anscombe, F. J. (1961). Examination of residuals. in J. Neyman (ed.), *Proc. 4th Berkeley Symp. Mathematics, Probability and Statistics*, Vol. 1 (University of California Press, Berkeley), pp. 1–36.

Bahadur, R. R. and Ranga Rao, R. (1960). On deviations of the sample mean. *Annals of Mathematical Statistics* **31**, pp. 1015–1027.

Barndorff-Nielsen, O. E. (1980). Conditionality resolutions. *Biometrika* **67**, pp. 293–310.

Barndorff-Nielsen, O. E. (1983). On a formula for the distribution of the maximum likelihood estimator. *Biometrika* **70**, pp. 343–365.

Barndorff-Nielsen, O. E. (1986). Inference on full or partial parameters, based on the standardized signed likelihood ratio. *Biometrika* **73**, pp. 307–322.

Barndorff-Nielsen, O. E. (1987). Discussion of "Parameter orthogonality and approximate conditional inference" by D. R. Cox and N. Reid. *Journal of the Royal Statistical Society: Series B* **49**, pp. 18–20.

Barndorff-Nielsen, O. E. and Chamberlin, S. R. (1991). An ancillary invariant modification of the signed log likelihood ratio, *Scandinavian Journal of Statistics* **18**, pp. 341–352.

Barndorff-Nielsen, O. E. and Chamberlin, S. R. (1994). Stable and invariant adjusted directed likelihoods. *Biometrika* **81**, pp. 485–499.

Barndorff-Nielsen, O. E. and Cox, D. R. (1994). *Inference and Asymptotics.* (Chapman and Hall, London).

Barndorff-Nielsen, O. E. and Wood, A. T. A. (1998). On large deviations and choice of ancillary for p^* and r^*. *Bernoulli* **4**, pp. 35–63.

Berg, R. and Ditlevsen, S. (2003). Synaptic inhibition and excitation estimated via the time constant of membrane potential fluctuations. *Journal of Neurophysiology* **110**, pp. 1021–1034.

Bhattacharya, R. N. and Ghosh, J. K. (1978). On the validity of the formal Edgeworth expansion. *Annals of Statistics* **6**, pp. 434–451.

Bhattacharya, R. N. and Rao, R. R. (1976). *Normal Approximations and Asymptotic Expansions.* (Wiley, New York).

Bickel, P. J. and Chernoff, H. (1993). Asymptotic distribution of the likelihood ratio statistic in a prototypical non-regular problem. in J. Ghosh, S. Mitra, K. Parthasarathy, and B. P. Rao (eds.), *Statistics and Probability: A Raghu Raj Bahadur Festschrift* (Wiley, New York), pp. 83–96.

Bickel, P. J. and Ghosh, J. K. (1990). A decomposition for the likelihood ratio statistic and the Bartlett correction – a Bayesian argument. *Annals of Statistics* **18**, pp. 1070–1090.

Bleistein, N. (1966). Uniform asymptotic expansions of integrals with stationary point near algebraic singularity. *Communications on Pure and Applied Mathematics* **19**, pp. 353–370.

Bohman, H. (1963). What is the reason that Esscher's method of approximation is as good as it is? *Scandinavian Actuarial Journal* **1963**, pp. 87–94.

Brereton, R. G. (2014). A short history of chemometrics: a personal view. *Journal of Chemometrics* **28**, pp. 749–760.

Brillinger, D. R. (1972). On the number of solutions of systems of random equations. *Annals of Mathematical Statistics* **43**, pp. 534–540.

Brockhoff, P. B. (2011). Sensometrics. in M. Lovric (ed.), *International Encyclopedia of Statistical Science* (Springer, Berlin).

Brockhoff, P. B., Schlich, P., and Skovgaard, I. M. (2015). Taking individual scaling differences into account by analyzing profile data with the mixed assessor model. *Food Quality and Preference* **39**, pp. 156–166.

Brockhoff, P. B. (1994). *Statistical Analysis of Sensory Data.*, Ph.D. thesis, Dept. of Mathematics and Physics, Royal Veterinary and Agricultural University, Copenhagen, Denmark.

Brockhoff, P. B. and Skovgaard, I. M. (1994). Modelling individual differences between assessors in sensory evaluations. *Food Quality and Preference* **5**, pp. 215–224.

Cai, Z. and Sun, Y. (2003). Local linear estimation for time-dependent coefficients in Cox's regression models. *Scandinavian Journal of Statistics* **30**, pp. 93–111.

Carstensen, B. (2004). Comparing and predicting between several methods of measurement. *Biostatistics* **5**, pp. 399–413.

Cheah, P., Fraser, D. A. S., and Reid, N. (1994). Multiparameter testing in exponential models: third order approximations from likelihood. *Biometrika* **81**, pp. 271–278.

Cook, R. D. (1977). Detection of influential observations in linear regression. *Technometrics* **19**, pp. 15–18.

Cox, D. R. (1958). Some problems connected with statistical inference. *Annals of Mathematical Statistics* **29**, pp. 357–372.

Cox, D. R. (1980). Local ancillarity. *Biometrika* **67**, pp. 279–286.

Cox, D. R. and Reid, N. (1987). Parameter orthogonality and approximate conditional inference (with discussion). *Journal of the Royal Statistical Society: Series B* **49**, pp. 1–39.

Cox, D. R. and Snell, E. J. (1968). A general definition of residuals (with discussion). *Journal of the Royal Statistical Society: Series B* **30**, pp. 248–275.

Cox, D. R. and Snell, E. J. (1971). On test statistics constructed from residuals. *Biometrika* **58**, pp. 589–594.

Daniels, H. E. (1954). Saddlepoint approximations in statistics. *Annals of Mathematical Statistics* **25**, pp. 631–650.

Daniels, H. E. (1987). Tail probability approximations. *International Statistical Review* **55**, pp. 37–48.

Davison, A. C., Fraser, D. A. S., Reid, N., and Sartori, N. (2014). Accurate directional inference for vector parameters in linear exponential families. *Journal of the American Statistical Association* **109**, pp. 302–314.

Davison, A. C. and Hinkley, D. V. (1997). *Bootstrap Methods and their Application.* (Cambridge University Press, Cambridge).

Davison, A. C. and Snell, E. J. (1991). Residuals and diagnostics. in D. V. Hinkley, N. Reid, and E. J. Snell (eds.), *Statistical Theory and Modelling: In Honour of Sir David Cox, FRS* (Chapman and Hall, London), pp. 83–106.

Dawid, A. P. (1975). Contribution to the discussion of "Curvature of a statistical problem" by B. Efron. *Annals of Statistics* **3**, pp. 1231– 1234.

DiCiccio, T. J. and Martin, M. A. (1993). Simple modifications for signed roots of likelihood ratio statistics. *Journal of the Royal Statistical Society: Series B* **55**, pp. 305–316.

Dolby, G. R. (1976). The ultrastructural relation: A synthesis of the functional and structural relations. *Biometrika* **63**, pp. 39–50.

Durbin, J. (1980). Approximations for densities of sufficient estimators. *Biometrika* **67**, pp. 311–333.

Edgeworth, F. Y. (1904). The law of error. *Transactions of the Cambridge Philosophical Society* **20**, pp. 36–65 and 113–141.

Efron, B. (1975). Defining the curvature of a statistical problem. *Annals of Statistics* **3**, pp. 1189–1242.

Efron, B. and Hinkley, D. V. (1978). Assessing the accuracy of the maximum likelihood estimator: Observed versus expected Fisher information. *Biometrika* **65**, pp. 457–482.

Esscher, F. (1932). On the probability function in the collective theory of risk. *Scandinavian Actuarial Journal* **1932**, pp. 175–195.
Esscher, F. (1963). On approximate computation of distribution functions when the corresponding characteristic functions are known. *Scandinavian Actuarial Journal* **1963**, pp. 78–86.
Fan, J. and Gijbels, I. (1996). *Local Polynomial Modelling and Its Applications.* (Chapman and Hall, London).
Ferrari, S. L. P. and Pinheiro, E. C. (2011). Improved likelihood inference in beta regression. *Journal of Statistical Computation and Simulation* **81**, pp. 431–443.
Fisher, R. A. (1925). Theory of statistical estimation. *Mathematical Proceedings of the Cambridge Philosophical Society* **122**, pp. 700–725.
Fisher, R. A. (1934). Two new properties of mathematical likelihood. *Proceedings of the Royal Society of London* **144**, pp. 285–307.
Fisher, R. A. (1936). Uncertain inference. *Proceedings of the American Academy of Arts and Sciences* **71**, pp. 245–258.
Fisher, R. A. (1938). The statistical utilization of multiple measurements. *Annals of Eugenics* **8**, pp. 376–386.
Fisher, R. A. (1956). On a test of significance in Pearson's Biometrika Table No. 11. *Journal of the Royal Statistical Society: Series B* **18**, pp. 56–60.
Fraser, D. A. S. (1967). Data transformations and the linear model. *Annals of Mathematical Statistics* **38**, pp. 1456–1465.
Fraser, D. A. S. (1968). *The Structure of Inference.* (Wiley, New York).
Fraser, D. A. S. (1993). Directional tests and statistical frames. *Statistical Papers* **34**, pp. 213–236.
Fraser, D. A. S. (2003). Likelihood for component parameters. *Biometrika* **90**, pp. 327–339.
Fraser, D. A. S. (2016). Definitive testing of an interest parameter: using parameter continuity. *Journal of Statistical Research* **47**, to appear.
Fraser, D. A. S., Fraser, A. M., and Staicu, A. M. (2010). The second order ancillary: A differential view with continuity. *Bernoulli* **16**, pp. 1208–1223.
Fraser, D. A. S. and Massam, H. (1985). Conical tests: observed levels of significance and confidence regions. *Statistische Hefte* **26**, pp. 1–17.
Fraser, D. A. S. and Reid, N. (1989). Adjustments to profile likelihood. *Biometrika* **76**, pp. 477–488.
Fraser, D. A. S. and Reid, N. (1995). Ancillaries and third order significance, *Utilitas Mathematica* **47**, pp. 33–53.
Fraser, D. A. S. and Reid, N. (2006). Assessing a vector parameter. *Student* **5**, pp. 247–256.
Fraser, D. A. S., Reid, N., and Sartori, N. (2016). Accurate directional inference for vector parameters. *Biometrika* **103**, pp. 625–639.
Gervini, D. and Gasser, T. (2005). Nonparametric maximum likelihood estimation of the structural mean of a sample of curves. *Biometrika* **92**, pp. 97–128.
Götze, F. and Hipp, C. (1983). Asymptotic expansions for sums of weakly dependent random variables. *Zeitschrift für Wahrscheinlichkeitstheorie und Verwandte Gebiete* **64**, pp. 211–239.

Guihenneuc-Jouyaux, C. and Rousseau, J. (2005). Laplace expansions in Markov chain Monte Carlo algorithms. *Journal of Computational and Graphical Statistics* **14**, pp. 75–94.

Hall, P. and Stewart, M. (2005). Theoretical analysis of power in a two-component normal mixture model. *Journal of Statistical Planning and Inference* **134**, pp. 158–179.

Halmos, P. R. (1948). The range of a vector measure. *Bulletin of the American Mathematical Society* **54**, pp. 416–421.

Hartigan, J. A. (1985). A failure of likelihood asymptotics for normal mixtures. in L. LeCam and R. Olshen (eds.), *Proceedings of the Berkeley Conference in Honor of Jerzy Neyman and Jack Kiefer* (Wadsworth, Inc., New York), pp. 807–810.

He, H. and Severini, T. A. (2007). Higher-order asymptotic normality of approximations to the modified signed likelihood ratio statistic for regular models. *Annals of Statistics* **35**, pp. 2054–2074.

Hinkley, D. V. (1980). Likelihood as approximate pivotal distribution. *Biometrika* **67**, pp. 287–292.

Hougaard, P. (1985). The appropriateness of the asymptotic distribution in a nonlinear regression model in relation to curvature. *Journal of the Royal Statistical Society: Series B* **47**, pp. 103–114.

James, G. M. (2007). Curve alignment by moments, *Annals of Applied Statistics.* **1**, pp. 480–501.

Jensen, J. L. (1992). The modified signed likelihood statistic and saddlepoint approximations. *Biometrika* **79**, pp. 693–703.

Jensen, J. L. (1995). *Saddlepoint Approximations.* (Clarendon Press, Oxford).

Jensen, J. L. and Wood, A. T. A. (1998). Large deviations results for minimum contrast estimators. *Annals of the Institute of Statistical Mathematics* **50**, pp. 673–695.

Jensen, S. P., Kristensen, K., and Brockhoff, P. B. (2016). *Multiplicative Mixed Models using Template Model Builder*, `https://github.com/sofpj/mumm`, R package version 0.1. (Presented at the use R conference, Stanford, June 2016).

Johnson, R. A. (1970). Asymptotic expansions associated with posterior distributions, *Annals of Mathematical Statistics* **41**, pp. 851–864.

Jones, M. C., Marron, J. S., and Park, B. U. (1991). A simple root n bandwidth selector. *Annals of Statistics* **19**, pp. 1919–1932.

Kline, P. and Santos, A. (2012). Higher order properties of the wild bootstrap under misspecification. *Journal of Econometrics* **171**, pp. 54–70.

Kristensen, K., Nielsen, A., Berg, C. W., Skaug, H. J., and Bell, B. (2016). Tmb: Automatic differentiation and Laplace approximation. *Journal of Statistical Software* **70**, pp. 1–21.

Kuznetsova, A., Brockhoff, P. B., and Christensen, R. H. B. (2015a). *lmerTest: Tests in Linear Mixed Effects Models*, `http://CRAN.R-project.org/package=lmerTest`, R package version 2.0-29.

Kuznetsova, A., Brockhoff, P. B., and Christensen, R. H. B. (2015b). *SensMixed: Analysis of Sensory and Consumer Data in a Mixed Model Framework*, http://CRAN.R-project.org/package=SensMixed, R package version 2.0-8.

Kuznetsova, A., Christensen, R. H. B., and Brockhoff, P. B. (2017). lmerTest package: Tests in linear mixed effects models. *Journal of Statistical Software*, to appear.

Lieberman, O., Rousseau, J., and Zucker, D. M. (2003). Valid asymptotic expansions for the maximum likelihood estimator of the parameter of a stationary, Gaussian, strongly dependent process. *Annals of Statistics* **31**, pp. 586–612.

Linton, O. (1996). Edgeworth approximation for MINPIN estimators in semiparametric regression models. *Econometric Theory* **12**, pp. 30–60.

Liu, X. and Shao, Y. (2004). Asymptotics of the likelihood ratio test in a two-component normal mixture model. *Journal of Statistical Planning and Inference* **123**, pp. 61–81.

Lugannani, R. and Rice, S. (1980). Saddlepoint approximation for the distribution of a sum of independent variables. *Advances in Applied Probability* **12**, pp. 475–490.

Martinussen, T., Scheike, T. H., and Skovgaard, I. M. (2002). Efficient estimation of fixed and time-varying covariate effects in multiplicative intensity models. *Scandinavian Journal of Statistics* **28**, pp. 57–74.

McCullagh, P. (1987). *Tensor Methods in Statistics.* (Chapman and Hall, London).

Munkholt, N. (2012). *Likelihood ratio tests – When a parameter disappears under the hypothesis*, Master's thesis, University of Copenhagen.

Murphy, S. A. and Sen, P. K. (1991). Time-dependent coefficients in a Cox-type regression model. *Stochastic Processes and their Applications* **39**, pp. 153–180.

Murphy, S. A. and van der Vaart, A. W. (2000). On profile likelihood. *Journal of the American Statistical Association* **95**, pp. 449–465.

Næs, T., Tomic, O., and Brockhoff, P. B. (2010). *Statistics for Sensory and Consumer Science.* (Wiley, New York).

Nelder, J. A. (1961). The fitting of a generalization of the logistic curve. *Biometrics* **17**, pp. 89–110.

Pace, L. and Salvan, A. (1997). *Principles of Statistical Inference: From a Neo-Fisherian Perspective.* (World Scientific, Singapore).

Peltier, C., Brockhoff, P. B., Visalli, M., and Schlich, P. (2013). The mam-cap table: a new tool for monitoring panel performances. *Food Quality and Preference* **32**, pp. 24–27.

Petrov, V. V. (1975). *Sums of Independent Random Variables.* (Springer, Berlin).

Pfanzagl, J. (1980). Asymptotic expansions in parametric statistical theory. in P. R. Krishnaiah (ed.), *Developments in Statistics*, Vol. 3 (Academic Press, New York), pp. 1–97.

Pierce, D. and Peters, D. (1992). Practical use of higher order asymptotics for multiparameter exponential families (with discussion). *Journal of the Royal Statistical Society: Series B* **54**, pp. 701–738.

Pierce, D. A. and Bellio, R. (2016). Modern likelihood-frequentist inference. *Submitted manuscript*, Available from the authors.

Pierce, D. A. and Schafer, D. W. (1986). Residuals in generalized linear models. *Journal of the American Statistical Association* **81**, pp. 977–986.

Qi, L., Sun, Y., and Gilbert, P. B. (2016). Generalized semiparametric varying-coefficient model for longitudinal data with applications to treatment switching. *Biometrics*, Early view online DOI:10.111/biom.12626.

R Core Team (2016). *R: A Language and Environment for Statistical Computing*, R Foundation for Statistical Computing, Vienna, Austria, `https://www.R-project.org`.

Ramsay, J. O. and Silverman, B. W. (2005). *Functional Data Analysis.* (Springer, Berlin).

Reid, N. (1988). Saddlepoint methods and statistical inference. *Statistical Science* **3**, pp. 213–227.

Reid, N. (1995). The roles of conditioning in inference. *Statistical Science* **10**, pp. 138–199.

Reid, N. (2005). Asymptotics and the theory of statistics. in A. C. Davison, Y. Dodge, and N. Wermuth. (eds.), *Celebrating Statistics: Papers in Honour of D. R. Cox* (Oxford University Press, Oxford), pp. 73–88.

Ritz, C. and Skovgaard, I. M. (2005). Likelihood ratio tests in curved exponential families with nuisance parameters present only under the alternative. *Biometrika* **92**, pp. 507–517.

Robinson, J. (1982). Saddlepoint approximations for permutation tests and confidence intervals. *Journal of the Royal Statistical Society: Series B* **44**, pp. 91–101.

Rønn, B. B. (2001). Nonparametric maximum likelihood estimation for shifted curves. *Journal of the Royal Statistical Society* **63**, pp. 243 –259.

Rue, H., Martino, S., and Chopin, N. (2009). Approximate Bayesian inference for latent Gaussian models by using integrated nested Laplace approximations. *Journal of the Royal Statistical Society* **71**, pp. 319 –392.

Saporta, G. (2015). A conversation with Jean-Louis Bodin. *International Statistical Review* **83**, pp. 2–16.

Schmidt, W. H. and Zwanzig, S. (1986). Second order asymptotics in nonlinear regression. *Journal of Multivariate Analysis* **18**, pp. 187–215.

Scott, D. and Dong, C. Y. (2015). *VarianceGamma: The Variance Gamma Distribution*, `http://CRAN.R-project.org/package=VarianceGamma`, R package version 0.3-1.

Severini, T. A. (2000). *Likelihood Methods in Statistics.* (Oxford University Press, Oxford).

Sharma, G., Mathew, T., and Bebu, I. (2014). Combining multivariate bioassays: accurate inference using small sample asymptotics. *Scandinavian Journal of Statistics* **41**, pp. 152–166.

Skovgaard, I. M. (1981a). Transformation of an Edgeworth expansion by a sequence of smooth functions. *Scandinavian Journal of Statistics* **8**, pp. 207–217.

Skovgaard, I. M. (1981b). Edgeworth expansions of the distributions of maximum likelihood estimators in the general (non-i.i.d.) case. *Scandinavian Journal of Statistics* **8**, pp. 227–236.

Skovgaard, I. M. (1985a). Large deviation approximations for maximum likelihood estimators. *Probability and Mathematical Statistics* **6**, pp. 89–107.

Skovgaard, I. M. (1985b). A second-order investigation of asymptotic ancillarity. *Annals of Statistics* **13**, pp. 534–551.

Skovgaard, I. M. (1986). On multivariate Edgeworth expansions. *International Statistical Review* **54**, pp. 169–186.

Skovgaard, I. M. (1988). Saddlepoint expansions for directional test probabilities. *Journal of the Royal Statistical Society: Series B* **50**, pp. 269–280.

Skovgaard, I. M. (1989a). Amendment to "On multivariate Edgeworth expansions, Internat. Statist. Rev. (1986) 54, 169–186.". *International Statistical Review* **57**, p. 183.

Skovgaard, I. M. (1989b). *Analytical Statistical Models.* (Institute of Mathematical Statistics, Hayward).

Skovgaard, I. M. (1990). On the density of minimum contrast estimators. *Annals of Statistics* **18**, pp. 779–789.

Skovgaard, I. M. (1996). An explicit large-deviation approximation to one-parameter tests. *Bernoulli* **2**, pp. 145–165.

Skovgaard, I. M. (2001). Likelihood asymptotics. *Scandinavian Journal of Statistics* **28**, pp. 3–32.

Skovgaard, I. M., Martinussen, T., and Andersen, S. W. (2001). On the use of a parameter dependent group structure for inference. in Saleh (ed.), *Data Analysis and Statistical Foundations* (Nova Scientific), pp. 53–63.

Smith, A., Cullis, B., Brockhoff, P. B., and Thompson, R. (2003). Multiplicative mixed models for the analysis of sensory evaluation data, *Food Quality and Preference* **14**, pp. 387–395.

Smith, R. L. (1989). A survey of non-regular problems. *Proceedings of the 47th session of the International Statistical Institute* **3**, pp. 353–372.

Sørensen, H. (2004). Parametric inference for diffusion processes observed at discrete points in time: a survey. *International Statistical Review* **72**, pp. 337–354.

Sørensen, M. (2012). Estimating functions for diffusion-type processes. in M. Kessler, A. Lindner, and M. Sørensen (eds.), *Statistical Methods for Stochastic Differential Equations* (Chapman Hall/CRC Press), pp. 1–107.

Stapelfeldt, H., Björn, H., Skovgaard, I. M., Skibsted, L. H., and Bertelsen, G. (1992). Warmed-over flavour in cooked sliced beef chemical analysis in relation to sensory evaluation. *Zeitschrift für Lebensmittel-Untersuchung und Forschung* **195**, pp. 203–208.

Stone, A. H. and Tukey, J. W. (1942). Generalized sandwich theorems. *Duke Mathematical Journal* **9**, pp. 356–359.

Stuart, A. and Ord, J. K. (1987). *Kendall's Advanced Theory of Statistics*, Vol. 1, 5th edn. (Charles Griffin, London).

Sun, J., Loader, C., and McCormick, W. P. (2000). Confidence bands in generalized linear models. *Annals of Statistics* **28**, pp. 429–460.

Sun, Y., Hyun, S., and Gilbert, P. (2008). Testing and estimation of time-varying cause-specific hazard ratios with covariate adjustment. *Biometrics* **64**, pp. 1070–1079.

Temme, N. M. (1982). The uniform asymptotic expansion of a class of integrals related to cumulative distribution functions. *SIAM Journal on Mathematical Analysis* **13**, pp. 239–253.

Tierney, L., Kass, R. E., and Kadane, J. B. (1989). Approximate marginal densities of nonlinear functions. *Biometrika* **76**, pp. 425–433.

Wallace, D. L. (1958). Asymptotic approximations to distributions. *Annals of Mathematical Statistics* **29**, pp. 635–654.

Williams, E. J. (1967). The analysis of association among several variates (with discussion). *Journal of the Royal Statistical Society: Series B* **29**, pp. 199–242.

Yu, Z. and Lin, X. (2010). Semiparametric regression with time-dependent coefficients for failure time data analysis. *Statistica Sinica* **20**, pp. 853–869.

Zhou, H. and Wang, C.-Y. (2000). Failure time regression with continuous covariates measured with error. *Journal of the Royal Statistical Society: Series B* **62**, pp. 657–665.

Zucker, D. M. and Karr, A. F. (1990). Nonparametric survival analysis with time-dependent covariate effects: A penalized partial likelihood approach. *Annals of Statistics* **18**, pp. 329–353.

Printed in the United States
By Bookmasters